GALECTINS

GALECTINS

Edited by

ANATOLE A. KLYOSOV
ZBIGNIEW J. WITCZAK
DAVID PLATT

WILEY

Published by John Wiley & Sons, Inc., Hoboken, New Jersey
Published simultaneously in Canada

Library of Congress Cataloging-in-Publication Data:

Galectins / [edited by] Anatole A. Klyosov, Zbigniew J. Witczak, David Platt.
p. ; cm.
Includes bibliographical references and index.
ISBN 978-0-470-37318-7 (cloth)
1. Lectins . 2. Galactose. I. Klesov, A. A. (Anatolii Alekseevich)
II. Witczak, Zbigniew J., 1947– III. Platt, David, 1953–
[DNLM: 1. Galectins–physiology. 2. Galectins–metabolism. 3. Galectins–pharmacology.
QU 55 G152 2008]
QP552.L42G35 2008
572'.69–dc22

2008003753

Printed in the United States of America
10 9 8 7 6 5 4 3 2 1

CONTENTS

PREFACE

This book emerged from topics discussed at the symposium "Galectins: Structures, Functions, and Therapeutic Targets" that was a part of the 234th American Chemical Society meeting held in Boston, MA, from August 19th to August 23rd, 2007. The reasons for organizing this symposium were to (1) bring together leading researchers in the field, (2) better understand mechanisms underlining galectins structures and modes of action, (3) further identify therapeutic targets for galectins, and (4) outline innovative ways for new drug designs based on newly acquired knowledge.

It was announced at the opening of the symposium that the number of galectin publications recorded in the SciFinder database reached almost 3000 since the word "galectin" first appeared in 1994; a new peer-review article on galectins is now published—on average—every 48 h. Of those publications, 2541 were journal articles, besides 69 conference abstracts, 335 patents, 44 dissertations, and no books. These publications were authored and coauthored by almost 10,000 individuals, contributors to the field.

Galectins play a paramount role in a variety of different pathologies, from triggering cell death to inducing skeletal muscle differentiation, to muscle regeneration and proliferation of stem cells. Galectins can also suppress or, conversely, induce tumor cell growth, depending on which galectin is activated. Galectins can also kill pathogenic fungi, increase or decrease parasite adhesion to extracellular matrix, increase or decrease cell survival, regulate cell migration, and affect tumor angiogenesis. Galectins might affect a predisposition of people to cancer depending on their diet. The mechanism of action of galectins is still under investigation. Further research is ongoing to delineate the cause-and-effect relation of these molecules in various disease states including oncology, fibrosis, and many infectious and inflammatory diseases. We have limited data thus far but current research indicates that the biological action of these molecules is quite complex and includes a multiplicity of chemical and biochemical reactions.

What makes the whole field even more complicated from a scientific standpoint is the word "carbohydrates." Lectins in general and, galectins in particular, are proteins specific to certain carbohydrates. Vocabulary, language, and methodologies related to carbohydrates have been building lately so fast, in fact, that most chemists and biochemists, who are not among the few who actually work in the area of carbohydrates, often feel a bit overwhelmed. Moreover, it has become increasingly clear that carbohydrates, which can be biological tools for — or targets of — specific diseases, represent a whole new dimension in medical biochemistry and drug design. Characterized by a variety of terms—"specific recognition," "lectins," and "molecular diversity"—this new dimension, based on carbohydrates, has essentially introduced a new "language" into chemistry, biochemistry, and other related disciplines.

This book will assist chemists, biochemists, and many other scientists to better understand the current state of the art and the challenges that "remain," or better said, "overwhelm" the study of galectins. The translation of these study results, inventions, and discoveries into practical medicine will provide a better understanding of the biological and biochemical bases of many pathologies and the respective drug design.

This is the first book about galectins. It opens with some of the early history of galectin research and an explanation of the nature of galectins. The next part moves to the structure and functions of galectins, their ligand specificity, molecular mechanisms of action, and localization of galectins in the cell, along with the roles galectins play in tumor growth and cancer, fibrosis, inflammation, and immunity. The last part focuses on efforts to better understand the effect of galectins on cell migration, angiogenesis, and chemoresistance, as well as describe new approaches to designing galectin inhibitors.

We thank our wives—Gail, Wanda, and Naomi—who had to tolerate our spending so many evenings and weekends in order to prepare this book.

Anatole A. Klyosov
Zbigniew J. Witczak
David Platt

October 25, 2007

CONTRIBUTORS

Editors

Anatole A. Klyosov
Pro-Pharmaceuticals, Inc.
7 Wells Avenue
Newton, MA 02459, USA

David Platt
Pro-Pharmaceuticals, Inc.
7 Wells Avenue
Newton, MA 02459
USA

Zbigniew J. Witczak
Department of Pharmaceutical Sciences
Nesbitt School of Pharmacy
Wilkes University
84 W. South Street
Wilkes-Barre, PA 18766
USA

Authors

Samuel H. Barondes
Center for Neurobiology and Psychiatry
UCSF
401 Parnassus Avenue
San Francisco, CA 94143-0984, USA

Robert S. Bresalier
Department of Gastroenterology, Hepatology, and Nutrition
The University of Texas MD Anderson Cancer Center
Houston, TX 77030-4009, USA

James C. Byrd
Department of Gastroenterology, Hepatology, and Nutrition
The University of Texas MD Anderson Cancer Center
Houston, TX 77030-4009, USA

Huan-Yuan Chen
Department of Dermatology
University of California, Davis
School of Medicine
3301 C Street, Suite 1400
Sacramento, CA 95816, USA

Ruud P.M. Dings
Department of Biochemistry, Molecular Biology & Biophysics
University of Minnesota
Minneapolis, MN 55455, USA

Hans-Joachim Gabius
Institut fur Physiologische Chemie
Ludwig-Maximilians-Universität
Tierärztliche Fakultät
Veterinärstr. 13
D-80539 München, Germany

Vladislav V. Glinsky
Department of Biochemistry
University of Missouri
Columbia, MO 65211, USA

Kevin C. Haudek
Department of Biochemistry and
Molecular Biology
Michigan State University
East Lansing, MI 48824
USA

Neil C. Henderson
The Queen's Medical Research Institute
47 Little France Crescent
Edinburgh EH16 4TJ
Scotland, UK

Anna Hoffmann
Department of Neurology
Medical School
University of Göttingen
Robert Koch Strasse 40
D-37099 Göttingen
Germany

Daniel K. Hsu
Department of Dermatology
University of California, Davis
School of Medicine
3301 C Street, Suite 1400
Sacramento, CA 95816
USA

Robert Kiss
Laboratory of Toxicology
Institute of Pharmacy
Université Libre de Bruxelles
B-1050 Brussels
Belgium

Anatole A. Klyosov
Pro-Pharmaceuticals, Inc.
7 Wells Avenue
Newton, MA 02459
USA

Florence Lefranc
Laboratory of Toxicology
Institute of Pharmacy
Université Libre de Bruxelles
B-1050 Brussels, Belgium;
Department of Neurosurgery
Erasme University Hospital
Université Libre de Bruxelles
Brussels, Belgium

Marie Le Mercier
Laboratory of Toxicology
Institute of Pharmacy
Université Libre de Bruxelles
B-1050 Brussels
Belgium

Fu-Tong Liu
Department of Dermatology
University of California, Davis
School of Medicine
3301 C Street, Suite 1400
Sacramento, CA 95816
USA

Véronique Mathieu
Laboratory of Toxicology
Institute of Pharmacy
Université Libre de Bruxelles
B-1050 Brussels
Belgium

Thomas P. Mawhinney
Department of Biochemistry
University of Missouri
Columbia, MO 65211
USA

Kevin H. Mayo
Department of Biochemistry, Molecular
Biology & Biophysics
University of Minnesota
Minneapolis, MN 55455, USA

Valeri V. Mossine
Department of Biochemistry
University of Missouri
Agriculture Bldg, Rm 4
Columbia, MO 65211, USA

Irina V. Nesmelova
Department of Biochemistry, Molecular
Biology & Biophysics
University of Minnesota
Minneapolis, MN 55455, USA

Ronald J. Patterson
Department of Microbiology
and Molecular Genetics
Michigan State University
East Lansing, MI 48824, USA

Gabriel A. Rabinovich
Laboratory of Immunopathology
Institute of Biology
and Experimental Medicine
IBYME/CONICET
Argentina

Jun Saegusa
Department of Dermatology
University of California, Davis
School of Medicine
3301 C Street, Suite 1400
Sacramento, CA 95816, USA

Sachiko Sato
Glycobiology Laboratory
Research Centre for Infectious Diseases
Faculty of Medicine
Laval University
CHUL Research Centre (CHUQ)
2705 boul, Laurier, Ste-Foy
Québec, Canada G1V 4G2

Tariq Sethi
MRC Centre for Inflammation
Research, The Queen's Medical
Research Institute
University of Edinburgh Medical
School
Royal Infirmary
51 Little France Crescent
Old Dalkieth Road
Edinburgh EH16 4SA
UK

Herwig M. Strik
Department of Neurology
Medical School
University of Göttingen
Robert Koch Strasse 40
D-37099 Göttingen
Germany

Patricia G. Voss
Department of Biochemistry and
Molecular Biology
Michigan State University
East Lansing, MI 48824
USA

John L. Wang
Department of Biochemistry and
Molecular Biology
Michigan State University
East Lansing, MI 48824, USA

Albert M. Wu
Glyco-Immunochemistry
Research Laboratory
Institute of Molecular and Cellular
Biology
Chang-Gung University
Kwei-san, Tao-yuan 333
Taiwan

Ri-Yao Yang
Department of Dermatology
University of California, Davis
School of Medicine
3301 C Street, Suite 1400
Sacramento, CA 95816, USA

1

STUMBLING ON GALECTINS

SAMUEL H. BARONDES
Center for Neurobiology and Psychiatry, UCSF, 401 Parnassus Ave, San Francisco, CA 94143-0984, USA

INTRODUCTION

This symposium about galectins has given me the opportunity to look back on the early research that led to their discovery and to take pleasure in reviewing the many fascinating developments since I moved on to other things. My interest in carbohydrate-binding proteins grew out of speculations about the possible role of protein–glycoconjugate interactions in the formation of specific synaptic connections in the brain (1), which explains how I—a psychiatrist and neuroscientist—stumbled into this field.

I first became aware of the potential biological importance of protein–glycoconjugate interactions in conversations with Victor Ginsburg, whom I met in 1961 at the National Institutes of Health in Bethesda, Maryland. I was, at the time, a post-doctoral fellow in Marshall Nirenbergs's lab and had the good fortune to participate in his studies of synthetic polynucleotides and the genetic code that soon brought him a Nobel prize. Vic, who was in the lab next door, was convinced that the next important code to solve was the one that determined how cells selectively adhere to each other to form complex tissues, and that protein–glycoconjugate interactions must play a role. He based his argument on the presence of complex glycoconjugates on cell surfaces and his hunch that their specific structures must have some biological significance. I found this idea very persuasive and tucked it away until I could explore it.

Over the next 10 years, I completed my clinical training in psychiatry, studied axoplasmic transport and the role of brain protein synthesis in memory storage,

Galectins, Edited by Anatole A. Klyosov, Zbigniew J. Witczak, and David Platt

and helped found the new School of Medicine at the University of California in San Diego. My chance to explore Vic's idea came in 1972 when Steven Rosen joined my laboratory as a postdoctoral fellow. While in graduate school at Cornell, Steve began to study cell adhesion in cellular slime molds, and we decided to continue this project in the hope that it would teach us something about synapses in the brain. This led to a program of research that paved the way for my work with galectins. To explain the early history of galectin research, I will begin with our studies of cell adhesion in slime molds.

LECTINS IN CELLULAR SLIME MOLDS

Research on cell adhesion of the cellular slime mold, *Dictyostelium discoideum*, was pioneered by Gunther Gerisch in the 1960s. It was already known that slime mold cells exist as individual amebas when food is available but form multicellular aggregates when food is gone, and Gerisch had raised antisera that block aggregation (2) to search for the relevant adhesion molecules. Rosen then made antisera of his own and tested their reaction with slime mold extracts bound to sheep erythrocytes—a common immunoassay of that period. He was surprised to find that the extracts from aggregating cells agglutinated the sheep erythrocytes without the addition of antibodies, whereas the extracts from the growing cells did not.

When Rosen arrived in my laboratory, we decided to test the possibility that the extract from aggregating slime mold cells agglutinated the erythrocytes by binding to glycoconjugates on their surface, just like lectins such as concanavalin-A. This was, at the time, a wild idea because lectins were supposed to be restricted to plants. Nevertheless, it was easy to test by adding various sugars to see if they blocked agglutination. The very first experiment worked: *N*-acetylgalactosamine, galactose, and lactose blocked agglutination whereas other sugars did not (3).

Our first attempt to purify the active material by gel filtration on Sepharose also gave a dramatic result: none of the protein fractions that came through the column had any agglutination activity. After trying to recombine various fractions to recover the activity, we figured out that the active material must have bound to the galactose residues in the Sepharose column. Realizing that we had unwittingly used Sepharose as a matrix for affinity chromatography, we eluted the bound protein from the column with a lactose solution. When we examined the eluate by gel electrophoresis, we found a major protein that we named discoidin I, and a minor one that we named discoidin II (4, 5).

Subsequent work in my lab established that discoidin I is mainly located in the cytoplasm of aggregating slime mold cells, that it is secreted by an atypical mechanism (6), and that it plays a role in aggregation (7). When the discoidin I gene was sequenced in Rick Firtel's lab (8) at U.C.S.D., we learned that the protein contains an arg-gly-asp (RGD) sequence similar to that found in cell adhesion proteins such as fibronectin. This suggested that the RGD in discoidin I is involved in the formation of aggregates, and we confirmed this in experiments in which we inhibited aggregation of slime mold cells with RGD-containing peptides (7).

Ironically, a functional site (RGD) that had already been discovered in studies with a human protein (fibronectin) helped us understand how aggregation works in our so-called "simple model system," the cellular slime mold.

FINDING GALECTINS

Encouraged by the discovery of the discoidins, Tom Nowak, a graduate student, began looking for agglutinins of erythrocytes in extracts of vertebrate tissues. He started with extracts of mouse brains because of the hope that a lectin like discoidin might play a role in synapse formation. He also tested extracts of embryonic chick muscle because of the possibility that mammalian lectins might be made at a specific developmental stage as they are in slime molds. In initial experiments, Tom found some agglutination activity in both the brain and muscle extracts. But it was not blocked by sugars.

Scientists at the Weizmann Institute had better luck. Also interested in the possible role of lectins in synapses, and encouraged by our work with slime molds (3, 4), Vivian Teichberg and his colleagues published a classic paper in 1975 about a lactose-inhibiting hemagglutinin in extracts of the electric organ of an electric eel. When they applied the extracts to a modified Sepharose column and eluted it with beta-galactosides, they obtained a pure protein that they named electrolectin (9)—the first purified galectin. They also found a similar lectin activity in extracts of several vertebrate tissues including chick muscle, which shares some properties with the electric organ.

Why did Teichberg et al. succeed while we had failed? The answer is that their buffers contained a reducing agent that is essential for maintaining the lectin activity, whereas ours did not. When we subsequently prepared chick muscle extracts in buffers that contained dithiothreitol or beta-mercaptoethanol, we too found a lactose-inhibitable lectin. Furthermore, when we studied chick muscle extracts prepared at various stages of embryogenesis we found that, like discoidin, the chick lectin was developmentally regulated (10): it was virtually undetectable in 8-day-old embryos but abundant by day 16.

Stimulated by the pioneering work of Teichberg et al., several groups began purifying related lectins from other vertebrate tissues. The first, which we now know as galectin-1, was purified in 1976 in Stuart Kornfeld's laboratory from calf heart and lung (11). This was followed in 1977 by the purification of the lectin from embryonic chick muscle by Halina Den and David Malinzack (12), and in my lab (13). Like electrolectin, those from the other sources were dimers with subunit molecular weights of about 15,000; and in each case, activity was only retained in buffers with sulfhydryl-containing compounds. This led Kurt Drickamer (14) to call these lectins "S-type" (for sulfhydryl-dependent type), a term that was later abandoned in favor of the term galectins when it became clear that other members of this family retain activity in the absence of reducing agents.

The first evidence of other members of this family came in 1980 from Eric Beyer, a student in my laboratory, who found an abundant lectin in chick intestine with a subunit molecular weight of 14,000 and other differences from the muscle lectin.

To distinguish them, we named the one from muscle chicken-lactose-lectin I (CLL-I), and the one from intestine chicken-lactose-lectin II (CL-II) (15).

Further evidence that there are multiple members of this family came in the 1980s from work on mammals. First, a group led by John Wang showed that cultured mammalian fibroblasts have three lectins and went on to purify the one now known as galectin-3 from mouse lung (16, 17). Then, Robert Cerra, Michael Gitt, and I purified three lectins from rat lung, now known to be galectin-1, -3, and -5 (18, 19). Later, Hakon Leffler, Frank Masiarz, and I found yet another member in extracts of rat intestine, now known to be galectin-4 (20).

Subsequent discovery of new galectins was greatly facilitated by cloning and sequencing of cDNA and genomic DNA. In the first attempt that Michael Gitt and I made to clone the cDNA for galectin-1 from a human hepatoma cDNA expression library, we found a related cDNA for a hitherto unknown galectin, now called galectin-2 (21). Many of the other galectins discussed in this symposium were initially identified on the basis of nucleotide sequencing and were later confirmed as galectins by demonstrating that the purified proteins bind beta-galactosides. The availability of cDNAs for galectins also made possible the synthesis of large quantities of recombinant protein for crystallography. In fact, the conserved structure of the carbohydrate-binding domain of galectins was first identified in Jim Rini's lab in 1993 (22, 23) by studying crystals of recombinant galectin-2 that we made with Gitt's cDNA.

NAMING GALECTINS

As this information accumulated, it became clear that a standard nomenclature was needed. In 1994, I contacted Kurt Drickamer, who agreed that "S-type lectins" was no longer appropriate. I then contacted the other main workers in the field and we all decided to give up the various names we were using in favor of the general term galectin. In the joint statement we published in *Cell* (24), we indicated that membership in the galectin family required "fulfillment of two criteria: affinity for beta-galactosides and significant sequence similarity in the carbohydrate-binding site, the relevant amino acid residues of which have been determined by X-ray crystallography." For mammalian galectins, numbers were assigned for those then known and continued to be assigned for new ones (e.g., (25, 26)) as they were discovered: there were 4 when we agreed on these criteria and now there is a pretty good agreement that there are at least 15. Because Gitt and I had already worked with a nomenclature committee to name the galectin-1 gene LGALS-1 and the galectin-2 gene LGALS-2 (27–29), the designation LGALS for "lectin, galactoside-binding, soluble" was used for all the others as they were identified.

LOOKING BACKWARD AND FORWARD

When Steve Rosen and I discovered discoidin I in *Dictyostelium discoideum*, we took this as a support for the general idea that protein–carbohydrate binding plays a role in cell–cell and cell–matrix interactions during tissue development not only in slime molds but also in vertebrates. Looking back at discoidin I, it is remarkable how many molecular and biological properties it shares with galectins such as galectin-1: (a) they both bind galactosides but with relatively low affinity; (b) they are both developmentally regulated, often rising to very high levels at a particular stage of development; (c) they are both concentrated in the cytosol and may constitute as much as 1% of the soluble cytosolic proteins; (d) like other cytosolic proteins they both lack a signal peptide; (e) despite the absence of a signal peptide they both are secreted onto the cell surface and into the extracellular matrix by atypical secretory mechanisms (6, 30–33); (f) they both influence cell–matrix interactions and cell migration. These striking similarities suggest that soluble lectins (34) fill a special biological niche that has remained important through vast periods of cellular evolution (35, 36).

Another remarkable thing about discoidin I is that since 1993 many human proteins that contain discoidin domains have been identified. The first to be discovered (37), named discoidin domain receptor-1 (DDR-1), is an integral membrane protein with an extracellular N-terminal domain that resembles discoidin I. It also has a tyrosine kinase domain that is activated when specific ligands such as collagen bind the discoidin domain, and this interaction activates intracellular signaling pathways (38, 39).

The discovery of discoidin domain proteins is of personal interest to me because several have important roles in the brain. For example, DDR-1 has been implicated in axon migration (40), and neurexin, another discoidin domain protein, has been found in synapses where it interacts with neurologins (41). Even more tantalizing is the finding that patients with rare forms of autism have mutations in neurexins and neuroligins (41). Having been teased by some of my psychiatric colleagues when I devoted so much time to discoidin I, I find it amusing that this slime mold protein is related to a human one that appears to play a role in a mental disorder.

Also of interest to psychiatrists and neuroscientists is the identification of galectin-1 and -3 in specific populations of neurons (42–44). But most galectin research is now focused elsewhere, and their role in the nervous system is not yet well established.

In the 10 years since I left galectin research, and turned my attention to psychiatric genetics and psychopharmacology, there has been a great deal of progress, as described in the course of this volume. One idea that continues to guide this work is that galectins influence cell adhesion, cell signaling, and other functions by interaction with glycoconjugates on and around cells. But, as you will see, there is a growing awareness that these proteins also have other domains (45), have important effects inside the cells that make them, and have been adapted for functions that no one imagined at the time they were discovered.

REFERENCES

1. Barondes SH. Brain glycomacromolecules and interneuronal recognition. In:Schmitt FO, editor. The Neurosciences: A Second Study Program. New York: Rockefeller University Press;1970. p747–760.

2. Beug H, Gerisch G, Muller E. Cell dissociation: univalent antibodies as a possible alternative to proteolytic enzymes. Science 1971;173:742–743.
3. Rosen SD, Kafka JA, Simpson DL, Barondes SH. Developmentally regulated, carbohydrate-binding protein in *Dictyostelium discoideum*. Proc Natl Acad Sci USA 1973;70:2554–2557.
4. Simpson DL, Rosen SD, Barondes SH. Discoidin, a developmentally regulated carbohydrate-binding protein from *Dictyostelium discoideum*: purification and characterization. Biochemistry 1974;13:3487–3493.
5. Frazier WA, Rosen SD, Reitherman RW, Barondes SH. Purification and comparison of two developmentally regulated lectins from *Dictyostelium discoideum*: discoidin I and II. J Biol Chem 1975;250:7714–7721.
6. Barondes SH, Haywood-Reid PL, Cooper DNW. Discoidin I, an endogenous lectin, is externalized from *Dictyostelium discoideum* in multilamellar bodies. J Cell Biol 1985;100:1825–1833.
7. Springer WR, Cooper DNW, Barondes SH. Discoidin I is implicated in cell–substratum attachment and ordered cell migration of *Dictyostelium discoideum* and resembles fibronectin. Cell 1984;39:557–564.
8. Poole S, Firtel RA, Lamar E, Rowekamp W. Sequence and expression of the discoidin I gene family in *Dictyostelium discoideum*. J Mol Biol 1981;153:273–289.
9. Teichberg VI, Silman I, Beisch DD, Resheff G. A beta-D-galactoside binding protein from electric organ tissue of *Electrophorus electricus*. Proc Natl Acad Sci USA 1975;72: 1383–1387.
10. Nowak TP, Haywood PL, Barondes SH. Developmentally regulated lectin in embryonic chick muscle and a myogenic cell line. Biochem Biophys Res Commun 1976;68: 650–657.
11. De Waard A, Hickman S, Kornfeld S. Isolation and properties of beta-galactoside binding lectins of calf heart and lung. J Biol Chem 1976;251:7581–7587.
12. Den H, Malinzak DA. Isolation and properties of beta-D-galactoside-specific lectin from chick embryo thigh muscle. J Biol Chem 1977;252:5444–5448.
13. Nowak TP, Kobiler D, Roel L, Barondes SH. Developmentally regulated lectin from embryonic chick pectoral muscle: purification by affinity chromatography. J Biol Chem 1977;252:6026–6030.
14. Drickamer K. Two distinct carbohydrate-recognition domains in animal lectins. J Biol Chem 1988;263:9557–9560.
15. Beyer EC, Zweig SE, Barondes SH. Two lactose binding lectins from chicken: purified lectin from intestine is different from those in liver and muscle. J Biol Chem 1980;255: 4236–4239.
16. Roff CF, Wang JL. Endogenous lectins from cultured cells. Isolation and characterization of carbohydrate-binding proteins from 3T3 fibroblasts. J Biol Chem 1983;258: 10657–10663.
17. Crittenden SL, Roff CF, Wang JL. Carbohydrate-binding protein 35: identification of the galactose-specific lectin in various mouse tissues. Mol Cell Biol 1984;4: 1252–1259.
18. Cerra RF, Gitt MA, Barondes SH. Three soluble rat β-galactoside-binding lectins. J Biol Chem 1985;260:10474–10477.

19. Leffler H, Barondes SH. Specificity of binding of three soluble rat lung lectins to substituted and unsubstituted mammalian β-galactosides. J Biol Chem 1986;261: 10119–10126.
20. Leffler H, Masiarz FR, Barondes SH. Soluble lactose-binding vertebrate lectins: a growing family. Biochemistry 1989;28:9222–9229.
21. Gitt MA, Barondes SH. Evidence that a human soluble β-galactoside-binding lectin is encoded by a family of genes. Proc Natl Acad Sci USA 1986;83:7603–7607.
22. Lobsanov YD, Gitt MA, Leffler H, Barondes SH, Rini JM. Crystallization and preliminary X-ray diffraction analysis of the human dimeric S-Lac lectin (L-14-II). J Mol Biol 1993;233:553–555.
23. Lobsanov YD, Gitt MA, Leffler H, Barondes SH, Rini JM. X-ray crystal structure of the human dimeric S-Lac lectin, L-14-II, in complex with lactose at 2.9 Å resolution. J Biol Chem 1993;268:27034–27038.
24. Barondes SH, Castronovo V, Cooper DNW, Cummings RD, Drickamer K, Feizi T, Gitt MA, Hirabayashi J, Hughes C, Ken-ichi K, Leffler H, Liu FT, Lotan R, Mercurio AM, Monsigny M, Pillai S, Poirer F, Raz A, Rigby PWJ, Rini JM, Wang JL. Galectins: a family of animal β-galactoside-binding lectins. Cell 1994;76:597–598.
25. Gitt MA, Wiser MF, Leffler H, Herrmann J, Xia Y, Massa SM, Cooper DNW, Lusis AJ, Barondes SH. Galectin-5: sequence and mapping of a β-galactoside-binding lectin found in rat erythrocytes. J Biol Chem 1995;270:5032–5038.
26. Gitt MA, Xia Y, Atchison RE, Lusis AJ, Barondes SH, Leffler H. Sequence, structure, and chromosomal mapping of the mouse *Lgals6* gene, encoding galectin-6. J Biol Chem 1998;273:2961–2970.
27. Gitt MA, Massa SM, Leffler H, Barondes SH. Isolation and expression of a gene encoding L-14-II, a new human soluble lactose-binding lectin. J Biol Chem 1992;267: 10601–10606.
28. Gitt MA, Barondes SH. Genomic sequence and organization of two members of a human lectin gene family. Biochemistry 1991;30:82–89.
29. Mehrabian M, Gitt MA, Sparkes RS, Leffler H, Barondes SH, Lusis AJ. Two members of the S-Lac lectin gene family, *LGALS1* and *LGALS2*, reside in close proximity on human chromosome 22q12–q13. Genomics 1993;15:418–420.
30. Barondes SH, Haywood-Reid PL. Externalization of an endogenous chicken muscle lectin with *in vivo* development. J Cell Biol 1981;91:568–572.
31. Beyer EC, Barondes SH. Secretion of endogenous lectin by chicken intestinal goblet cells. J Cell Biol 1982;92:28–33.
32. Cooper DNW, Barondes SH. Evidence for export of a muscle lectin from cytosol to extracellular matrix and for a novel secretary mechanism. J Cell Biol 1990;110: 1681–1691.
33. Lindstedt R, Apodaca G, Barondes SH, Mostov KE, Leffler H. Apical secretion of a cytosolic protein by Madin–Darby canine kidney cells. J Biol Chem 1993;268: 11750–11757.
34. Barondes SH. Soluble lectins: a new class of extracellular proteins. Science 1984;223: 1259–1264.
35. Barondes SH, Cooper DNW, Gitt MA, Leffler H. Galectins: structure and function of a large family of animal lectins. J Biol Chem 1994;269:20807–20810.

36. Cooper DNW, Barondes SH. God must love galectins; He made so many of them. Glycobiology 1999;9:979–984.
37. Johnson JD, Edman JC, Rutter WJ. A receptor tyrosine kinase found in breast carcinoma cells has an extracellular discoidin-I domain. Proc Natl Acad Sci USA 1993;90: 5677–5681.
38. Vogel W. Discoidin domain receptors: structural relations and functional implications. FASEB J 1999;13 (Suppl):577–582.
39. Kiedzierska A, Smietana K, Czepczynska H, Otlewski J. Structural similarities and functional diversity of eukaryotic discoidin-like domains. Biochim Biophys Acta 2007; 1774:1069–1078.
40. Bhatt RS, Tomoda T, Fang Y, Hatten ME. Discoidin domain receptor 1 functions in axon extension of cerebellar granule neurons. Genes Dev 2000;14:2216–2228.
41. Garber K. Autism's cause may reside in abnormalities at the synapse. Science 2007;317:190–191.
42. Regan LJ, Dodd J, Barondes SH, Jessell TM. Selective expression of endogenous lactose-binding lectins and lactoseries glycoconjugates in subsets of rat sensory neurons. Proc Natl Acad Sci USA 1986;83:2248–2252.
43. Hynes MA, Buck LB, Gitt M, Barondes S, Dodd J, Jessell TM. Carbohydrate recognition in neuronal development: structure and expression of surface oligosaccharides and β-galactoside-binding lectins. In:Carbohydrate Recognition in Cellular Function. CIBA Foundation Symposium 145. Chichester, UK: John Wiley & Sons, Ltd.;1989. p189–218.
44. Hynes MA, Gitt MA, Barondes SH, Jessell TM, Buck LB. Selective expression of an endogenous lactose-binding lectin gene in subsets of central and peripheral neurons. J Neurosci 1990;10:1004–1013.
45. Massa SM, Cooper DNW, Leffler H, Barondes SH. L-29, an endogenous lectin, binds to glycoconjugate ligands with positive cooperativity. Biochemistry 1993;32:260–267.

2

GALECTINS AND THEIR FUNCTIONS IN PLAIN LANGUAGE

Anatole A. Klyosov
Pro-Pharmaceuticals, Inc., 7 Wells Avenue, Newton, MA 02459, USA

This is the farmer sowing his corn,
That kept the cock that crowed in the morn,
That waked the priest all shaven and shorn,
That married the man all tattered and torn,
That kissed the maiden all forlorn,
That milked the cow with the crumpled horn,
That tossed the dog,
That worried the cat,
That killed the rat,
That ate the malt,
That lay in the house that Jack built.

INTRODUCTION

Galectins are galactose-binding lectins. Lectins are proteins or glycoproteins (glycosylated proteins) possessing a relatively high affinity to specific (for a given lectin) carbohydrate moieties and triggering—upon the tight, specific binding—certain biological functions.

A number of confusions start right here. Or, if not outright confusions, then things that need clarifications.

Galectins, Edited by Anatole A. Klyosov, Zbigniew J. Witczak, and David Platt

LECTINS VERSUS ENZYMES

The first item concerns lectins themselves. A number of definitions of lectins emphasize that lectins are "proteins except enzymes and antibodies." However, there is no reasonable ground to dismiss enzymatic functions of lectins, except that earlier studies of lectins have not revealed their enzymatic activities. In other words, known lectins have not catalyzed chemical conversions of their ligands. They rather bind to their ligands, forming with them multivalent (i.e., multipoint, multicontact) noncovalent complexes, which, as a result, precipitated (with glycoconjugates) or agglutinated (with human erythrocytes), if to use historical terms. The ligands in the precipitated complexes remained chemically intact and could be recovered by the following dissolution, for example, in detergents.

Hence, by definition, enzymes chemically modify their ligands (substrates), while lectins do not chemically modify their ligands. However, lectins still can be enzymes with respect to other ligands or substrates. For example, the earliest (1860) identified lectin, ricin, turned out to be an enzyme, RNA-N-glycosidase, and cleaves the specific glycosidic bond of an adenine residue of rat liver 28S rRNA. That is why ricin is an RIP (ribosome-inactivating protein) and inhibits cellular protein synthesis (1, 2).

To be exact, enzymes also do not modify their ligands, because very often, if not always, a reactive bond in their substrate is just adjacent to the ligand, which in fact interacts with the enzyme binding site. For example, in trypsin the ligand is an electrostatically charged lysine or arginine side chain, while the enzyme splits an ester or an amido group. In chymotrypsin, the ligand is a hydrophobic tyrosine or tryptophan side chain, which the enzyme does not modify. In an alcoholdehydrogenase, the ligand is an alcohol primary (typically) hydrocarbon group, not an alcohol itself, which is oxidized by the enzyme. Hence, it would be proper to say that enzymes chemically modify their substrates, which typically represent a combination of a ligand and a reactive group chemically attached to it.

Another example of a lectin possibly possessing an enzymatic activity is galectin-10, which was initially found to be lysophospholipase, or 2-lysophosphatidylcholine acylhydrolase, catalyzing hydrolysis of the ester bond in 2-lysophosphatidylcholine into glycerophosphocholine with a liberation of a free acid. An enzymatic activity of galectin-10 was questioned, though (3). However, the same enzymatic activity of galectin-13 was confirmed not only for the purified galectin, but also for its bacterially expressed variant (4).

Four more additional galectin candidates, analogues of galectin-10, have been suggested using the GenBank databases, and all four were suggested to be also candidates for lysophospholipase activity (5). It remains to be verified. From these four, at least one was isolated and named galectin-13. It had a weak lysophospholipase activity (4).

Therefore, a current view on lectins does not deny their enzymatic activity. A lectin can tightly bind its specific ligand that in turn initiates (or further conveys) a signal resulting in a biological function, and, in some cases, the same lectin can catalyze a chemical conversion of another substrate, taking part in the same or a separate biological function.

Overall, it might be a bit superficial to talk on weak (and disputable) enzymatic activities of some lectins; however, these data are mentioned here just to show that lectins might be more complex systems then we typically think about them.

WHAT ARE GALECTINS, ANYWAY?

This is a galectin,
That is a lectin,
That is a protein,
That may be a glycoprotein,
That often is obtained in a recombinant form,
That often does not have a carbohydrate component,
That may be important for a possible biological function,
That may include—or start with—recognition, binding specificity, etc.,
That makes us wonder if we lose some critical information by studying recombinant galectins.

All right, let us start again:

This is a galectin,
That is a lectin,
That does not chemically split or covalently fuse its ligands,
That can be both carbohydrates and glycoproteins,
That often situate on surface of a cell,
That therefore provides a network of ligands to the galectin,
That enters into noncovalent cross-linking with its ligands,
That often aggregates the whole formation of cells along with galectins,
That therefore sends a mysterious "biological signal" to other components,
That are typically unknown,
That makes us wonder what is a biological meaning of a job that galectins do.

There are 15 galectins in mammals identified and isolated up to date.
Common things between different galectins are

- Carbohydrate recognition domains (CRD), consisting of about 130 amino acids, arranged in a tightly folded structure (so-called β-sheets).
- Binding of β-galactose-containing carbohydrates and glycoconjugates, though this criterion nowadays is not an absolute one, particularly when polymeric ligands are considered.
- Absence of the signal sequence required for protein secretion through the usual secretory pathway (though galectins manage to get secreted into the extracellular space anyhow, apparently via some "unorthodox pathway").
- A very old origin; it is believed that galectins must have evolved before the split of organisms to the Kingdoms, because there are many galectin-like sequences

in genomes of organisms from all five Kingdoms: animal, plant, fungi, protists, and (apparently) prokaryotes.

- Galectins are found in a variety of cell types, including fibroblasts, ovary cells, epithelial cells, endothelial cells, dendritic cells, macrophages, bone marrow cells, T and B cells, and so on.

Besides, common things about galectins are plenty of conjectures regarding their important biological functions and mechanisms of action of galectins in terms of "signal transduction." Apparently, not in a single case that "signal transduction" was completely deciphered and relayed to the final biological function; however, nobody (including the author) has any doubt that galectins are extremely important for something related to many pathologies and their either development or cure. Or both, in different circumstances. It should be mentioned here, however, that in the case of induction of apoptosis in T cells by galectin-1, the signaling pathway engaged by the lectin is partially elucidated (6–10). The same can be said on the induction of p21 transcription and selective increasing of p27 protein stability as a result of interaction of galectin-1 with the $\alpha 5\beta 1$ fibronectin receptor (integrin), which in turn restricts carcinoma cell growth. The antiproliferative effect results from inhibition of the canonical Ras-MEK-ERK cascade (ERK—extracellular signal-regulated kinase, MEK—mitogen-activated protein kinase/ERK kinase, Ras—an oncogene) by galectin-1 and consecutive transcriptional induction of p27 (11).

Different things between different galectins are

- Some of them contain only one CRD (carbohydrate recognition domain) and are biologically active as monomers (galectin-5 and -10).
- Some of them contain only one CRD and are biologically active as homodimers (galectin-1, -2, -11, -13, -14, and -15). Some of them are dimerized by noncovalent interactions, some (as galectin-13) are held together by disulfide bonds (4). Galectin-1 from Chinese hamster ovary cells can exist in a reversible monomer–dimer equilibrium (12); however, it is not known whether other galectin-1 proteins exist in a similar equilibrium (with a dissociation constant of 7 μM in one example (13)) or behave differently.

  The above two types of galectins, containing one CRD, are called "prototype galectins." Evolutionary, they seem to be quite distinct from each other (14).
- One galectin found so far (galectin-3) contains a "nonlectin domain," through which it aggregates into oligomers, and a single CRD, fused to an N-terminal sequence. The N-terminal domain is a polypeptide of about 100–150 amino acids, made of tandem repeats of short proline–glycine rich stretches of amino acids. This galectin is called "chimera type galectin."
- Some of them contain two CRDs, connected by a short linker peptide (galectin-4, -6, -8, -9, and -12). These five galectins are called "tandem-repeat type galectins."
- They show different affinities toward simple carbohydrates, such as lactose (Galβ4Glc, or (β-D-galactosido)-D-glucose) and *N*-acetyllactosamine

(Gal-β1,4-GlcNAc), and more so toward complex carbohydrates and glycoconjugates, though it is not generally known which affinity is related to their primary biological functions, and which prevents those functions. Galectin-1, for example, shows preference in binding with *N*-acetyllactosamine, while others often show different preferences.

- Some differences exist even between the same galectins but from different sources. For example, human galectin-1 has a different binding preference compared to bovine galectin-1 toward long chains of poly-*N*-acetyllactosamine (15). Human galectin-1 does not bind to poly-*N*-acetyllactosamine sequences terminating in nonreducing GlcNAc residue, while bovine galectin-1 does. At the same time, human galectin-1 shows a preference to polyacetyllactosamine with terminal Galβ4-GlcNAc (15).

Galectins are widely distributed throughout the animal kingdom (mammals, fishes, frogs, etc.) and are expressed in a variety of organs and tissues such as the central and peripheral nervous systems, lung, liver, and intestinal system. Some galectins have more limited tissue distribution, though. Galectins have been shown (or are thought otherwise) to play a number of important roles in cell–cell and cell–matrix interactions, neuronal cell differentiation and survival, embryonic development, growth and development, malignant progression, metastasis, cell adhesion, angiogenesis, proliferation, anoikis/apoptosis, pre-RNA splicing, and other crucial functions.

However, removing of galectins from experimental animals (and resulting in Gal-1−/− null mutants) left the animals viable and they could grow into adults without any obvious developmental or morphological abnormalities (except some minor apparent aberrations). It has been noticed, though, that upon certain challenges they demonstrated phenotypes, such as reduced inflammatory response (16–19).

It is commonly said that "galectins have different ligand specificities," though it is commonly not defined what is "specificity" in this context. Unlike "affinity," "specificity" is supposed to be related to biological activity, which is usually not known. These questions will be considered below.

"NEAT" LECTINS VERSUS GLYCOSYLATED LECTINS

There are very few, if any data pointing out at the role of glycosylated parts of lectins in their specificity of binding and signaling. There are very few data on glycosylated parts of lectins in the first place.

One example is a human HL-60 lectin, specific for β-D-galactosides and having a molecular weight of 14 kDa in a recombinant form and 17–20 kDa in a glycosylated form (20). A glycosyl part of 3–6 kDa can be approximated with 18–35 sugar residues per lectin molecule.

Another example is lectin CBP70, isolated from the same HL60 and having an affinity for *N*-acetylglucosamine. It has different types of glycosylation in its cytoplasmic and nuclear forms, but the protein part is identical (21, 22). Besides, both have different binding specificities: the nuclear form of the lectin preferentially binds with galectin-3, and the cytoplasmic form of the lectin preferentially binds with Bcl-2 (23).

Yet another example of a different pattern of glycosylation of lectins in their cytoplasmic and nuclear forms is HSP70 lectin (HSP stands for heat shock protein), a GlcNAc-binding lectin from both the cytosol and the nucleus, or a "nucleocytosolic" lectin (24–28).

It might well be that glycosylation of a lectin changes its conformation, ligand specificity, affinity, mode of contacts with ligands, and so on. These questions have not apparently been raised in the literature as yet, except maybe in Reference 22. Furthermore, specificity and apparent biological function(s) of galectins are commonly studied using recombinant galectins, which likely do not have their glycosyl moieties, particularly when expressed in *E. coli* (a common procedure for galectins). In this context, the suggestion (22) that a different glycosylation pattern of a lectin might act as a different complementary signal for cellular targeting calls for a potential reconsideration of data obtained for recombinant galectins, as the native forms of at least some galectins contain glycosyl moieties.

There are some indications that galectins are characterized by the "absence of glycosylation" (29); however, the generality of such a statement is not apparently proven as yet. Galectin-13, isolated from human placenta and purified, contained 0.6% of carbohydrate content (4, 30), which corresponds to a bound monosugar residue.

"CROSS-LINKING" OF GALECTINS WITH THEIR SPECIFIC LIGANDS

"Cross-linking" in general chemistry, as well as in polymer sciences, protein chemistry, and enzymology, to name just a few, is related to covalent intra- or intermolecular bonds. In lectinology, "cross-linking" is related to noncovalent interactions between a lectin and its specific ligands. Some definitions of lectins include statements such as they "specifically cross-link carbohydrates." "Cross-linking" in lectinology results in aggregation, precipitation, agglutination of lectins along with their ligands and attached to them high molecular formations (cells, proteins, polysaccharides, and glycoconjugates).

What the "cross-linking" does, besides physically removing the ligands and attached to them formations from the "reaction zone"—it is not known. However, conjectures abound.

Lectins in general and galectins in particular are described as having multivalent binding properties. It is essentially referred to their two carbohydrate recognition domains, hence, "bivalent" interactions with their ligands. Again, "bivalent" here has nothing to do with covalent bonds. It means two noncovalent binding sites, each one can provide several contact points (hydrogen bonds, electrostatic interactions, and hydrophobic interactions). The resulting noncovalent complex, or formation, exceeds limits of its water solubility, or an ability to be a stable suspension, and precipitates, coagulates, and aggregates.

A known example is an aggregation of galectin-1 with the human melanoma cells (line A375) through binding to 90K/MAC-2BP glycoprotein. The latter is a tumor-secreted antigen in the culture medium of human breast cancer cells. Each subunit of this glycoprotein contains a number of cysteins and N-glycosylation sites. Since galectin-1 binds to 90K in a dose-dependent manner, and the binding is blocked by lactose, it suggests that the galectin recognizes carbohydrate moieties of 90K (31). It is interesting that the authors described this interaction as the galectin "induces" cell aggregation or "mediates" the aggregation (see also Reference 29), while the galectin apparently just takes part in it. The galectin is an active participant in the aggregation, not a "mediator" or an "inducer," so it might be looked upon in different ways. It cannot be excluded, however, that the galectin can activate cells as well as participate in cell aggregation (32). So, it may not just bridge cells.

Galectin-1 plays the same role as 90K glycoprotein, namely, the two form a complex. Apparently, galectin-1, having two carbohydrate recognition domains, binds to 90K displayed on surface of two or several cells, or to 90K and another cell-surface high molecular ligand.

It is of certain interest that not only galectin-1, but galectin-3 also aggregates along with the same 90K-carrying human melanoma cells A375 (33). The authors have shown that human recombinant galectin-3 bound immobilized 90K, and this binding was also inhibited by lactose. Since galectin-1 and -3 did not affect the binding of each other, they appeared to bind to separate sites on 90K (31). It was evaluated that the binding constant (in the dimension of the dissociation constant) of galectin-1 to 90K was on the order of 100 nM, which is reasonably tight. It can be compared with a concentration of galectin-1, which decreases human SK-N-MC neuroblastoma cells number by 50% after treatment and 48-h incubation with the galectin (34). This concentration was approximately equal to 50 μg/mL, that is 1.7 μM, or 17 times less effective, in terms of strength of binding and/or precipitation.

Galectin-13 (from human term placenta) agglutinates human erythrocytes (4) at small amounts of the galectin (strong agglutination is detected at and above 50 μg/mL). Some other galectins were also observed to be potent as agglutinating agents. *N*-Acetyllactosamine at and above concentrations of 1 mM abolished the agglutination activity of galectin-13 (4).

What is an alleged structure of a network of linkages between a galectin and its ligands when they are precipitating?

In fact, that network typically involves multicell or other multiligand aggregates along with a number of galectin molecules linked to their ligands on cell surface. In principle, those aggregates can be formed between bivalent (or oligovalent) galectin molecules and bivalent (or oligovalent) carbohydrate-carrying ligands, such as glycoproteins and glycolipids.

In a simple case of large complex formations, a bivalent lectin forms a linear, "one-dimensional" chain such as

...–lectin–ligand–lectin–ligand–lectin–ligand–...

In principle, other lectins (with a similar specificity) and other ligands (of a similar structure) can build into such a chain. If a lectin or a ligand (or both) possesses a valency greater than 2, they can form two-dimensional (e.g., on a cell surface) or three-dimensional linked formations that commonly precipitate from solution.

Such complexes and their experimental models were considered in Reference (35). For example, "biantennary" 2,3-, 2,4-, 3,6-, and 2,6-pentasaccharides formed 3D complexes with tetrameric soybean agglutinin (SBA), in which one tetramer was linked by four pentasaccharide molecules to four other SBA monomers that in turn were parts of adjacent tetramers (35).

By using a series of synthetic analogues of multivalent (divalent) carbohydrates, containing two LacNAc residues separated by a various number of methylene groups, as well as biantennary analogs possessing two LacNAc residues, Ahmad et al. (36, 37) have shown that each bivalent analogue bound to the two galectins, galectin-1 or galectin-3, though in a differed manner. Galectin-1 was indifferent to length of the straight methylene chain and branched chain analogues, while galectin-3 showed the highest affinity for lacto-*N*-hexaose, a naturally occurring branched chain carbohydrate (36). Furthermore, galectin-3 precipitated as a pentamer with a series of divalent pentasaccharides with terminal LacNAc residues (37).

Apparently, in the same manner galectins form "cross-linked" complexes with specific glycoproteins. For example, galectin-1 forms 3D complexes with asialofetuin, a 48-kDa monomeric glycoprotein possessing three triantennary N-linked complex carbohydrates with terminal galactose residues (38, 39). By doing so, galectins may "cross-link" cell surface receptors, carrying glycoligands specific for the galectins.

There is a belief, common in the literature, which is based, however, mainly on conjectures, that cross-linking of cell surface receptors is involved in apoptotic signals and other biological signal transductions leading to metastasis, control of cell growth, and control of inflammatory phenomena, though respective biochemical mechanisms are not known as yet. Those conjectures are in turn based on observations such as inhibition of the apoptosis by lactose (a weak ligand of all galectins), the presence of galectin-1 at sites of T-cell death by apoptosis during normal T-cell development (but not in resting T cells), and an observation that cells that express galectin-1 induced apoptosis of T cells (6). Such a possible function of galectin-1 is supported by the observation that several specific T-cell surface glycoproteins, called CD7, CD43, and CD45, are specific receptors for galectin-1 binding, and that binding results in a striking redistribution of these glycoproteins into segregated membrane microdomains on the cell surface, which apparently triggers apoptosis (8).

It is remarkable that before addition of galectin-1, the distribution of CD43 and CD45 were practically random across the cell surface (that was shown using color-labeled antibodies), but after treating the cells with galectin-1 for 20 min, removing the galectin with lactose, and fixing the treated cells and their membrane glycoproteins, CD43 and CD45 were segregated and clustered into patches (8). Similarly, CD7 and CD43 were colocalized and distributed diffusely over the cell surface prior to galectin-1 addition, but coaggregated, clustered, and segregated from CD45, which in turn was colocalized with CD3, following galectin-1 treatment (8). True biological significance of this phenomenon remains to be clarified.

In addition to it, we do not have enough data to conclude that cross-linking by galectins on the cell surface proceeds in the same manner as that in solutions.

BINDING STRENGTH AND SPECIFICITY OF GALECTIN LIGANDS

What makes binding with galectins strong and/or specific?

Strong binding is characterized by a low dissociation constant for a ligand–galectin noncovalent complex (association constants are not in common use in biochemistry; however, strangely enough, dissociation constants in biochemistry are called "binding constants"). What is "low," though? According to a rule of thumb, poor binding constants are those in the range of 1 M to 1 mM, good binding constants are around 1 μM, and excellent, outstanding binding constants are around 1 nM.

To make sense out of these figures, let us consider the simplest binding in the following form:

$$G + L \leftrightarrow GL$$

where G = galectin, L = ligand, and GL = galectin–ligand noncovalent complex.

Dissociation constant for such a complex can be described as

$$K = [G][L]/[GL]$$

where concentrations (molar concentrations in this case) are shown in brackets. When the initial concentration of a ligand is much higher than the initial concentration of a galectin, then concentrations of free ligand are always higher than concentrations of free or bound galectin, and within a reasonable accuracy $[L] = [L]_0$. In these symbols,

$$K = [G][L]_0/[GL]$$

and when a concentration of free galectin [G] is equal to concentration of bound galectin [GL], the following equation holds:

$$K = [L]_0$$

That is, if the binding constant is equal to the initial concentration of the ligand, the galectin is evenly distributed between its free, unbound galectin and the bound galectin.

In other words,

- If only at 1 M (mol/L) of the ligand the galectin is 50% free and 50% bound, the binding constant $K = 1$ M.
- If it takes 10^{-3} M of the ligand to bind 50% of the galectin in a complex, $K = 1$ mM.
- If it takes only 10^{-9} M of the ligand to bind half of the galectin, $K = 1$ nM.

This, of course, is valid when the initial concentration of the ligand is much higher than that of the galectin.

This describes strength of binding. However, strength of the binding does not necessarily reflect specificity of the binding. In enzyme mechanisms, for example, specific substrate is not the one that binds tightly, but the one that leads to the greatest acceleration of the enzymatic catalysis. "Specificity of binding" mobilizes the catalytic machinery of the enzyme in such a way that the enzyme employs it as fully as possible in the given catalytic process. Many substrates bind tightly but result in a poor conversion. It is often called "nonproductive binding" and is not specific by all means.

Specific inhibitors, on the contrary, are those that just bind very tightly and therefore block the enzymatic reaction at low concentration of the inhibitor compared to other inhibitors.

Apparently, specific binding of galectins with their ligands is that which triggers the specific biological signal most effectively, at lowest amounts (concentrations) of the ligand. In some cases, tight ligands can result in nonproductive binding. They can be specific blockers of the galectin's biological function, not specific promoters of it.

Examples of such specific blockers of a noncarbohydrate nature are given in Reference 40. Some synthetic peptides bound to purified recombinant galectin-3 and to "cell surface galectin-3" with binding constants of 70 ± 30 and 90 ± 20 nM, and have been shown to bind to the C-terminal CRD of galectin-3 (but not to the other lectins tested) and to block both the interaction of the galectins with β-galactosides and metastasis-associated tumor cell adhesion.

Another example of a tight blocker of sugar nature was shown to be thiodigalactosides bearing benzamido substituents at the C-3 position of both the sugar units. They inhibited galectin-3 on a nanomolar level, between 33 and 61 nM (41).

An introduction of a large hydrophobic substituent into a carbohydrate ligand for galectin-3 not necessarily leads to a sharp increase in the binding capacity. For example, a substitution of the *N*-acetyl group at the GlcNAc moiety in lacto-*N*-biose (β-D-Gal-(1-3)-β-D-GlcNAc) with the *N*-naphthoyl group resulted in only sevenfold increase of the binding capacity of the new compound (the dissociation constant decreased from 73 to 10.6 μM) (42). However, a difference in hydrophobicity of the two compounds could have resulted in an energy gain of 3.5 kcal/mol; that is, the dissociated constant decreased by 370 times:

$$\Delta F = 0.59 \ln K/K' = 3.5\ \text{kcal/mol}$$

(see below). This could have led to $K = 0.2$ μM (not 10.6 μM), provided that the binding site accommodating the naphthalene group was entirely hydrophobic.

To sum it up, "tight" ligands for galectins can either effectively trigger their biological function or effectively block that function. Without that knowledge, it is rather misleading to call the binding "specific." This area of science on galectins is still awaiting its researchers. Currently, X-ray data on tightly binding ligands, nicely fitting the galectin-binding sites, or data on "specific" binding of galectins with some glycoproteins on cell surface resulting in rearranging of the glycoproteins, or precipitation of galectins by some ligands, all (or some of them) could be indications and examples of "nonproductive" binding, interfering with true biological functions of galectins. At any rate, these effects are indeed not linked as yet, as a rule, with actual biological functions of galectins.

NATURAL AND ARTIFICIAL LIGANDS FOR GALECTINS

Let us consider tightness of binding of galectins to their known ligands, natural or artificial ones.

The most known ligands (of a nonglycoprotein nature) for galectins are lactose, *N*-acetyllactosamine, and poly-*N*-acetyllactosamine. Their chemical structures, in main different notations, or nomenclatures, which are in the use in the literature on galectins, are as follows:

- For lactose:
  - Galβ4Glc
  - Gal-β1,4-Glc
  - (β-D-Galactosido)-D-glucose
  - 4-*O*-β-D-Galactopyranosyl-D-glucose
- For *N*-acetyllactosamine:
  - Gal-β1,4-GlcNAc
- For poly-*N*-acetyllactosamine:
  - Poly-Gal-β1,4-GlcNAc

These three simple β-galactosides and their derivatives cannot be considered as "specific" ligands, because their binding constants are around 1 mM, and qualified for only "poor binding." In fact, for such a poor binding, a ligand can have only one contact with the binding site in the ligand–lectin complex, such as one hydrogen bond. The hydrogen bond energy usually ranges from 1 to 4 kcal/mol, and it will bring an increase in the binding from 5- to 1000-fold (compared to a reference compound not having such a bond). This can be calculated using a formula

$$\Delta F = RT \ln K/K'$$

where ΔF is a free energy difference (in kcal/mol) between two binding modes, characterized by constants K and K', for the same galectin (in this context). If two ligands differ with their binding constants with the same lectin by 1000 times, the above formula gives

$$\Delta F = 0.59 \ln 1000 = 4.08 \text{ kcal/mol}$$

Just one proper electrostatic interaction (not a hydrogen bond) in a ligand–lectin complex with an energy of 4 kcal/mol can also bring a similar 1000-fold increase in the affinity of the ligand to the lectin. A simple hydrophobic interaction of a methyl (CH_3) group or a methylene (CH_2) group in a ligand with a proper hydrophobic site in a lectin can give an energy gain in the complex of 0.65 kcal/mol and improve the binding by three times.

It was suggested that both the hydrogen bonding and hydrophobic interactions are equally important for binding of carbohydrates (exemplified with five Gal(1-4),

Gal(1-3), Glc(1-4) disaccharides, and a Gal(1-3) tetrasaccharide) to galectin-3 (43). Of course, such suggestions are not based, as a rule, on direct observations, but rather aim at interpretations of values of binding constants.

One more comment in this regard. It was noticed that small saccharides such as (β1,3-galactosido)-*N*-acetylglucosamine (Gal-β1,3-GlcNAc), (β1,3-galactosido)-arabinose (Gal-β1,3-Ara), and (β1,4-galactosido)-mannose (Gal-β1,4-Man) inhibit galectins two to four times more effectively than lactose (Gal-β1,4-Glc) (44, 45). In fact, such an increase in the binding potency reflects some small changes in energy of the binding, equal to about 0.4–0.8 kcal/mol, and might correspond to one weak (or significantly distorted) hydrogen bond. As it was noted, "all these small natural ligands have low levels of inhibitory potency *in vitro*. ... Nevertheless, when administered ... every 8 h at 2 mg/g body weight, D-galactose completely inhibits the liver metastasis ... in mice" (45). However, simple calculations show that 2 mg/g of the plain sugar in mice' blood corresponds to about 6 mg/mL, that is 35 mM, which is much higher than the binding constant of 1-methyl-β-D-galactoside equal to 4.4 mM (with respect to galectin-3) (46). This concentration is enough to overcome the "low inhibitory potency" of the inhibitor *in vivo*. In other words, there were no contradictions between the inhibitory effect of the compounds *in vitro* and *in vivo*, as the authors (45) implied.

It was noticed that binding sites for small sugars in galectins are extended and can be hydrophobic in nature. Therefore, they can be suited not only for sugars, but also for hydrophobic molecules, which can be principally instrumental for "biological signals," we do not know about as yet. For example, 1-methyl-*N*-acetyllactosamine (1-Me-Lac-NAc) has the binding constant with galectin-3 of 70 μM, while for its 3-OH Gal-substituted benzamide (C_6H_5-CO-NH-) product the binding constant is 10 times better, 7 μM (47–49). A 10-fold increase in affinity corresponds to 1.4 kcal/mol gain in the energy of binding. This is less than the benzyl group alone could provide by binding with a proper hydrophobic site (about 2.1 kcal/mol) but could be just right considering the accompanying polar (hydrophilic) groups (CO and NH). However, it could be just one or two hydrogen bonds (with CO and/or NH) negatively compensated by an unfavorable benzyl hydrophobic group.

These variants cannot be resolved by X-ray data and practically always are subjects of conjectures and speculations. The following replacement of the benzyl group with naphthalyl, which is supposed to provide an additional binding energy of about 1.9 kcal/mol via pure hydrophobic interactions, would have increased the binding (in terms of the dissociation constant) by about 25 times. In reality, the increase was from 7 to 0.5 μM, that is by 14 times (50). Again, either the site is not entirely hydrophobic or the improvement of the binding is due to a more favorable orientation of a hydrogen bond.

Overall, X-ray data may be wonderful in terms of a general pattern and an overall fit, but they never directly show us "hydrogen bonds" or "hydrophobic interactions" in carbohydrates and proteins. These all are matters of interpretations and conjectures based on distances between atoms, molecules, and groups in question and their (often assumed) orientations.

Below are some examples of binding constants for lactose to various galectins:

Galectin-3:	0.14 mM (as I_{50})–0.22 mM (as 1-methyl ether) (46, 51); 0.6 mM (52); 1 mM (53)
Galectin-4:	0.86 mM–2.0 mM (as I_{50}) (51)
Galectin-7:	nearly equal affinities for lactose and *N*-acetyllactosamine, but both were much weaker compared to those with galectin-1 and -3 (6- and 11-fold, respectively) (54)
Galectin-10:	very little or no β-galactoside-binding activity (55)
Galectin-13:	preference to *N*-acetyllactosamine compared to lactose, with a difference at least 50-fold in affinity (4).

A variant of galectin-1 from a nematode binds asialofetuin 32 times better than lactose, laminin—200 times better, and asialolaminin—533 times better than lactose (44).

Some examples of binding constants for *N*-acetyllactosamine to various galectins (data for galectin-3 and -4 are listed in Reference 51):

Galectin-1:	0.050 mM (36, 56)
Galectin-3:	0.034 mM (as I_{50}) for Gal-β1,4-GlcNAc
	0.052 mM (as I_{50}) for Gal-β1,3-GlcNAc
	0.2 mM (57) for Gal-β1,4-GlcNAc
	0.34 mM (as I_{50}) for Gal-β1,3-GalNAc
Galectin-4:	2 mM (as I_{50}) for Gal-β1,4-GlcNAc
	1 mM (as I_{50}) for Gal-β1,3-GlcNAc
	0.4 mM (as I_{50}) for Gal-β1,3-GalNAc

These data illustrate strength of binding of the ligands to the galectins, but not necessarily their "specificity," as it was defined above.

α-GALACTOSIDE-CONTAINING LIGANDS FOR GALECTINS

It is commonly assumed that galectins recognize the terminal nonreducing β4Gal residues in their ligands. Crystallographic and kinetic data often support that assumption, or at least are interpreted that way. However, those ligands are typically restricted with short-chain lactose, *N*-acetyllactosamine, and oligosaccharides. When ligands are more complex, the recognition of β-Gal residues is much less certain, probably, because other binding sites of the galectin are involved as well. Surprisingly enough, it becomes inclusive also with α-Gal residues. For instance, bovine galectin-1 shows even enhanced recognition toward terminal nonreducing alpha-galactosyl residue, such as in the sequence Galα3Galβ4GlcNAc-R in glycoproteins (58). Specificity to terminal alpha-galactose residues is also shown by human galectin-3 that binds with α-D-GalNAc-(1-3)-[α-L-Fuc-(1-2)]-β-D-Gal-(1-4)-D-Glc-NAc almost 100-fold better compared to galectin-1 (53). A variant of galectin-1 from nematode binds Galα1,4Gal with almost the same affinity as Galβ1,6Gal (44). Besides, galectin-8 binds NeuAcα2,3Lac very tight, with a dissociation constant of 50 nM (59).

BINDING OF SOME GALACTOSE-CONTAINING POLYSACCHARIDES TO GALECTINS: ELISA 101

Let us consider strength of binding of an alpha- and beta-galactose-containing polysaccharides, such as

- 1,4-β-D-galactomannan [(1 → 6)-α-D-galacto-(1 → 4)-β-D-mannan]

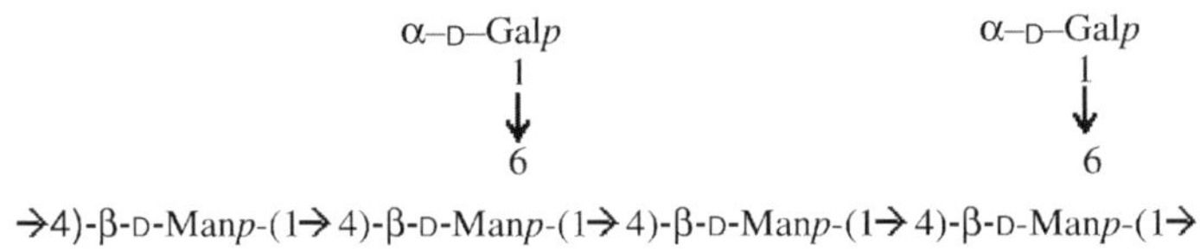

and

- polymers of α-L-rhamnose-α-1,4-galacturonic acid with 1,4-β-D-galacto-arabinogalactan and 1,6-β-D-galactan side chains, attached to O-4 of the rhamnosyl residues, such as

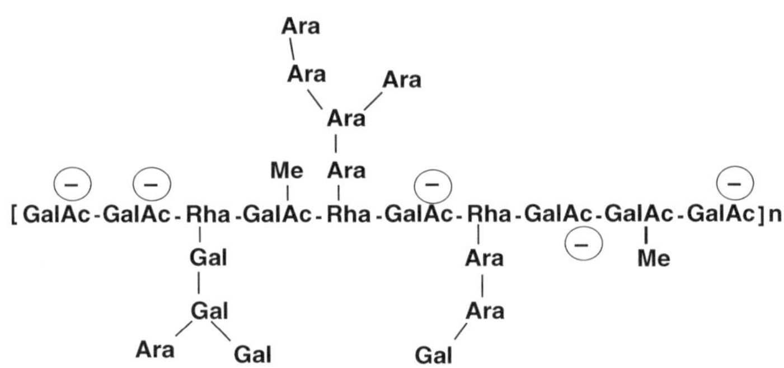

to galectin-1 and galectin-3. The binding was studied using sandwich ELISA (the enzyme-linked immunosorbent assay). The wells of a microtiter ELISA plate were

- layered with anti-h-galectin-1 or -3 (capture antibodies),
- blocked with bovine serum albumin and sucrose to prevent nonspecific binding of galectins and/or the polysaccharides to the plates, then
- added various amounts of galectin-1 or -3, respectively (serial dilutions, from 200 to 0.4 ng/mL; from 20 to 0.04 ng of either galectin was added to each

plate, that was from 1.4 to 0.0028 pmol of galectin-1 or from 0.67 to 0.0013 pmol of galectin-3). After it

- 0.1 mg of a polysaccharide was added into each plate (approximately 2 nmol, that was a thousand to a million times higher in molar equivalents compared to amounts of the galectins). Then
- anti-h-galectin-1 or -3 (detection antibodies) was added to each plate, then
- a secondary antibody, conjugated with peroxidase (an "amplifier"), was added to each plate, and finally
- hydrogen peroxide along with a chromogenic substrate (3,3′,5,5′-tetramethylbenzidine) was added to each plate for color developing and reading. The color indicated the quantity of the galectin bound to the plate and available to the detection antibody. If a polysaccharide binds to the galectin, the latter would be masked and color would be less developed (see figures below).

Each step was followed with a thorough multiple washing with a mild detergent solution to remove proteins, antibodies, or polysaccharides that were not specifically bound. If a polysaccharide does not bind to the galectin, it would be washed away, and the color readout would not change (see figures below).

Note: Capture antibodies—anti-human galectin-1 (PeproTech, #500-P210) and anti-human galectin-3 (PeproTech, #500-P246) were each produced from sera of rabbits preimmunized with highly purified (98%) recombinant hGalectin-1 and hGalectin-3 and purified by affinity chromatography on immobilized hGalectin-1 and hGalectin-3 matrix, respectively.

Detection antibodies—anti-human galectin-1 polyclonal antibody (R&D Systems, #AF1152) and anti-human galectin-3 monoclonal antibody (R&D Systems, #MAB 1164) were produced as follows:

- Anti-human galectin-1—in goats immunized with purified, *E. coli*-derived, recombinant galectin-1 (rhGalectin-1), and purified by human galectin-1 affinity chromatography,
- Anti-human galectin-3—in a hybridoma resulting from the infusion of a mouse myeloma with B cells obtained from a mouse immunized with purified, *E. coli*-derived, recombinant human galectin-3 (rhGalectin-3), and purified by Protein G affinity chromatography.

As it was expected, the polysaccharide, having 1,4-β-D-mannose backbone and single 1,6-α-D-galactose side residues (DAVANAT®), did not bind to galectin-1 or galectin-3 (Figs. 2.1 and 2.2).

However, the same galectin-1 binds, albeit loosely, the polysaccharide with beta-galactoside side chain (Fig. 2.3).

Galectin-3 binds this polysaccharide with beta-galactoside side residue much stronger (Fig. 2.4).

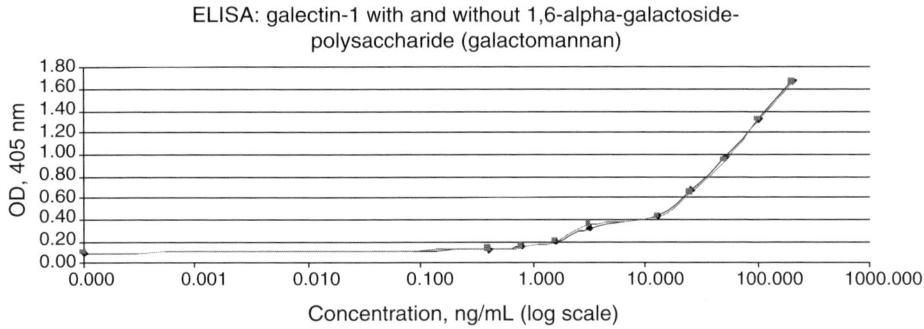

FIGURE 2.1 Effect (rather, lack of it) of 1,6-α-galactoside-polysaccharide (galactomannans) on binding of galectin-1 to its antibodies (ELISA test). Obviously, the polysaccharide does not compete appreciably with the antibodies. The calculated binding constant exceeds 40 nM.

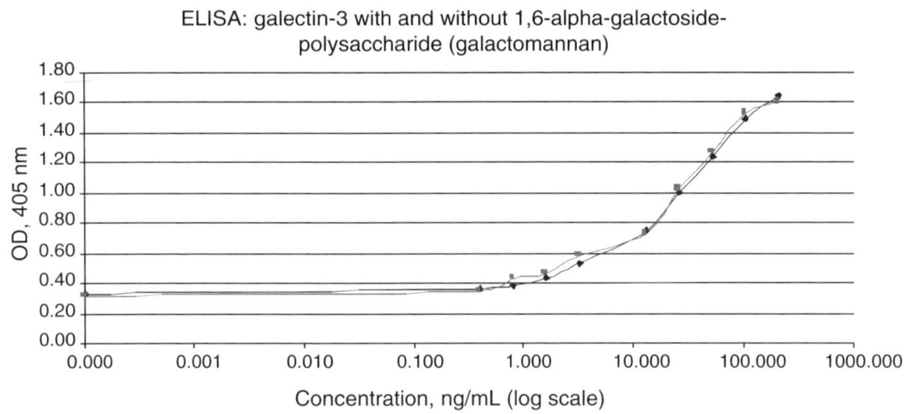

FIGURE 2.2 Effect (rather, lack of it) of 1,6-α-galactoside-polysaccharide (galactomannans) on binding of galectin-3 to its antibodies (ELISA test). Obviously, the polysaccharide does not compete appreciably with the antibodies. The calculated binding constant exceeds 40 nM.

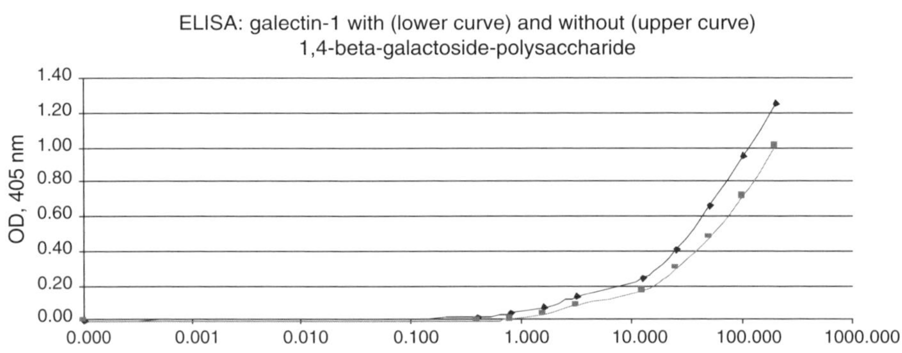

FIGURE 2.3 Effect of 1,4-β-galactoside-polysaccharide (GR-300) on binding of galectin-1 to its antibodies (ELISA test). The calculated binding constant is equal to 2.9 nM.

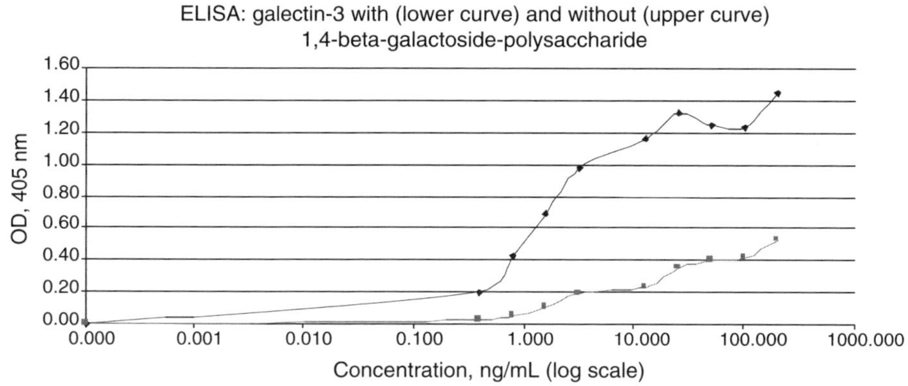

FIGURE 2.4 Effect of 1,4-β-galactoside-polysaccharide (GR-300) on binding of galectin-3 to its antibodies (ELISA test). The calculated binding constant is equal to 0.5 nM.

Interpretation and Calculation of the Data

The galectins (G) bind to the capture antibodies (A) as shown below:

$$G + A \leftrightarrow GA$$

where GA is a complex of the galectin with the antibody, which eventually gives a color readout. Free galectin is in an equilibrium with a bound antibody, and the equilibrium is governed by a binding (or an equilibrium, or a dissociation) constant K_a.

$$K_a = [G] \cdot [A]/[GA]$$

where [G] is the concentration of free galectin, [A] is the "concentration" of the bound antibody, and [GA] is the concentration of the galectin–antibody complex.

At small concentrations of G, added to the well of a microtiter plate, $[G]_0 \ll [A]$, where $[G]_0$ is a total concentration of the galectin (free and bound to the antibody). At a high concentration of the galectin (a 500-fold difference in concentrations was employed in the described experiment), $[G]_0$ can be close to [A] or even exceed it, that is, $[G]_0 \approx [A]$ or $[G]_0$ [A].

Hence,

$$[G]_0 = [G] + [GA]$$

and

$$[GA] = [G]_0/(1 + K_a/[A])$$

For a very tight binding of a galectin for its antibody, $K_a \ll [A]$ and $[GA] = [G]_0$. Indeed, there is practically no free galectin in the well, as it would have been washed away. As a result, optical density (a readout) in the above figures is proportional to the initial concentration of the galectin, at least for low amount of it (at $[G]_0 \ll [A]$). For a high amount of added galectins, the readout (OD) curves become hyperbolic ones and level off at a saturation of the antibody with the galectin.

In the presence of a competitive inhibitor—a polysaccharide in our case, which binds to a galectin and prevents a color readout—optical density will be again proportional to the initial concentration of the galectin, at least for low amount of it, but the slope of that straight line will be lower compared to that without the competitive inhibitor:

$$[\mathrm{GA}] = [\mathrm{G}]_0/(1 + [\mathrm{I}]/K_\mathrm{i})$$

where [I] is the inhibitor concentration and K_i is the binding (dissociation) constant of the inhibitor. This is, of course, for the inhibitor concentration much higher compared to the galectin. Indeed, as described above, a molar amount of a polysaccharide in the experiment was a thousand to a million times higher than that of the respective galectin.

Data of the above figures plotted as (OD versus galectin amounts in the wells) graphs for initial amounts of galectins (0.04 to 0.30–1.25 ng/well) showed that 0.1 mg of a polysaccharide (2 nmol) added to a well decreased the slope by 1.7 times (galectin-1 with 1,4-β-galactoside-polysaccharide, GR-300) and by 5.1 times (galectin-3 with the same 1,4-β-D-galactoside-polysaccharide). This corresponds to the binding constants of 1,4-β-D-galactoside-polysaccharide with galectin-1 equal to 2.9 nM, and that with galectin-3 equal to 0.5 nM.

As it was indicated above, 1,6-α-galactoside-polysaccharide (galactomannan) did not reveal any detectable binding to galectin-1 and galectin-3. Its binding constant, if anything, would be higher than 40 nM, that is, at least 15–80 times higher compared to that for 1,4-β-D-galactoside-polysaccharide with galectin-1 and galectin-3.

BINDING OF THE POLYSACCHARIDES (WITH 1,6-α-D-GALACTOSIDE SIDE RESIDUE AND WITH 1,4-β-D-GALACTOSIDE SIDE CHAINS) TO ANTIBODIES TO GALECTIN-1 AND -3

In the course of the above study it was observed that both polysaccharides bound directly to antibodies to galectin-1 and galectin-3, showing an obvious preference of the binding compared to each other and to the antibodies. In other words, the antigalectins recognized the galactose-containing polysaccharides and responded with the same color readout.

The lowest degree of binding was shown by the 1,6-α-D-galactoside-mannan with anti-h-galectin-1. The readout was 0.12 ± 0.03 OD, which corresponded to 0.18 ng (0.25 ng) of galectin-1 in the well. Compared to that, 1,4-β-D-galactoside-rhamnogalacturonan showed the readout of 0.38 ± 0.01 OD; that is, the binding with the same anti-h-galectin-1 was three times better.

The preference was the same for the binding with anti-h-galectin-3. In case with 1,6-α-D-galactoside-mannan, the readout was 0.34 ± 0.01 OD, and with 1,4-β-D-galactoside-rhamnogalacturonan the readout was 1.08 ± 0.05 OD, again about three times better.

It seems that the polysaccharides can in a way mimic galectins with respect to the antibodies, and antigalectins can recognize them. The binding of the polysaccharides to antigalectin-1 and -3 was so tight that they could not be washed away by multiple treatment with mild detergents. The binding cannot be considered as nonspecific, because the polysaccharides showed preference to one antigalectin compared to another and formed sandwich complexes with antigalectins (capture and detection antibodies).

A possible biological significance of the "galectin" properties of the polysaccharides with respect to the antibodies and the ability of them being recognized by antibodies is not clear as yet. However, the galactomannan was shown to reduce toxicity of a chemotherapeutic agent 5-fluorouracil in clinical trials, and both the galactomannan and the galacto-rhamnogalacturonan showed a rather significant effect in decreasing levels of fibrotic markers in stellate cells *in vitro*, with the latter polysaccharide being more efficient (marker levels was decreased by 50–80% compared to control).

ACKNOWLEDGMENTS

I am indebted very much to Dr. David Platt (Pro-Pharmaceuticals, Inc., Newton, MA) for fruitful collaboration and discussions, and to Professors Fu-Tong Liu (University of California, Davis) and Hans-Joachim Gabius (Institute fur Physiologische Chemie, Ludwig-Maximilians Universitat, Munchen, Germany) for valuable comments and suggestions.

REFERENCES

1. Stirpe F, Hudges RC. Specificity of ribosome-inactivating proteins with RNA N-glycosidase activity. Biochem J 1989;262:1001–1002.
2. Pallanca A, Mazzaracchio R, Brigotti M, Carnicelli D, Alvergna P, Sperti S, Montanaro L. Uncompetitive inhibition by adenine of the RNA-N-glycosidase activity of ribosome-inactivating proteins. Biochem Biophys Acta 1998;1384:277–284.
3. Ackerman SJ, Liu L, Kwatia MA, Savage MP, Leonidas DD, Swaminathan GJ, Acharya KR. Charcot–Leyden crystal protein (galectin-10) is not a dual-function galectin with lysophospholipase activity, but binds a lysophospholipase inhibitor in a novel structural fashion. J Biol Chem 2002;277:14859–14868.
4. Than NG, Pick E, Bellyei S, Szigeti A, Burger O, Berente Z, Janaky T, Boronkai A, Kliman H, Meiri H, Bohn H, Than GN, Sumegi B. Functional analyses of placental protein 13/galectin-13. Eur J Biochem 2004;271:1065–1078.
5. Cooper DNW, Barondes SH. God must love galectins; He made so many of them. Glycobiology 1999;9:979–984.
6. Perillo NL, Pace KE, Seilhamer JJ, Baum LG. Apoptosis of T-cells mediated by galectin-1. Nature 1995;378:736–739.
7. Yang RY, Hsu DK, Liu F-T. Expression of galectin-3 modulates T-cell growth and apoptosis. Proc Natl Acad Sci USA 1996;93 (13):6737–6742.

8. Pace KE, Lee C, Stewart PL, Baum LG. Restricted receptor segregation into membrane microdomains occurs on human T cells during apoptosis induced by galectin-1. J Immunol 1999;163:3801–3811.
9. Rabinovich GA, Daly G, Dreja H, Tailor H, Riera CM, Hirabayashi J, Chernajovsky Y. Recombinant galectin-1 and its genetic delivery suppress collagen-induced arthritis via T cell apoptosis. J Exp Med 1999;190 (3):385–398.
10. Pace KE, Hahn HP, Pang M, Nguyen JT, Baum LG. Cutting edge: CD7 delivers a pro-apoptotic signal during galectin-1-induced T cell death. J Immunol 2000;165:2331–2334.
11. Fischer C, Sanchez-Ruderisch H, Welzel M, Wiedenmann B, Sakai T, Andr S, Gabius H-J, Khachigian L, Detjen KM, Rosewicz S. Galectin-1 interacts with the $\alpha5\beta1$ fibronectin receptor to restrict carcinoma cell growth via induction of p21 and p27. J Biol Chem 2005;280 (44):37266–37277.
12. Cho M, Cummings RD. Galectin-1, a beta-galactose-binding lectin in Chinese hamster ovary cells. II. Localization and biosynthesis. J Biol Chem 1995;270:5207–5212.
13. Cho M, Cummings RD. Galectin-1, a beta-galactose-binding lectin in Chinese hamster ovary cells. I. Physical and chemical characterization. J Biol Chem 1995;270: 5198–5206.
14. Houzelstein D, Concalves IR, Fadden AJ, Sidhu SS, Cooper DNW, Drickamer K, Leffler H, Poirier F. Phylogenetic analysis of the vertebrate galectin family. Mol Biol Evol 2004;21:1177–1187.
15. Stowell SR, Dias-Baruffi M, Penttila L, Renkonen O, Nyame AK, Cummings RD. Human galectin-1 recognition of poly-*N*-acetyllactosamine and chimeric polysaccharides. Glycobiology 2004;14:157–167.
16. Poirier F, Robertson EJ. Normal development of mice carrying a null mutation in the gene encoding the L 14 S-type lectin. Development 1993;119:1229–1236.
17. Tenne-Brown J, Puche AC, Key B. Expression of galectins-1 in the mouse olfactory system. Int J Dev Biol 1998;42:791–799.
18. McGraw J, Gaudet AD, Oschipok LW, Steeves JD, Poirier F, Tetzlaff W, Ramer MS. Altered primary afferent anatomy and reduced thermal sensitivity in mice lacking galectins-1. Pain 2005;114:7–18.
19. Hsu DK, Yang R-Y, Pan Z, Yu L, Salomon DR, Fung-Leung W-P, Liu F-T. Targeted disruption of the galectin-3 gene results in attenuated peritoneal inflammatory responses. Am J Pathol 2000;156:1073–1083.
20. Nedwin G, Bringman T, Conraud P-O. US Patent 5,633,148. 1997.
21. Rousseau C, Muriel MP, Musset M, Botti J, Seve AP. Glycosylated nuclear lectin CBP70 also associated with endoplasmic reticulum and the Golgi apparatus: does the "classic pathway" of glycosylation also apply to nuclear glycoproteins? J Cell Biochem 2000;78: 638–649.
22. Rousseau C, Felin M, Doyennette-Moyne MA, Seve AP. CBP70, a glycosylated nuclear lectin. J Cell Biochem 1997;66:370–385.
23. Rousseau C, Felin M, Seve AP. Stable expression of functional CBP70 lectin during heat shock. J Cell Biochem 2000;77:615–623.
24. Facy P, Seve AP, Hubert M, Monsigny M, Hubert J. Analysis of nuclear sugar-binding components in undifferentiated and *in vitro* differentiated human promyelocytic leukemia cells (HL60)Exp Cell Res 1990;190:151–160.

25. Felin M, Doyennette-Moyne MA, Hadj-Sahraoui Y, Aubery M, Hubert J, Seve AP. Identification of two nuclear *N*-acetylglucosamine-binding proteins. J Cell Biochem 1994;56:527–535.
26. Lefebvre T, Cieniewski C, Lemoine J, Guerardel Y, Leroy Y, Zanetta JP, Michalski JC. Identification of *N*-acetyl-D-glucosamine-specific lectins from rat liver cytosolic and nuclear compartments as heat-shock proteins. Biochem J 2001;360:179–188.
27. Rondanino C, Bousser M-T, Monsigny M, Roche A-C. Sugar-dependent nuclear import of glycosylated proteins in living cells. Glycobiology 2003;13 (7):509–519.
28. Guinez C, Losfeld M-E, Cacan R, Michalski J-C, Lefebvre T. Modulation of HSP70 GlcNAc-directed lectin activity by glucose availability and utilization. Glycobiology 2006;16:22–28.
29. Elola MT, Chiesa ME, Alberti AF, Mordoh J, Fink NE. Galectin-1 receptors in different cell types. J Biomed Sci 2005;12:13–29.
30. Bohn H, Kraus W, Winckler W. Purification and characterization of two new soluble placental tissue proteins (PP13 and PP17). Oncodev Biol Med 1983;4:343–350.
31. Tinari N, Kuwabara I, Huflejt ME, Shen PF, Iacobelli S, Liu F-T. Glycoprotein 90K/MAC-2BP interacts with galectin-1 and mediates galectin-1-induced cell aggregation. Int J Cancer 2000;91:167–172.
32. Kuwabara I, Liu F-T. Galectin-3 promotes adhesion of human neutrophils to laminin. J Immunol 1996;156:3939–3944.
33. Inohara H, Akahani S, Koths K, Raz A. Interactions between galectin-3 and Mac-2-binding protein mediate cell–cell adhesion. Cancer Res 1996;56:4530–4534.
34. Kopitz J, von Reitzenstein C, Andre S, Kaltner H, Uhl J, Ehemann V, Cantz M, Gabius H-J. Negative regulation of neuroblastoma cell growth by carbohydrate-dependent surface binding of galectin-1 and functional divergence from galectin-3. J Biol Chem 2001;276:35917–35923.
35. Brewer CF. Binding and cross-linking properties of galectins. Biochim Biophys Acta 2002;1572:255–262.
36. Ahmad N, Gabius H-J, Sabesan S, Oscarson S, Brewer CF. Thermodynamic binding studies of bivalent oligosaccharides to galectin-1, galectin-3, and the carbohydrate recognition domain of galectin-3. Glycobiology 2004;14:817–825.
37. Ahmad N, Gabius H-J, Andre S, Kaltner H, Sabesan S, Roy R, Liu B, Macaluso F, Brewer CF. Galectin-3 precipitates as a pentamer with synthetic multivalent carbohydrates and forms heterogeneous cross-linked complexes. J Biol Chem 2004;279 (12):10841–10847.
38. Mandal DK, Brewer CF. Cross-linking activity of the 14-kilodalton β-galactose-specific vertebrate lectin with asialofetuin: comparison with several galactose-specific plant lectins. Biochemistry 1992;31:8465–8472.
39. Gupta D, Kaltner H, Dong X, Gabius H-J, Brewer CF. Comparative cross-linking activities of lactose-specific plant and animal lectins and a natural lactose-binding immunoglobulin G fraction from human serum with asialofetuin. Glycobiology 1996;6:843–849.
40. Zou J, Glinsky VV, Landon LA, Matthews L, Deutscher SL. Peptides specific to the galectin-3 carbohydrate recognition domain inhibit metastasis-associated cancer cell adhesion. Carcinogenesis 2005;26:309–319.
41. Cumpstey I, Sundin A, Leffler H, Nillson UJ. C2-symmetrical thiodigalactoside bis-benzamido derivatives as high-affinity inhibitors of galectin-3: efficient lectin

inhibition through double arginine–arene interactions. Angew Chem Int Ed Engl 2005;44:5110–5112.

42. Fort S, Kim H-S, Hindsgaul O. Screening for galectin-3 inhibitors from synthetic lacto-*N*-biose libraries using microscale affinity chromatography coupled to mass spectrometry. J Org Chem 2006;71:7146–7154.
43. Mandal TK, Mukhopadhyay C. Binding free energy calculations of galectin-3–ligand interactions. Protein Eng 2002;15:979–986.
44. Ahmed H, Bianchet MA, Amzel LM, Hirabayashi J, Kasai K-I, Giga-Hama Y, Tohda H, Vasta GR. Novel carbohydrate specificity of the 16-kDa galectin from *Caenorhabditis elegans:* binding to blood group precursor oligosaccharides (type 1, type 2, Tα, and Tβ) and gangliosides. Glycobiology 2002;12:451–461.
45. Ingrassia L, Camby I, Lefranc F, Mathieu V, Nshimyumukiza P, Darro F, Kiss R. Anti-galectin compounds as potential anti-cancer drugs. Curr Med Chem 2006;13: 3513–3527.
46. Tejler J, Leffler H, Nilsson UJ. Synthesis of *O*-galactosyl aldoximes as potent LacNAc-mimetic galectin-3 inhibitors. Bioorg Med Chem Lett 2005;15:2343–2345.
47. Salameh BA, Sundin A, Leffler H, Nilsson UJ. Thioureido *N*-acetyllactosamine derivatives as potent galectin-7 and 9N inhibitors. Bioorg Med Chem 2006;14: 1215–1220.
48. Sörme P, Qian Y, Nyholm P-G, Leffler H, Nilsson UJ. Low micromolar inhibitors of galectin-3 based on 3′-derivatization of *N*-acetyllactosamine. ChemBioChem 2002;3: 183–189.
49. Sörme P, Kahl-Knutsson B, Wellmar U, Magnusson B-G, Leffler H, Nilsson UJ. Design and synthesis of galectin inhibitors. Methods Enzymol 2003;363:157–169.
50. Sörme P, Arnoux P, Kahl-Knutsson B, Leffler H, Rini JM, Nilsson UJ. Structural and thermodynamic studies on galectin-3 in complex with synthetic inhibitors: carbohydrate–protein affinity enhancements through fine-tuning of an arginine–arene interaction. J Am Chem Soc 2005;127:1737–1743.
51. Ideo H, Seko A, Ohkura T, Matta KL, Yamashita K. High-affinity binding of recombinant human galectin-4 to $SO_3^- \rightarrow 3Gal\beta1 \rightarrow 3GalNAc$ pyranoside. Glycobiology 2002;12: 199–208.
52. Hsu DK, Zuberi RI, Liu FT. Biochemical and biophysical characterization of human recombinant IgE-binding protein, an S-type animal lectin. J Biol Chem 1992;267: 14167–14174.
53. Sparrow CP, Leffler H, Barondes SH. Multiple soluble β-galactoside-binding lectins from human lung. J Biol Chem 1987;262:7383–7390.
54. Ahmad N, Gabius HJ, Kaltner H, Andre S, Kuwabara I, Liu F-T, Oscarson S, Norberg T, Brewer CF. Thermodynamic binding studies of cell surface carbohydrate epitopes to galectins-1, -3, and -7: evidence for differential binding specificities. Can J Chem 2002;80: 1096–1104.
55. Dyer KD, Rosenberg HF. Eosinophil Charcot–Leyden crystal protein binds to beta-galactoside sugars. Life Sci 1996;58:2073–2082.
56. Schwarz FP, Ahmed H, Bianchet MA, Amzel LM, Vasta GR. Thermodynamics of bovine spleen galectins-1 binding to disaccharides: correlation with structure and its effect on oligomerization at the denatured temperature. Biochemistry 1998;37: 5867–5877.

57. Leffler H, Barondes SH. Specificity of binding of three soluble rat lung lectins to substituted and unsubstituted mammalian beta-galactosides. J Biol Chem 1986;261: 10119–10126.

58. Appukuttan PS. Terminal alpha-linked galactose rather than *N*-acetyl lactosamine is ligand for bovine heart galectin-1 in N-linked oligosaccharides of glycoproteins. J Mol Recognit 2002;15:180–187.

59. Carlsson S, Oberg CT, Carlsson MC, Sundin A, Nilsson UJ, Smith D, Cummings RD, Almkvist J, Karlsson A, Leffler H. Affinity of galectin-8 and its carbohydrate recognition domains for ligands in solution and at the cell surface. Glycobiology 2007;17:663–676.

3

UNDERSTANDING GALECTIN STRUCTURE–FUNCTION RELATIONSHIPS TO DESIGN EFFECTIVE ANTAGONISTS

Irina V. Nesmelova, Ruud P.M. Dings, and Kevin H. Mayo

Department of Biochemistry, Molecular Biology & Biophysics, University of Minnesota, Minneapolis, MN 55455, USA

INTRODUCTION

Galectins are a subfamily of lectins that bind β-galactoside and have a significant sequence similarity in their overall carbohydrate-binding site (1). Although some galectins have been recognized to have intracellular functions, they are best known for their extracellular activities, being found on the cell surface and even deposited in the extracellular matrix. Galectin expression, which varies considerably from cell type to cell type, depends upon the activation state of a certain cell type. All cells appear to express at least one galectin, and each galectin tends to be expressed at high concentration in a few, but different cell types (2).

Galectins can be translocated to the nucleus or to other subcellular sites after being synthesized on cytosolic ribosomes. Galectins have several features in common with cytosolic proteins, such as being deficient in a secretion signal peptide or typical transmembrane segments, and they can have acetylated N-termini. This diversity in their occurrence is also reflected in the multimodal biological roles they exhibit in controlling cell–cell and cell–matrix interactions, adhesion, proliferation, apoptosis,

Galectins, Edited by Anatole A. Klyosov, Zbigniew J. Witczak, and David Platt

pre-mRNA splicing, immunity, and inflammation (3). The underlying principle of all these functions is most often, but not always, carbohydrate recognition, as will be discussed below.

Galectin-1 has been the most studied and well-characterized galectin (4), since it was the first galectin discovered in 1975 through its display of hemagglutinating activity in electric eels (5). Currently at least 14 mammalian galectins have been reported, and many more are found in different organisms, for example, vertebrates, invertebrates, and protists (1, 6, 7). Early on following the discovery of galectins, it was proposed that they be divided into three groups (8, 9): prototype (galectins 1, 2, 5, 7, 10, 11, 13, and 14), chimera (galectin-3), and tandem repeat (galectins 4, 6, 8, 9, and 12). Prototype galectins consist of a single core domain, usually referred to as the carbohydrate recognition domain (CRD). Galectin-3 is the only chimera galectin known, and it has a CRD and a nonlectin N- or C-terminal domain with different properties. Tandem-repeat galectins have two homologous, yet distinct, CRDs that are connected to each other via a linker polypeptide chain.

In this chapter, we will focus on structure–function relationships of galectins, particularly on those whose X-ray and/or NMR three-dimensional structures are known, as well as on open questions to help understand their mechanisms of action on the molecular level, and on the use of structure-based design and library-directed screening of antagonists of galectin function.

FOLDED STRUCTURES

Three-dimensional structures of CRDs are primarily derived from X-ray crystallographic diffraction studies, and the remainder result from the use of heteronuclear NMR spectroscopy. While some galectin structures were done in the apoprotein state, most have been derived in complex with various carbohydrates, primarily with the disaccharide lactose, a carbohydrate recognition unit. So far, structures have been determined for human galectin-1 (10), galectin-2 (11), CRD of galectin-3 (12, 13), C-terminal CRD of galectin-4 (14), galectin-7 (15), galectin-10 (16–18), bovine galectin-1 (19, 20), mouse galectin-9 (21), toad ovary galectin (22, 23), chicken galectin-16 (24), galectins from conger eel: congerin I (25) and congerin II (26, 27), and fungal galectins from *Coprinopsis cinerea* (28) and *Agrocybe cylindracea* (29). We have summarized these in Table 3.1, along with references and PDB access codes. No structures of full-length tandem-repeat galectins are as yet known.

The Monomer Fold

The secondary structure of all CRDs is composed of multiple β-strands that fold as contiguous antiparallel β-sheets as shown in Fig. 3.1 with the amino acid sequence of galectin-1. Most is known about galectin-1, and its amino acid sequence has been reported for multiple species, including humans, mice, rats, bovine, chicken, and eels. In general, galectins have 11 β-strands. The amino acid sequences of all galectins (whose folded structures are known) are aligned in Fig. 3.2, with the β-strands

TABLE 3.1 Protein Database (PDB) Access Codes for Known X-Ray and NMR Derived Structures of Galectins

Human	
Galectin-1	1gzw (lactose)
	1w6m (C2S)
	1w6o (lactose + C2S)
	1w6q (R111H)
Galectin-2	1hlc (lactose)
Galectin-3	1a3k (free)
	1kjl (*N*-acetyl-D-glucosamine, D-galactose)
	1kjr (2,3,5,6-tetrafluoro-4-methoxy-benzamide)
Galectin-4	1 × 50 (free)
Galectin-7	1bkz (free)
	4gal (lactose)
	2gal (galactose)
	3gal (galactosamine)
	5gal (*N*-acetyllactosamine)
Galectin-10	1lcl (free)
	1qkq (mannose)
	1g86 (*N*-ethylmaleimide)
Bovine	
Galectin-1	1sla (*N*-acetyllactosamine 8-mer)
	1slb (8-mer, 7-mer)
	1slc (8-mer, 8-mer)
	1slt (*N*-acetyl-D-glucosamine, D-galactose)
Mouse	
Galectin-9	2d6l (free)
	2d6m (lactose)
	2d6n (*N*-acetyllactosamine)
	2d6o (*N*-acetyllactosamine dimer)
	2d6p (T-antigen)
	1d6k (cf1)*
Chicken	
Galectin-16#	1qmj (free)
Toad	
Bufo arenarum	1a78 (thiodigalactose)
	1gan (*N*-acetylgalactose)
Conger eel	
Congerin I#	1c1f (free)
	1c1l (lactose)
Congerin II	1is5 (free)
	1is4 (lactose)
	1is6 (2-(*N*-morpholino)-ethanesulfonic acid)
	1is3 (lactose, 2-(*N*-morpholino)-ethanesulfonic acid)
	1wld (T88I)
	1wlw (Y16S)
	1wlc (Y16S/T88I)

(continued)

Table 3.1 (*Continued*)

Fungi	
Agrocybe cylindracea	1ww7 (free)
	1ww6 (lactose)
	1ww4 (NeuAca2-3lactose)
	1ww5 (3′-sulfonyl lactose)
Coprinopsis cinerea##	1ul9 (free)
	1ulc (lactose)
	1ule (linear trisaccharide)
	1ulf (tetrasaccharide)
	1ulg (*N*-acetyl-D-galactosamine, D-galactose)

The state (free or indicated ligand bound) or the mutated amino acid(s) are specified between the parentheses.
Comparable to human #galectin-1 or ##galectin-2.

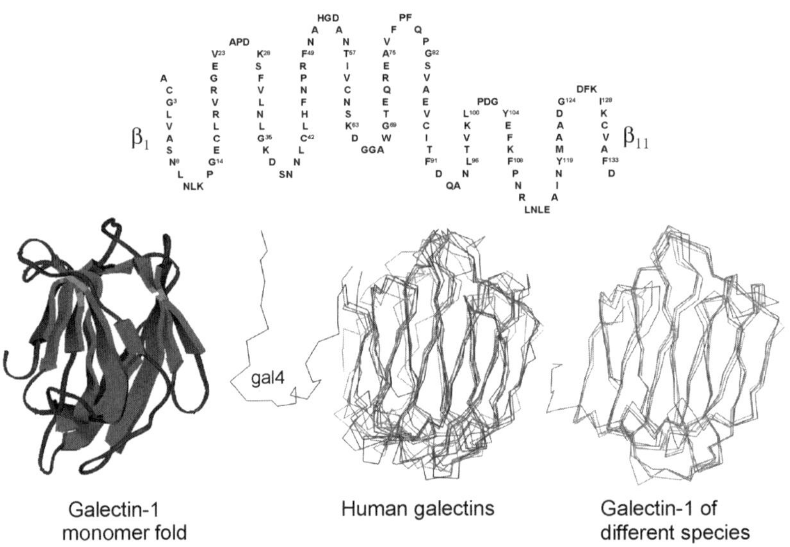

FIGURE 3.1 Superposition of CRDs from different galectin monomers. In top panel, a schematic representation of the 11 β-strands (β1 and β11 labeled) and β-sheet elements for human galectin-1 is illustrated using single letter codes for amino acid residues. Residues of the carbohydrate-binding site encompass β-strands 4–6. The structure at the bottom left shows the overall β-sheet fold of the galectin-1 monomer, with the carbohydrate-binding site at the upper left quadrant of the structure. The superposition of CRDs of all known human galectin structures (Table 3.1) is shown at the bottom middle of the figure, and the superposition of known 3D structures (Table 3.1) of galectin-1 from different species is shown at the bottom right. In general, RMSD values for backbone superimposed β-strands are less than 1.5 Å. (See color insert.)

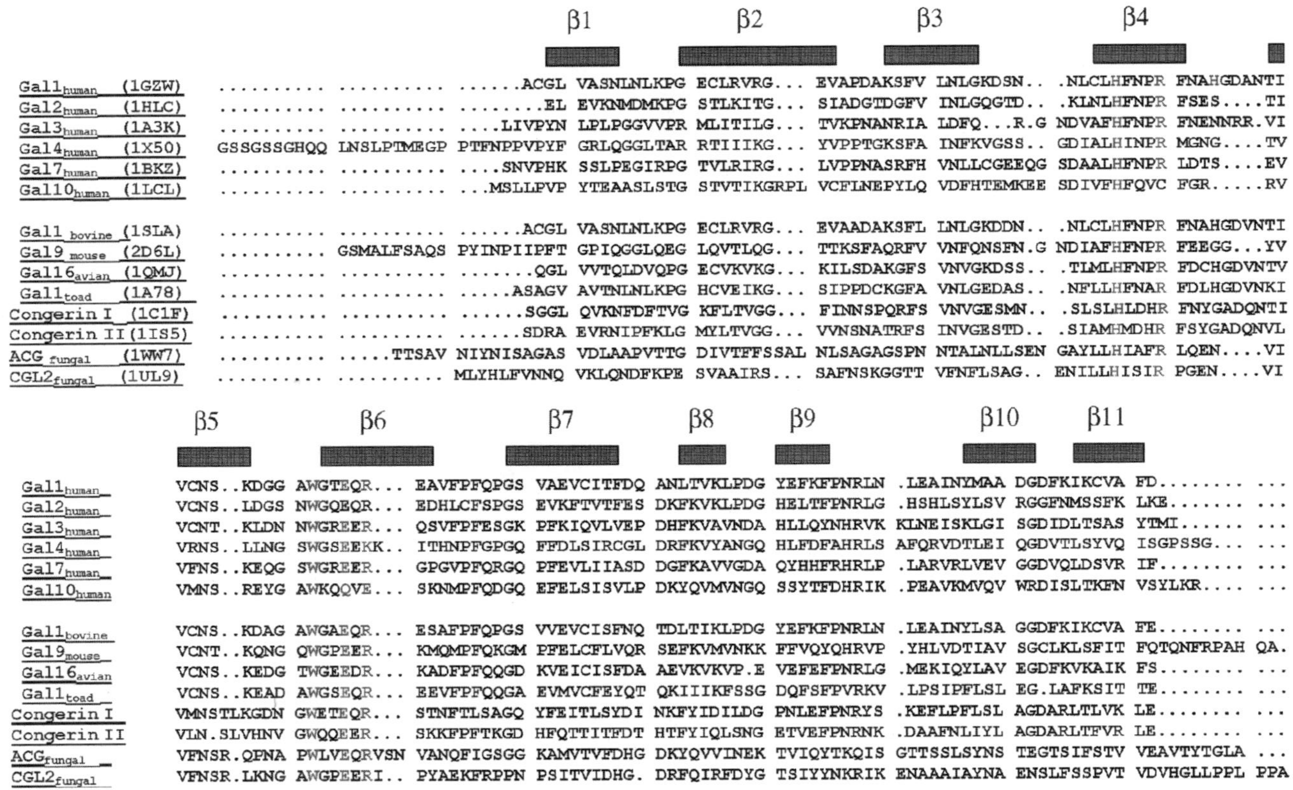

FIGURE 3.2 Amino acid sequences of galectins. The amino acid sequences of galectins, whose three-dimensional folded structures have been reported, are shown aligned. References for these sequences are provided in Table 3.1. The 11 β-strands common to these galectins are indicated at the top by hatched bars and labeled β1 through β11. Conserved residues that form the lactose-binding pocket are identified in red. (See color insert.)

identified by shaded bars and labeled β1 to β11. Note that these amino acid sequences are highly homologous (intra- and interspecies; e.g., 90% of the 134 amino acids are identical between humans and mice (30)), and generally conserved, particularly with respect to residues in the carbohydrate-binding site (highlighted in red) to be discussed later. Unlike other galectins, however, galectin-1 contains six cysteine residues and displays carbohydrate-binding activity only when the three intramolecular disulfide bonds (Cys2–Cys130, Cys16–Cys88, and Cys42–Cys60) are reduced (31, 32). This suggests that disulfide bond formation somehow distorts aspects of the galectin-1 monomer fold that are necessary for carbohydrate binding.

In general, the galectin CRD folded structure is composed of two antiparallel β-sheets of five (β1–β5) and six (β6–β11) β-strands, arranged in a β-sheet sandwich motif, with no helical segments. The galectin folded structure forms a more or less globular shaped, with a "jelly roll" topology, as illustrated in ribbon format in Fig. 3.1 (bottom left). The β-sheet "sandwich" is shown here edge on and is "filled" with mostly hydrophobic residues. The β-sheet sandwich has one concave and one convex surface, a structural feature that is more evident with the illustrations of galectin-2 and galectin-7 in Fig. 3.3. More interesting is the fact that the monomer folds of all CRDs are remarkably similar, not only between the same galectin from different species, but also between/among all galectins. This is shown in Fig. 3.1 (bottom middle), where the β-strands of all known human galectin structures have been superimposed. RMSD values for backbone atoms in the β-strands are about 1.5 Å. The loops are more variable in their average conformations determined by X-ray or NMR. This is due in part to increased flexibility in the turn/loop domains, as well as to variability in amino acid type and length of the loop.

The same thing is observed when one superposes known galectin-1 structures from different species (Fig. 3.1, bottom right). Some exceptions are certainly known. For example, galectin-4 has a relatively longer N-terminal sequence (identified in Fig. 3.1), whereas galectin-5 and galectin-7 have shorter N-terminal sequences.

Self-Association

In general, galectins are known to self-associate, most forming dimers (especially those from the prototype group), and in some cases, higher oligomeric states. In fact, despite the similarity of their monomer structures, galectins can form different types of dimers. The so-called "terminal" dimer is formed through hydrophobic interactions between N- and C-terminal residues of the two monomer subunits, with the subunits being related by a twofold rotation axis that runs approximately perpendicular to the plane of the β-sheets (e.g., galectin-1, Fig. 3.3).

The "nonsymmetric sandwich" dimer interface involves the β-strands β1 and β6 (see Fig. 3.1) from each subunit (11) and is stabilized by electrostatic interactions among charged residues at the convex surfaces of two monomers (e.g., fungal galectin-2, Fig. 3.3). The "symmetric sandwich" dimer (e.g., human galectin-7, Fig. 3.3), also formed by electrostatic interactions among charged residues on the convex surfaces of two monomers, is symmetric with its intersubunit contact surface being reduced compared to that of the nonsymmetric dimer.

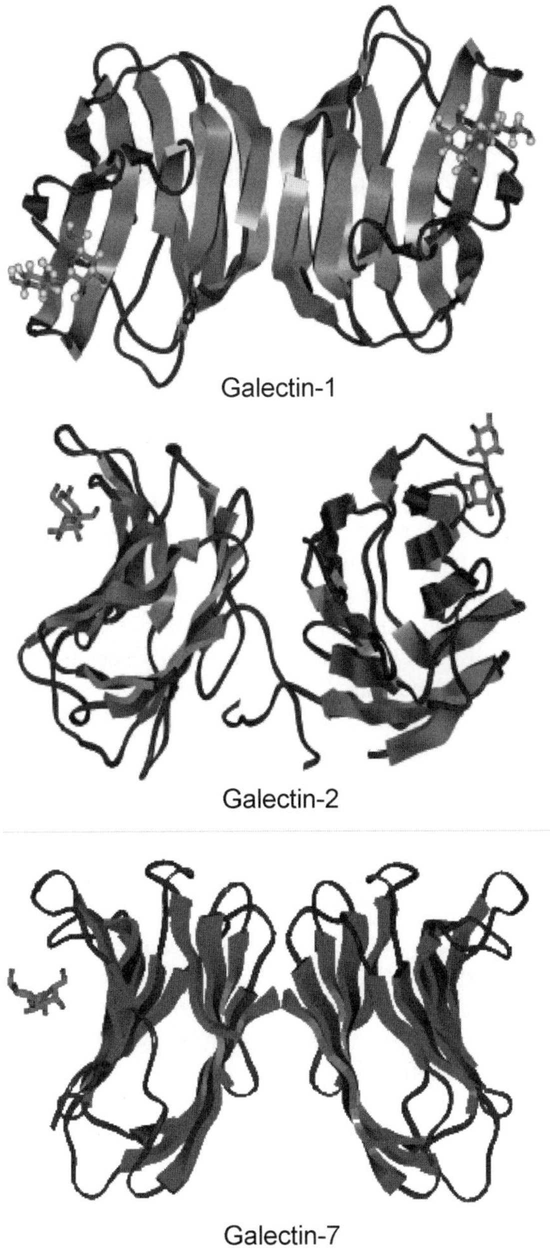

FIGURE 3.3 Galectin dimers. The different types of galectin dimers that can form are shown here. The CRD of the galectin-1 "terminal" dimer (PDB access code lgzw) (10) is shown at the top of the figure, followed by the "nonsymmetric sandwich" dimer of fungal galectin-2 (1ul9) (28) in the middle, and then the "symmetric sandwich" dimer of human galectin-7 (1bkz) (15) at the bottom. The carbohydrate-binding sites in each dimer are indicated by the lactose molecules shown as ball-and-stick structures. (See color insert.)

There are two reasons for the observations of these dimers. One is artifactual resulting from lattice packing forces in crystals used for X-ray structure determination, and the other is based on thermodynamic stability of one type of dimer over another. The former explanation appears to be the case at least for the crystal structure of the galectin-1 dimer, which is asymmetric, whereas the solution NMR structure of galectin-1 yields a symmetric dimer (Nesmelova et al., unpublished results), likely closer to that found *in situ*.

Since galectin monomer structures are so similar, the thermodynamic explanation for formation of a particular dimer state would be dictated by the specific amino acid residues within the intermonomer contact surface. If per residue free energies are summarily greater for one type of dimer, then that one will prevail in solution. In this regard, formation of dimer type (terminal, symmetric, and nonsymmetric) may be functionally important, because this may help differentiate how different galectins bind differently to complex glycans intra- or extracellularly on the surface of cells. These natural glycans are far more complex than simple disaccharides such as lactose that have been used to study galectin carbohydrate binding and function.

In solution, human galectin-1 most consistently is found as a dimer and exhibits an apparent dimer dissociation constant of 7×10^{-6} M (33), and even at low concentration (2 μM) running on a native gel (33, 34), human galectin-1 appears exclusively as a dimer. The situation with other prototype galectins is not always clear, and the use of different conditions clearly influences the galectin oligomeric state. In solution at intermediate concentrations, galectin-5 and -7 behave as monomers (35, 36), yet galectin-5 can agglutinate cells, indicative of some sort of cross-linking and therefore self-association.

In the crystal, galectin-7 appears to be a dimer (15), whereas in solution others have reported it to be either a monomer (6, 15, 37) or a dimer (15, 38, 39). At least over the range of concentrations examined, galectin-7 does not appear to form higher order oligomers. Another prototype galectin, galectin-10, can spontaneously form crystals, so-called Charcot–Leyden crystals, in tissue and in secretion (17). Interestingly, galectin-13 is the only prototype galectin to form a homodimer, wherein subunits are covalently linked together by disulfide bonds, which when reduced decrease carbohydrate binding and abolish agglutination activity. All other galectins dimerize via noncovalent interactions.

Chimera galectin-3, a 30-kDa molecule, consists of one C-terminal CRD linked to a proline, glycine, and tyrosine rich N-terminal domain having multiple homologue repeats that are important for higher order oligomerization (40). However, there is considerable confusion concerning the oligomeric state of galectin-3. From size-exclusion chromatography, it is either a monomer (41) or a dimer (42, 43).

Apparently, when binding to cell surface ligands galectin-3 induces glycan cross-linking, consistent with it behaving more as a dimer or a multimer (2). Moreover, even under nonreducing conditions and in the presence of sodium dodecyl sulfate (SDS) (44), galectin-3 forms dimers and higher order oligomers, reportedly by being chemically cross-linked (45) apparently through the action of transglutaminase (46). From these studies, it appears that galectin-3 can exist as a monomer, dimer, or even higher oligomer. In fact, Ahmad et al. reported that galectin-3 precipitates from solution in a pentameric state upon binding synthetic carbohydrates (47). Galectin-3 oligomerization occurs through its N-terminal nonlectin domain. Consequently, when CRD domains of galectin-3 bind to oligosaccharides, their N-termini associate forming oligomers, and thereby cross-link oligosaccharides on the cell surface.

Although tandem-repeat type galectin-4, -6, -8, and -9 are usually reported to be monomers in solution, they already possess two CRDs, and thereby mimic the dimer state of prototype galectins like galectin-1. This may indicate some level of biological control and/or evolutionary link, in that tandem-repeat type galectins cannot dissociate into single CRD monomers like all other galectins. With any galectins, the presence of two CRDs appears necessary to mediate full activity, at least in terms of promoting cell adhesion and migration. Since each CRD contains one carbohydrate-binding site, galectin dimerization can mediate intermolecular cross-linking by binding to more than one β-galactoside carbohydrate moieties on different glycoconjugates, especially in the extracellular environment.

In the case of tandem-repeat type galectins, two CRD domains are already present within their monomer folded structures. Nevertheless, two studies report that tandem-repeat type galectin-9 can self-associate as dimers (mouse galectin-9 (21)) or multimers (human galectin-9 (48)). In the latter paper, Miyanishi et al. actually used a galectin-9 mutant, which lacks a cleavable linker region, yet retains its biochemical and physiological activities (49). For the human galectin-9 intermolecular interaction, the K_d value determined by BIAcore analysis is about 1×10^{-6} M, which is close to that reported for prototype galectin-1 (7×10^{-6} M) determined by gel filtration analysis (33).

Galectin-9 oligomerization was enhanced upon increasing the temperature, indicating that hydrophobic interactions contribute significantly to thermodynamic stabilization of these oligomers. Moreover, galectin-9 oligomerization was abolished by the presence of saturating levels of lactose, suggesting that galectin-9 oligomers do not form when galectin-9 is bound to glycoconjugates *in situ*, for example, on the surface of cells.

In the following section, we will examine details of the binding of various carbohydrates to different galectins.

CARBOHYDRATE BINDING

In prototype galectins, the carbohydrate-binding site of the galectin CRD is located on the side of galectin opposite from the dimer interface and is formed from amino acid side chains of the six-stranded β-sheet (β6–β11, Fig. 3.1) (30). The actual amino acid residues that interact with the β-galactoside disaccharide, for example, lactose, are highly conserved among all galectins, as is evident from the amino acid sequence comparison shown in Fig. 3.2 (residues highlighted in red). Even though the β-galactoside disaccharide is the carbohydrate recognition element for galectin, the actual glycans present on various biomolecules, for example, cell surface proteoglycans, to which galectins bind, are far more complicated *in situ* than the simple recognition disaccharide. For example, it is likely that other saccharide units in those glycans interact to some extent with galectins, and this too may play a role in determining and/or differentiating the function of various galectins. The following sections discuss the binding of various carbohydrates (mostly simple ones) to galectins.

Binding Affinities

Lactose is the simplest carbohydrate to which galectins bind. Affinities for lactose binding to CRDs usually range from millimolar (weak binding affinity) to micromolar (moderate binding affinity), and these values can vary significantly depending on the analytic technique being used. In solution, human galectin-1 exhibits apparent dimer dissociation constant of 7×10^{-6} M (33), and even at low concentration (2 μM) running on a native gel [33, 34], human galectin-1 appears exclusively as a dimer.

Using isothermal titration calorimetry, the K_d value for lactose binding to bovine galectin-1 has been reported to be 64×10^{-6} M (50). Using affinity chromatography, K_d values range from 26×10^{-6} M for galectin-3 (strongest of all galectins using this technique) to 2×10^{-3} M for nematode galectin LEC-6 CRDs (51). In other reports, the K_d for galectin-3, for example, has been reported to be 1×10^{-3} M (52) and 0.6×10^{-3} M (45), whereas for galectin-2 it has been reported as 85×10^{-6} M (28), and for galectin-4 it is 0.9×10^{-3} M (53). In general, these K_d values indicate relatively weak binding of lactose to galectins. This may not be too surprising because lactose is known to be the minimal unit required for carbohydrate recognition by and binding to galectins.

Carbohydrate-binding affinity can be increased by modifying the lactose disaccharide. For example, *N*-acetyllactosamine binds about five-fold better to galectin-3, with a K_d of 0.2×10^{-3} M [52, 54]. A similar improvement is noted for binding to other galectins. Moreover, chimera galectin-3 has a distinct profile for binding more strongly to larger oligosaccharides [52, 54], like polyNAc-lactosaminoglycan, a polymer of β(1,3)-linked LacNAc units found on many extracellular matrix and cell surface molecules. This is important because much larger, more complex carbohydrates are more reflective of those found *in situ*.

Hirabayashi et al. have performed a thorough analysis of the binding of 13 galectins from various species to 41 pyridylaminated oligosaccharides, including linear, repeating, branched, and substituted ones (51). In general, they reported that binding affinities may be explained by some common carbohydrate recognition features (51), as follows:

(1) Galh1-4GlcNAc is the basic disaccharide unit recognized by all galectins, even though affinities (K_d values) vary greatly from galectin to galectin, for example, 1.7×10^{-6} M for galectin-3 (highest affinity) to 1.4×10^{-3} M for sponge galectin GC2. Isomers of Galh1-4GlcNAc modify these affinities, for example, with Galh1-3GlcNAc, the K_d value goes to 2.6×10^{-6} M for galectin-3 and to 0.13×10^{-6} M for GC2. Structural analysis (see the next section) shows that the 3-OH configuration (equivalent in Galh1-3GlcNAc and Galh1-4GlcNAc) is essential for carbohydrate recognition by galectins, which likely explains reduced affinity of galectin-3 for lactose, where the reducing end glucose pyranose ring is opened upon pyridylamination.

(2) Substitution at the 4-OH and 6-OH groups on the galactose ring generally abolishes binding. In support of this, galectins generally do not bind to nonreducing terminal mannosides or glucosides, and 2,6-sialylation completely abolishes binding to galectins, whereas 2,3-sialylation does not.
(3) Substitution at the 3(4)-OH of the penultimate Glc(NAc) of Galh1-4(3) GlcNAc also abolishes binding to most galectins.

Lactose-Bound State

From the structural perspective, we first need to understand simple carbohydrate binding to galectins, particularly because biophysical approaches generally necessitate the use of more simplified systems with which to work and because actual glycans are large, linear/branched, and heterogeneous in nature and difficult to prepare in quantities often required for high resolution structural analysis, either for X-ray crystallography or NMR spectroscopy.

Lactose is the minimal carbohydrate ligand necessary for binding to galectins, and most reported structures of galectins have been performed on the lactose-bound state. Figure 3.4 illustrates C_α traces of four CRD monomer structures: human galectin-1, mouse galectin-9, congerin II, and fungal galectin. Lactose is shown as a stick-and-ball structure within the carbohydrate-binding site of each of these galectins, and the side chains of key amino acid residues are indicated. Lactose-binding sites from each of these structures have been expanded in the insets shown at the right of each C_α trace, with key amino acid residues labeled. Note that most of these residues are conserved arginines and histidines, as well as a conserved tryptophan that appears under the lactose molecule (see Fig. 3.4). Lactose is effectively "grabbed" by the peptide loop above the lactose molecule and the relatively large and flat tryptophan side chain at the bottom of the disaccharide. NMR structural studies indicate that this loop is relatively flexible when the disaccharide is absent and is more firmly positioned when the disaccharide is bound (Nesmelova et al., unpublished data).

To illustrate that the carbohydrate-binding site is essentially the same for all galectins, Fig. 3.5a (left) shows the superposition of carbohydrate-binding sites of human galectin-1 in complex with lactose (magenta) and bovine galectin-1 in complex with an octasaccharide (green). The binding sites include all atoms within 4.5 Å from the saccharide molecule. Note that the tryptophan side chain is nearly invariant, whereas all other key residues are essentially in the same positions, especially considering that side chains can be relatively flexible and these are taken from different structures and with quite different saccharide moieties.

Since the carbohydrate-binding site is essentially the same in all galectins, we show the lactose-binding site of the most studied galectin-1 in greater detail in Fig. 3.5b (right). On the basis of distances between acceptor and donor groups, proposed H-bonds are illustrated as thin lines. These interactions with lactose are similar for all galectin CRDs (26). Three hydroxyl groups of lactose (or *N*-acetyllactosamine), that is, the C4-OH and C6-OH of Gal and the C3-OH of GlcNAc (or Glc), are required to form hydrogen bonds with side chains of hydrophilic residues from galectins. The galactose ring in particular forms

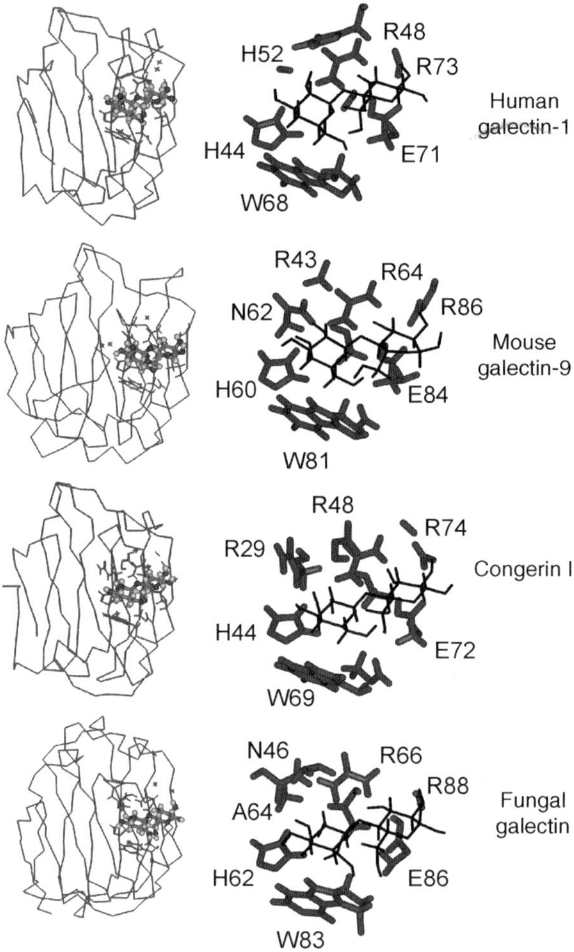

FIGURE 3.4 Lactose recognition site is similar in all galectins. The individual structures of lactose-bound galectins are shown for human galectin-1, mouse galectin-9, conger eel congerin II, and fungal galectin from *Agrocybe cylindracea* are illustrated at the left column in the figure. Atomic resolution expansions of these structures showing atoms within a distance of 4.5 Å around the lactose molecule are shown in the right column directly opposite from each of these structures. Lactose molecules are shown as ball-and-stick models. (See color insert.)

several H-bonds between its oxygen atoms O4, O5, and O6, and H44, E71, and N61 of galectin-1. The O3 of the glucose ring forms H-bonds with residues R48 and E71. H52 and W68 are proposed to have van der Waals interactions with both the glucose and galactose rings, respectively.

These conserved amino acid residues from other galectins are identified in Fig. 3.2. The nonpolar side of the galactose ring (composed primarily of its C1, C3, and C5 hydrogen atoms and distinguished by its axial C4-OH on the more polar side) interacts primarily with the tryptophan residue, which is conserved in all galectin carbohydrate-binding domains. This hydrophobically mediated interaction contributes favorably to the binding free energy and plays a role in the binding site discrimination between the galactose and glucose rings. In contrast to galactose, glucose has its C4-OH group in an equatorial position, which decreases the apolar nature of the corresponding surface and attenuates hydrophobic interactions with the conserved tryptophan (55).

Bovine heart galectin-1 binds the *syn* conformation of lactose and of galactosyl-xylosides as shown by NMR spectroscopy (transferred NOE) (56). However, if the side chains of H52 and W68 were rotated somewhat, H-bonds could also form with these residues. This description is essentially the same for all galectin CRDs. For example, orientation of the lactose molecule in the galectin-3 CRD bound state derived from NMR data (57) essentially agrees with that from the X-ray structure (PDB 1a3k) and as described above. There are van der Waals contacts between the homologous tryptophan residue and the more hydrophobic sides of the galactose and

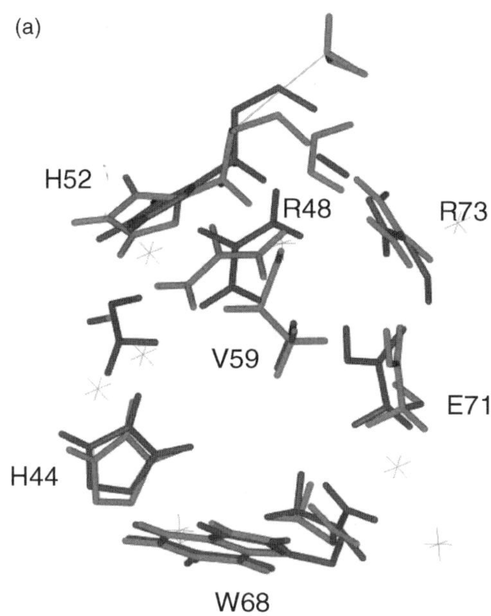

FIGURE 3.5 Carbohydrate-binding site of galectin-1. (a) Superpositon of the carbohydrate-binding sites of human galectin-1 in complex with lactose (1gzw, magenta) and bovine galectin-1 in complex with an octasaccharide (1sla, green) is shown. The binding sites include all atoms within a distance of 4.5 Å from the carbohydrate molecule. (b) The lactose-binding site of human galectin-1 (1gzw) is enlarged to illustrate the amino acid residues most responsible for binding to lactose. Potential hydrogen bonds formed with the hydroxyl groups O4, O6 of the galactose ring and the O3 of the glucose ring are shown by thin lines. The majority of contacts are formed between the side chains of H44, R48, N61, and E71, and the rings of H52 and W48 apparently provide stacking van der Waals interactions. The binding site shown was limited to a sphere with a radius of 4.5 Å about the lactose molecule. (See color insert.)

(b)

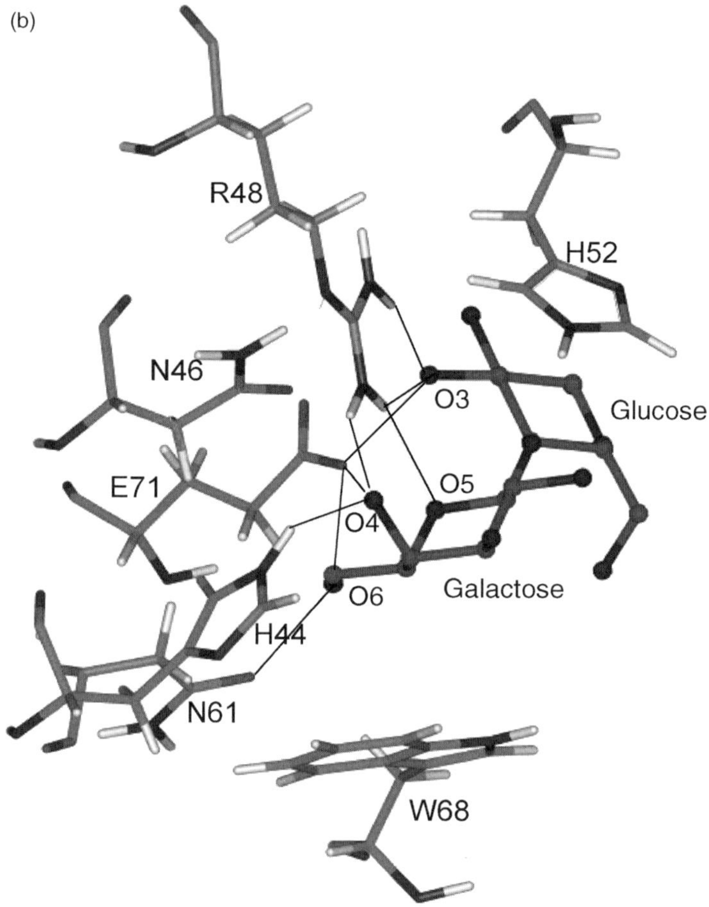

FIGURE 3.5 *(Continued)*

glucose rings, and H-bonding contacts with the 4-hydroxyl are also maintained. In both the galectin-3 crystal and NMR structures, the C-4 hydroxyl group is in a position to form hydrogen bonds with H158, R162, and D160.

Overall, however, the structure of the carbohydrate-binding site apparently varies little upon binding lactose, with the major change being the orientation of the tryptophan ring, which is disordered in the ligand-free state and highly ordered in the lactose-bound state. This lactose-induced orientation of the tryptophan ring occurs so that this site on the protein can interact optimally with the galactose ring. Moreover, this conformation seems to explain why in structures of galectin bound with longer carbohydrates, the remainder of the polysaccharide is oriented away from the protein surface and out into solution (19). In the water-bound and glycerol-bound galectin-3 crystallographic structures, the amino acid side chains within the carbohydrate-binding site and the loop regions on either side of the binding groove are found to be identical to those in the lactose-bound structures (58). This contrasts with the NMR results of the galectin-3 CRD, in which the loops around the binding site undergo a conformational change in the absence of ligand (59). This discrepancy is a consequence of crystal packing, as intermolecular contacts occur between the loop regions and symmetry-related molecules.

Other Carbohydrates

Aside from lactose, galectin structures in complex with *N*-acetyllactosamine (13, 15, 20, 21, 28), linear trisaccharide (28), and *N*-acetyllactosamine octasaccharide (19) have been reported. The resulting galectin structures are shown in Fig. 3.6. In all cases, a carbohydrate disaccharide moiety binds the galectin in a similar fashion.

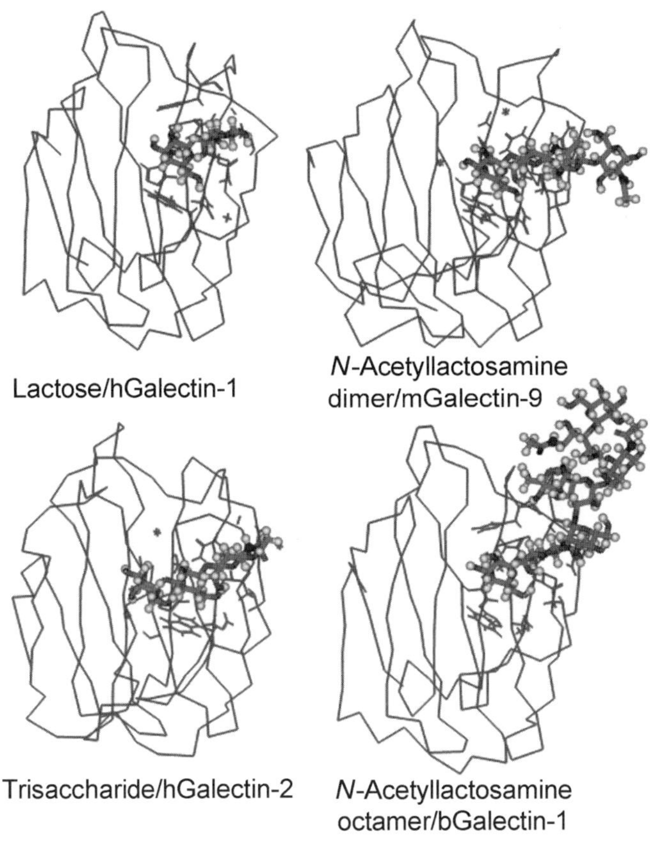

FIGURE 3.6 Carbohydrate-binding sites of some galectins. Human galectin-1 in complex with lactose (1gzw), mouse galectin-9 in complex with *N*-acetyllactosamine dimer (2d6o), human galectin-2 in complex with trisaccharide (1hlc), and bovine galectin-1 in complex with *N*- acetyllactosamine octamer (1sla) are shown. Binding site includes all atoms within the distance of 4.5 Å from saccharide molecule and is shown in magenta. (See color insert.)

Note in particular that the remainder of the *N*-acetyllactosamine octasaccharide then jets out from the galectin into solution. This binding state sets up the situation where another galectin dimer molecule can bind to the disaccharide at the free end of the octamer to allow for a scenario where multiple galectin–octasaccharides can polymerize. Because glycans to which galectin binds *in situ* are clearly much different structurally from this octasaccharide, this model may not occur *in vivo*.

In general, the number of ligand–CGL2 contacts is correlated with the binding affinities discussed in the previous section. For example, substitution of the hydroxyl group by deoxyacetamido in position 2 of the reducing sugar (glucose versus *N*-acetylglucose) increases binding affinity by about fivefold. This increase may be due to interaction of CGL2 with the acetyl group. Furthermore, derivatization of the β-galactoside in *N*-acetyllactosamine at the 3′ position with the charged substituent sialic acid increases binding affinity by twofold compared to that from LacNAc, and addition of the α1,2-fucoside further consolidates binding by reducing the K_d by fourfold over that of LacNAc. Perhaps the largest effect is observed with the linear B2 trisaccharide or Galili pentasaccharide, where the K_d value (1×10^{-6} M) is about 10-fold lower (i.e., better) than with LacNAc.

Working with more complex *N*-acetyllactosamine-based carbohydrates, Leppanen et al. reported that human galectin-1 binds to immobilized extended glycans (i.e., poly-*N*-acetyllactosamine, (-3Galβ1-4GlcNAcβ1-)$_n$ sequences, complex-type biantennary *N*-glycans, modified chitin-derived glycans, and even to native and desialylated human promyelocytic HL-60 cells) with similar K_d values of $2–4 \times 10^{-6}$ M (60). Interestingly, galectin-1-binding affinity fell when these glycans were free in solution (i.e., not immobilized), suggesting that the *N*-acetyllactosamine recognition element on surface-bound glycans is more optimally conformed for favorable interactions with galectin-1.

Working with even more complex carbohydrates, Hirabayashi et al. demonstrated that some galectins, like galectin-1 and galectin-7, showed increased binding affinity with an increase in branching number up to triantennary *N*-glycans (51). In general, binding affinity to branched *N*-glycans could be increased by up to about a factor of 5–10 compared to monovalent analogues. However, for other galectins like galectin-8, *N*-glycan branching leads to decreased affinity. Of further note, galectin-9 showed the greatest affinity for both repeated oligolactosamines and branched type *N*-glycans, with K_d values, for example, of 0.7×10^{-6} M for biantennary *N*-glycans and 0.16×10^{-6} M for both triantennary and tetraantennary *N*-glycans. These K_d values indicate about 100 times greater affinity for these branched glycans compared to analogous monovalent *N*-glycans.

Overall, these data suggest that binding of some galectins to glycans *in situ* will be much stronger than anticipated from results of routine galectin binding assays that employ standard disaccharides such as lactose. Moreover, this galectin-dependent variance of *N*-glycan binding (e.g., branched versus nonbranched) may be one way in which galectins could differentially modulate biological activity.

FUNCTIONAL RELATIONSHIPS

As pointed out by Barondes (1), galectins distinguish themselves by their selectivity to bind to β-galactosides. This is the essence of their function. Nevertheless, their functions can be quite different. For example, galectin-1 is known to induce T-cell apoptosis, whereas galectin-3 can suppress apoptosis and increase proliferation when overexpressed intracellularly in T cells (61, 62). Mechanistically, the function of any galectin on the molecular level may be quite complicated and multifaceted, and galectin self-association, along with the more complicated nature of glycans *in situ*, and their interactions with other biomolecules, both extracellularly and intracellularly, play crucial roles in determining and/or differentiating the functions of various galectins.

Functional Differentiation Through Self-Association

As discussed above, galectins can self-associate to form dimers, and in some cases, higher oligomers. But how does galectin self-association relate to galectin function? What is the oligomer state for galectins bound to actual cell surface glycans that are part of various glycoproteins or glycolipids? How does the presence of other carbohydrate groups adjacent (covalently or not) to the β-galactoside recognition element affect galectin binding to complex glycans? Little information is available to answer questions like these. One would think that self-association should account for at least some functional differentiation, primarily because monomer folds are essentially the same among all galectin CRDs and their ability to bind lactose, for example, is usually not affected by their oligomer state or quarternary structure.

As most is known about galectin-1, we shall start here. Galectin-1 is expressed in many tissues, predominantly of mesodermal origin, such as skeletal and smooth muscle, liver, lung, heart, skin, spleen, placenta, intestine, and kidneys. In addition, expression is also found on central and peripheral nervous tissues. During the development of an organism, galectin-1 is mainly restricted to the brain tissue and olfactory system, whereas with maturation it is limited to the peripheral nervous tissues. Overall, cell developmental dynamics, amounts of various binding partners such as glycans, and fluctuations in galectin concentration in different physiological states can dramatically affect galectin function.

Some information on a number of these variables is available in the literature. The influence of concentration of a galectin on its function actually reflects the effect of oligomer state on function. This is nicely exemplified by Adams et al., who reported that galectin-1-induced mitogenicity of human fibroblasts is reduced as the galectin-1 concentration is increased, whereas its effect on the growth of fibroblasts and HEP-2 carcinoma cells is enhanced at high concentrations (63). Although not mentioned in this paper by Adams et al., it appears that the key to understand this difference rests in the K_d value for the galectin-1 monomer–dimer equilibrium, that is, $1–7 \times 10^{-6}$ M (33, 34). From the concentration dependence (0.2–640 μg/mL) of their data (63), maximal cell growth inhibitory activity was observed at about 160 μg/mL (about 12×10^{-6} M), a concentration above the galectin-1 K_d, such that mostly galectin-1 dimers would be present. Moreover, the biological effect was negated at galectin-1 concentrations where monomers dominate (10 μg/mL and below), and the midpoint of the growth inhibition curve was about 40–80 μg/mL, where the concentration of monomers and dimers is about equal.

In the mitogenicity assay, however, the maximal effect from galectin-1 was observed at 1 μg/mL (0.07×10^{-6} M) (where monomers are dominant), and decreased as the galectin-1 concentration was increased, that is, as the population of galectin-1 dimers increased. This indicates that the monomer state mediates this mitogenic activity from galectin-1. While differences in the oligomer state may be the explanation, the situation may not be that simple. The presence of lactose only has an effect in the mitogenic assay and not on cell growth inhibition, implying that the effect of galectin-1 on cell growth is not dependent on its ability to bind cell surface glycans. Nevertheless, it is likely that both monomers and dimers can bind to cell surface glycans, as the CRD binding site is on the opposite side of galectin-1 compared to the prototype dimer interface. Furthermore, the presence/absence of various molecules with which galectin-1 (or any galectin) interacts can have significant effects in shifting binding in terms of monomer–dimer equilibrium. In this regard, the functional connection would be via mass action.

The functional importance of the galectin-1 oligomer state has been more clearly demonstrated (without the need for mass action explanations) by Nakabeppu's lab, who reported a naturally occurring form of galectin-1 (galectin-1β) that lacks the first 6 N-terminal residues (64) and remains monomeric (65) regardless of redox conditions. First of all, this observation substantiates the significance of the galectin-1 N-terminus in its ability to form dimers. More importantly, the galectin-1β monomer promotes axonal regeneration but not Jurkat cell death, unlike dimer-forming galectin-1, which promotes both (65). From this study, it is safe to conclude that the galectin-1 oligomer state (self-association) is important to function, in that galectin-1 monomers promote one activity (axonal regeneration) but not another (Jurkat cell death). This finding, however, raises other questions, for example, is the activity difference also mediated by β-galactoside interactions?

To address the question of which galectin oligomer state binds to cell surface glycans, Leppanen et al. examined the binding of a mostly monomeric form of a mutant galectin-1 and a chemically cross-linked galectin-1 dimer to poly-*N*-acetyllactosamine (-3Galβ1-4GlcNAcβ1-)$_n$, complex-type biantennary *N*-glycans, modified chitin-derived glycans, and native and desialylated human promyelocytic HL-60 cells (60). Both native galectin-1 and cross-linked galectin-1 dimers displayed similar carbohydrate-binding affinities that were almost fourfold greater than that for the mutant "monomer" galectin-1.

Even though this difference in binding affinities is somewhat small, these findings do suggest that native galectin-1 operates as a dimer to recognize *N*-acetyllactosamine units on complex glycans in its function to cross-link them on the surfaces of cells. For other galectins, the glycan-binding oligomer state certainly can differ. For example, Nieminen et al. reported that galectin-3 (via its N-terminal domain) oligomerizes, likely as pentamers (47), commensurate with its binding and cross-linking oligosaccharides (66).

In many papers on galectins, the oligomer state (in terms of the concentration of a given galectin used) is generally not considered (or not mentioned or thought about) when explaining function or functional differences. The first example of this is what we discussed above with galectin-1 (63). For another example, Karlsson et al. observed that neutrophils from peripheral blood were unresponsive to galectin-3, yet neutrophils harvested from a model inflammatory site were highly responsive (67). In this paper, the galectin-3 concentration used in their experiments was 20 μg/mL. Even though reported K_d values for galectin-3 (26×10^{-6} M (51), 1000×10^{-6} M (52), and 600×10^{-6} M (45)) vary considerably, they are all larger than the concentration of galectin-3 used in this study (about 1×10^{-6} M), such that galectin-3 monomers should be dominant. Therefore, activities observed at this or similar concentrations would mean that galectin-3 monomers would be the active state, and it would remain unknown whether galectin-3 dimers (or higher order oligomers) could elicit the same functional responses, as higher galectin-3 concentrations were not investigated.

Nieminen et al., however, have reported that various biological activities of galectin-3 do depend on its ability to form higher order oligomers. Based on fluorescence resonance energy transfer (FRET) studies, they proposed that galectin-3 oligomerization mediates cell activation/repression and cell adhesion via three different modes of action: receptor clustering, lattice formation, and cell–cell interactions (66). Nevertheless, the size of these oligomers remains unknown. Part of the problem here is that galectin-3 self-associates through its N-terminal nonlectin domain when the CRD domain binds to oligosaccharides, and a number of galectin-3 N-termini can associate to form these higher order oligomers, and thereby cross-link oligosaccharides on the cell surface.

There are other examples where the concentration of the galectin used has not really been taken into consideration when performing functional assays or interpreting functional data. For example, Henderson et al. used galectin-3 at a concentration of 30 μg/mL (about 1.5×10^{-6} M) to demonstrate that galectin-3 regulates myofibroblast activation and hepatic fibrosis (68). At this concentration, galectin-3 would be mostly monomeric, accepting its K_d values of $26–600 \times 10^{-6}$ M (51).

Certainly, the monomer–dimer equilibrium could be shifted *in situ* by binding to a glycoprotein, but this is unclear and was not investigated. However, Perillo et al. reported that galectin-1 at 20×10^{-6} M induced apoptosis in human thymocytes (69). In this case, galectin-1 should be mostly in the dimer state, although performing some variance on concentration would have helped address the question of whether the monomer state could be effective.

The point of this discussion is that the oligomer state of any galectin (aside from tandem-repeat galectin -4, -6, -8, and -9, which are monomers with two CRDs) could have significant consequences to biological function, and this should be investigated as one parameter when performing various functional assays with galectins or when designing galectin-based therapeutic agents. In addition, formation of dimer type (terminal, symmetric, and nonsymmetric) may be functionally important, as this may help differentiate how different galectins bind differently to complex glycans intra- or extracellularly on the surface of cells. These natural glycans are far more complex than simple disaccharides like lactose that have been used to study galectin carbohydrate binding and function.

Extracellular Function

Extracellular functions of galectins generally occur by their binding to and cross-linking of various glycan groups of glycoproteins and/or glycolipids on the surface of various cell types (70). In the past, galectins have been perhaps most generally associated with cells of the immune system, like leukocytes, as well as other cells in the cardiovascular system. More recently, some galectins have been associated with growth and differentiation of tumors. In general, galectins bind numerous glycoconjugates on the surfaces of various cells, and this binding must somehow relate to their function, in terms either of structural interactions that mediate cell–cell and/or cell–matrix adhesion and migration, or of signal transduction to effect intracellular changes necessary to, for example, a cell's growth and/or proliferation.

For example, prototype galectin-1 interacts with various glycoconjugate ligands of the extracellular matrix (e.g., laminin, fibronectin, β1 subunit of integrins, ganglioside GM1, and lysosomal membrane-associated proteins Lamp-1 and -2), as well as those on endothelial cells (e.g., integrins $\alpha_v\beta_3$ and $\alpha_v\beta_5$, ROBO4, CD36, and CD13) (71) and on T lymphocytes (e.g., CD7, CD43, and CD45) where it is known to induce apoptosis (72). Another prototype galectin, galectin-2, can induce exposure of cell membrane surface phosphatidylserine in activated neutrophils, but not in activated T cells (73), and has been associated with binding to lymphotoxin-α and myocardial infarction (74).

Little work has been done on galectin-5, which is expressed in bone marrow and erythrocytes, suggesting that it functions somehow in erythropoiesis (75). Galectin-7, specifically expressed by keratinocytes in the epidermis (76), is also associated with the induction of p53-induced apoptosis in keratinocytes (77), as well as in colon carcinoma (78). Galectin-10 is expressed at very high levels in eosinophil and basophil leukocytes and aggregates spontaneously to form Charcot–Leyden crystals (17). Galectin-13 specifically binds to annexin II and beta/gamma actin in placenta and fetal hepatic cells, which suggests that it is secreted to the outer cell surface by ectocytosis in microvesicles containing actin and annexin II.

Chimera galectin-3 is abundant in various epithelial cells and in macrophages, and can also bind to glycoconjugates of the extracellular matrix, such as laminin, fibronectin, vitronectin, elastin, N-CAM, LAMP1/2, and integrin $\alpha_3\beta_1$. Like galectin-1, galectin-3 can bind not only to CD43 and CD45 on leukocytes, but also to CD66, IgE, IgE receptor, and Mac-2-binding protein. Besides its constitutive expression, galectin-3 can be induced by inflammatory mediators (79), like CXC chemokine CXCL8 (interleukin-8, IL8) (80).

Interestingly, neutrophils from peripheral blood were found to be unresponsive to galectin-3, yet neutrophils harvested from a model inflammatory site were highly responsive (67), suggesting modification to their glycan groups. Moreover, whereas galectin-1 induces T-cell apoptosis, galectin-3 suppresses apoptosis and increases proliferation when overexpressed intracellularly in T cells (61, 62). Functionally, galectin-3 seems to be one of the most promiscuous galectins, as it displays a broad and diverse range of biological activities, including cell adhesion, apoptosis, immune regulation, as well as gene transcription regulation.

For tandem-repeat galectins (4, 6, 8, 9, 12), galectin-4 and galectin-6 (93% identical on a nucleotide level and 83% at the amino acid level (81)) tend to be restricted to epithelial cells of the alimentary tract from the tongue to the large intestine (76). Within the gastrointestinal tract, galectin-4 at least is known to bind to glycolipids in microvillar rafts, particularly of enterocytes, and to help to define and stabilize them (82). Galectin-8, the most abundant galectin in tumor cells of different origin (83), is closely related to the cell surface marker of prostate cancer, prostate carcinoma tumor antigen-1 (PCTA-1) (84). Galectin-8 binds to the cell surface ganglioside GM_3 (sialosyllactoseceramide), which associates with CD9 and CD82 to promote an antimetastatic effect [85, 86]. Galectin-9, which binds to extracellular matrix type IV collagen, Epstein-Barr virus latent membrane protein-1, and Tim-3 (a T helper type 1 specific cell surface protein), is an eosinophil-specific chemoattractant (87), functions as an urate transporter/channel UAT (88), and can induce apoptosis in T cells similar to galectin-1 and -3 (89).

The demonstration by Hirabayashi et al. (51) that stronger binding of some galectins (e.g., galectin-1, galectin-7, and galectin-9) to branched *N*-glycans, as well as weaker binding of others (e.g., galectin-8) to the same branched *N*-glycans, makes a more biologically relevant case for the situation that galectins will entertain *in situ*, that is, on the extracellular surface of cells. Certainly, when K_d values are in the 10×10^{-6} M to 100×10^{-6} M range (as is the case of galectin binding to lactose and related disaccharides), carbohydrate binding should be considered rather weak, and a relatively high concentration of a particular galectin may be required to elicit a solid biological response, especially when mass action of galectins to multiple glycan targets is taken into account. Therefore, having lower K_d values (stronger binding) with branched glycans is a biologically realistic finding, simply because this better reflects the nature of glycan groups on the surface of cells. Moreover, the galectin-dependent variance of *N*-glycan binding (e.g., branched versus nonbranched) may be one way in which galectins can differentially modulate their biological activity. Galectin function, therefore, may be differentiated by how one or another galectin interacts with specific sites on complex glycans found on the surfaces of various cell types mentioned above.

In this regard, structural similarity among all galectin CRDs raises the question of how this functional differentiation occurs. In other words, how can the functions of various galectins be differentiated given that their CRDs essentially have the same structural folds. Part of the answer lies in the fact that while most amino acid residues within the actual carbohydrate-binding domain are highly conserved among all galectins, other residues are not. Complex glycans found on the surface of cells may in fact interact with galectin amino acid residues and their formed surface patches that are not within, or are more distant from, the "lactose disaccharide defined" carbohydrate-binding domain. Moreover, differences in self-association profiles, organization of different galectins on the same glycoconjugate(s), and the possibility of galectin heterooligomerization (48) could all contribute to functional differentiation.

In an interesting study aimed at correlating galectin function and structure, Andre et al. combined computational modeling and the use of *in vitro* functional assays for insight into the functional overlap and divergence among some prototype galectins (primarily galectins 1, 2, and 5, and some cross-species comparisons), with respect to their folded structures and glycan/β-galactoside-binding properties (90). By *in silico* modeling galectin-1 interactions with the ganglioside GM_1 as a template, Andre et al. defined equivalent positions for substitutions in other prototype galectins, for example, Lys63 versus Leu60/Gln72 in galectin-2 and galectin-5, respectively. Briefly, they found from their *in silico* work and solid-phase assays that these galectins have affinity for the pentasaccharide of GM_1, whereas when the ganglioside was presented on the natural cell surface (human SK-N-MC neuroblastoma cells), galectin-5, for example, did not interact. In addition, they observed that a monomeric prototype galectin (CG-14) could impair galectin-1-dependent negative growth control by competitively blocking access to the shared ligand without acting as an effector. Overall, they concluded that the quaternary structure of prototype galectins can lead to functional divergence (90).

This conclusion is consistent with the findings of Nakabeppu et al. who reported that a monomeric form of galectin-1 (galectin-1β, lacking the first 6 N-terminal residues that are part of the dimer interface) promotes axonal regeneration but not Jurkat cell death, unlike dimer-forming galectin-1, which promotes both (65). Nevertheless, much more work needs to be done to delineate functional effects and correlate them with galectin structure (both tertiary and quarternary) and glycan binding on the molecular level.

Intracellular Function

Even though the multitude of reports on galectins is focused on their extracellular activities, there is growing evidence that at least some of them also play significant functional roles intracellularly. In this regard, information on at least two galectins, galectin-1 and galectin-3, has been reported, and it appears that this aspect of their function is mediated by interaction with various intracellular molecules via a galectin protein surface domain that is different from their β-galactoside-binding domain.

It has been known for a while now that galectin-1 can both interact primarily with H-Ras (and to some extent with K-Ras) to strengthen its association with the intracellular membrane and mediate H-Ras-GTP loading (91). Ras is certainly well known for its ability to transform cells, and this activity of Ras is dependent upon its intracellular membrane anchorage, which in turn is dependent upon covalent linkage of a C-terminal cysteine of Ras to a farnesyl moiety, the lipid part of which is anchored within the intracellular membrane. Because more structural information is available on Ras/galectin-1 interactions, we will focus our attention here on galectin-1.

Several reports are key to understand the structural interactions between Ras and galectin-1 (and galectin-3 by homology). First of all, Rotblat et al. identified a hydrophobic surface on galectin-1 (92) that is analogous to the Cdc42 geranylgeranyl-binding site on RhoGDI from the X-ray structure of the RhoGDI/Cdc42 complex (93). By comparing the Cdc42 geranylgeranyl-binding cavity on RhoGDI with the surface of homologously folded galectin-1, Rotblat et al. identified a homologous hydrophobic isoprenoid-binding site on galectin-1 (92). This proposed galectin-1 farnesyl-like binding site incorporates galectin-1 residues L9, L11, L17, F30, F32, and I128, which all lie within the β-sheet sandwich fold of galectin-1, that is, not on the solvent-exposed surface of galectin-1. Because a binding analysis suggested that L11 was most crucial for interactions with a farnesyl-like group, Rotblat et al. used site-directed mutagenesis to generate a L11A mutant of galectin-1 and provide proof for their model (92). Their work demonstrated that the L11A mutant destabilized Ras-GTP and its association with the membrane, and subsequently inhibited Ras function (fibroblast transformation and PC12-cell neurite outgrowth).

Moreover, the L11A mutant possessed normal galectin-1 carbohydrate-binding and dimerization properties, allowing them to conclude that intracellular galectin-1/Ras interactions are not dependent on galectin-1 β-galactoside (lactose) binding (92). Thus, independently of carbohydrate binding, native galectin-1 cooperates with Ras, whereas the galectin-1 L11A mutant inhibits it. Whether H-Ras/galectin-1 interactions indeed involve insertion of the farnesyl moiety of Ras into the described hydrophobic pocket in galectin-1 will need to be validated structurally. However, even at this point, it seems unlikely that the farnesyl group, while covalently attached to Ras, could interact in this way with galectin-1, as it would probably involve rather extreme conformational changes to galectin-1.

In a related study, Ashery et al. later confirmed that hydrophobic prenyl-binding surfaces on galectin-1 and galectin-3 interact with the farnesyl group on the C-terminus of Ras, thereby contributing to the prolongation of Ras signals in the plasma membrane (94). More recently, Gorfe et al. (95) modeled the structure of farnesyl-modified H-Ras protein in a DMPC bilayer and showed that part of the farnesyl group (the farnesyl cysteine carboxymethylester) in fact lies over a hydrophobic domain on the β-sheet surface of Ras, which in many respects resembles a part of the β-sheet domain in galectins, reminiscent of the Rotblat et al. study discussed above. It may be that one face of the β-sheet sandwich of galectin-1 interacts with the β-sheet surface of Ras, as well as with the farnesyl moiety, in a sort of hydrophobic ternary complex.

Aside from being found extracellularly, galectin-3 is also present in the nucleus and cytoplasm as a multifunctional oncogenic protein. In this regard, galectin-3 has been shown to associate intracellularly with Ras (94), as well as with other cytosolic molecules. Much earlier, Yang et al. demonstrated that galectin-3 interacts with Bcl-2, and concluded that, via this interaction, galectin-3 is a regulator of cell growth and apoptosis (61). Because the Bcl-2/galectin-3 interaction was inhibited by saccharide ligands of galectin-3, the interaction likely occurs through the galectin carbohydrate-binding domain, suggesting that certain intracellular glycoconjugates may have a profound effect through binding to galectin-3, or that carbohydrate binding to galectin-3 modifies its conformation such that the interaction with Bcl-2 is attenuated.

In any event, galectin-3 appears to be quite promiscuous. Paron et al. showed that galectin-3 also interacts with thyroid-specific transcription factor TTF-1 (albeit transiently in the presence of DNA to which the TTF-1 homeodomain binds), suggesting a role of galectin-3 in controlling proliferation and tumor progression in thyroid cancer (96). In yet another study, Yu et al. found that galectin-3 interacts with synexin (annexin VII, a Ca^{2+} and phospholipid-binding protein) that mediates galectin-3 translocation/trafficking to the perinuclear mitochondrial membrane, where it regulates mitochondrial integrity and cytochrome *c* release critical for apoptosis regulation (97).

Overall, the recognition of the roles that galectins in general play in controlling and mediating various biological processes is ever growing. Because of this, it is becoming more apparent that galectins should be good targets for therapeutic intervention in the clinic.

FUNCTIONAL ANTAGONISTS

The most obvious site on galectins to target when designing antagonists is the carbohydrate-binding site, and there are numerous compounds that have been synthesized to bind specifically at this site and compete with binding of natural, primarily lactose-based disaccharide, ligands to various galectins. However, antagonist binding to a specific galectin is paramount to develop an effective therapeutic, and this has been one of the major problems with discovering highly effective galectin antagonists, along with improving binding affinity of the designed therapeutic for efficient competition with the native glycan ligands *in situ.* Anti-galectin compounds could have therapeutic value as anti-inflammatory (98) and as anti-cancer agents (99).

Even though galectin-3, given its highly promiscuous nature, is a good target against which to develop a therapeutic antagonist, the observation that galectin-1 does not appear to be involved in normal bodily processes, such as wound healing (100), suggests that the increased expression of galectin-1 found in human tumors (101) renders this protein an excellent target for therapeutic purposes in oncology. In the following sections, various examples of galectin antagonists will be presented.

Targeting the Carbohydrate-Binding Site

Given the clear structural and functional significance of the galectin CRD carbohydrate-binding site, many investigators have focused their efforts on this site to design galectin antagonists. Moreover, the multivalent way in which galectins can cross-link glycans and induce aggregation has been exploited in the design of some of these antagonists. Most types of antagonists are based on derivatives of lactose, in one way or another.

The Leffler and Nilsson labs, in particular, have pioneered the design and synthesis of a number of lactose-based galectin antagonists. Salameh et al.

synthesized 3-(1,2,3-triazol-1-yl)-1-thio-galactosides that are small and hydrolytically stable inhibitors of galectin-3 (102). The best compounds in this class have K_d values down to 107×10^{-6} M, suggesting that they would be nearly as potent as the natural disaccharide inhibitors lactose and *N*-acetyllactosamine. Tejler et al. synthesized a panel of anomeric oxime ether derivatives of β-galactose (*O*-galactosyl aldoximes) as potent LacNAc-mimetic galectin-3 inhibitors, and using a competitive fluorescence polarization assay, they identified a number of galectin-3 inhibitors, with the best compound having a K_d value of 180×10^{-6} M (103).

Cumpstey et al. synthesized a small library (28 compounds) of phenyl thio-β-D-galactopyranoside analogues (104). Although any of these antagonists could bind to any of the galectins screened (galectins 1, 3, 7, 8, and 9), their best compound bound to galectin-7 with a significantly higher affinity (K_d as low as 140×10^{-6} M) than to the other galectins. Nevertheless, some compounds in this class could bind to all these five galectins with K_d values in the 0.4–2 mM range, raising the question of specificity. Salameh et al. also synthesized thioureido *N*-acetyllactosamine derivatives as galectin-7 and galectin-9 inhibitors (105). The best of these inhibitors displayed K_d values against galectin-7 and galectin-9 of 23×10^{-6} M and 47×10^{-6} M, respectively, or about one order of magnitude better than that for parent *N*-acetyllactosamine.

For structural insight into the affinity-enhancing effect from an aromatic-containing lactose-based analogue, Sorme et al. solved the crystal structure of the complex between galectin-3 and one of their lactose-based compounds having an aromatic 4-methoxy-2,3,5,6-tetrafluorobenzamido moiety (13). Basically, they found binding affinity of this compound to galectin-3 was improved by the presence of the aromatic (arene) group that stacked against Arg144 of galectin-3 (analogous to Arg73 of galectin-1 as illustrated in Fig. 3.4). Using structure-based design and fine-tuning the arginine–arene interactions, they then were able to synthesize another analogue with even greater affinity for galectin-3 (K_d of about 320×10^{-9} M).

In follow-up work, Cumpstey et al. (106) again used structure-based design to improve upon this initial arene compound (13) and synthesized double arene thiodigalactoside bisbenzamido analogues with arene groups on each end of the carbohydrate moiety. In this way, they demonstrated that their double arene antagonists could interact with two arginine residues (Arg144 and Arg186) within the carbohydrate-binding domain of galectin-3. In these double arene analogues, affinity was increased by about 10-fold, with K_d values as low as 33×10^{-9} M. Having double arginine–arene interactions clearly improved binding affinity and galectin inhibition.

The play on attaching hydrophobic groups to a lactose-based compound took a different slant when Fort et al. attached linear alkyl chains of varying length to the anomeric carbon of the glucose or *N*-acetylglucose ring of lactose (107). By screening their synthetic "lacto-*N*-biose" libraries for binding to galectin-3 using microscale affinity chromatography and mass spectrometry, they identified compounds with K_d values ranging from 11×10^{-6} M to 73×10^{-6} M.

As mentioned above, the normally inherent galectin multivalency for carbohydrate binding has been exploited in the design of various galectin antagonists. Tejler et al. functionalized unnatural amino acids (phenyl-bis-alanine and phenyl-tris-alanine) with 2-azidoethyl β-D-galactopyranosyl-(1–4)-β-D-glucopyranoside, effectively creating multivalent lactose derivatives as galectin antagonists (108). The resulting analogues were screened for binding to galectins 1, 3, 4, 7, 8, and 9. While binding specificity against these galectins was not impressive, the best analogue (a divalent "lactose" compound) did show a relatively good K_d value for galectin-1 of 3.2×10^{-6} M, a value that was about one order of magnitude higher affinity than for any other galectin tested. One of their trivalent "lactose" analogues also had reasonably good affinity (K_d of 22×10^{-6} M) against galectin-4.

The Kiss lab has also been active in this area, as they too synthesized a bilactosylated steroid-based compound with antigalectin-1 activity (109). The multivalent design approach was also exploited by Rabinovich et al. (110) who synthesized lactulose amine compounds (essentially polymethylene-spaced dilactoseamine derivatives) that demonstrated apparently selective effects in different events linked to tumor cell apoptosis, cell aggregation, and endothelial cell morphogenesis, suggesting that subtle differences in carbohydrate structures may be potentially useful to block tumor growth and metastasis.

To design glycomimetic inhibitors, Sirois et al. used a QSAR modeling approach by *in silico* docking a training set of 136 compounds taken from the literature to the 3D structure of galectin-3 (111). This was done to establish correlations between molecular properties and binding affinities (K_d values). The structured-based QSAR approach led to the conclusion that selective and potent inhibitors of galectin-3 could be designed by placing special emphasis on modification at both the carbohydrate C-3′ and O-3 positions.

At about the same time, Giguere et al. observed that compounds having the lowest O-3 electron density also have the highest inhibitory potency (112). The O-3 electron density calculations with galectin-1 showed that electron-withdrawing aglycons decreased the O-3 electron density and favored interaction with Glu71 of galectin-1 (see Fig. 3.4), which accounted for about 50% of the hydrogen bonding potential. They then synthesized aryl *O*- and *S*-galactosides and lactosides as inhibitors of galectin-1 and galectin-3, and identified compounds with K_d values in the 40×10^{-6} M to 2500×10^{-6} M range. Giguere et al. also synthesized carbohydrate triazoles and isoxazoles as inhibitors of galectin-1 and galectin-3 (113), similar to the work of Tejler et al. (108), but with a different linker scaffold. The compound with the best activity exhibited a K_d value of about 20×10^{-6} M, somewhat less potent than the Tejler divalent compound discussed above.

Numerous other papers have been published that report on various lactose-like-based compounds that target the carbohydrate-binding site of galectins. However, it does not appear that any of them, including those discussed above, have been tested in galectin-based functional assays *in vitro* or *in vivo*. In fact, most of these studies assume that if the binding affinity (K_d) to any given galectin is better than that of any other known compound(s) that inhibit(s) galectin-related biological function, then the new compound(s) will as well, but that argument is debatable.

In general, one might anticipate any number of problems with any of these galectin CRD-targeted lactose-like-based compounds when they are tested in an actual biological setting.

First of all, galectin-binding affinities for most, if not all, of these reported compounds in any of these chemical classes are generally relatively weak, with K_d values ranging from about 100×10^{-6} M to millimolar. Second, it is highly likely that specificity of any of these compounds will be a major issue. It seems that while K_d values may be some 10-fold better for one compound over another, there still remains relatively good cross-reactivity, bringing the question of specificity to the forefront. These compounds may be only pan-galectin specific, but this would still likely make them relatively poor therapeutic agents in the clinic, in addition to increasing the probability of eliciting various unwanted side effects. Finally, because of the issues mentioned above, lack of *in vivo* exposure is the overriding major concern. Unless such possible issues can be overcome, it appears that targeting the carbohydrate-binding site of any galectin with lactose-like-based compounds will remain an academic exercise.

Targeting the Oligomer State

As we have already noted functional differences resulting from shifts in the galectin oligomer state, another obvious design approach is to target the oligomer state and identify (a) compound(s) that can shift the equilibrium to one oligomer state or another, for example, monomer or dimer. Because each CRD contains one carbohydrate-binding site, galectin dimerization (and oligomerization) can mediate intermolecular cross-linking by binding to more than one carbohydrate moiety. Especially in the extracellular environment, galectin bioactivity is thought to be mediated by cross-linking β-galactosides found on different glycoconjugates. In fact, since Nakabeppu et al. demonstrated that the naturally occurring form of galectin-1 that lacks the first 6 N-terminal residues (galectin-1β) remains monomeric (65), galectin-1β could be such a therapeutic candidate, as it promotes axonal regeneration but not Jurkat (leukocyte) cell death, unlike dimer-forming galectin-1 that promotes both activities.

Targeting Other Sites

Recently, galectin-1 was identified as the molecular target of the designed antiangiogenic peptide anginex (101). Previously, anginex was known to inhibit the adhesion and migration of endothelial cells (EC) to the extracellular matrix (114), and galectin-1 had already been known as being integral to efficient EC adhesion and migration (115), particularly to highly proliferative, angiogenically-activated EC within tumors (101). BIAcore analysis yielded a K_d value of about 90×10^{-9} M for the binding of anginex to galectin-1 (101). Since galectin-1 is crucial to several processes required for tumor growth, this may explain why anginex displays multimodal activities (inhibition of tumor EC proliferation (114) and promotion of leukocyte infiltration into tumors (116, 117) that lead to tumor growth inhibition in mouse models (118–120). For instance, interfering with galectin-1 function (i) prevents tumor angiogenesis (101), (ii) abrogates tumor escape from immunity through blockade of galectin-1-induced apoptosis in activated T lymphocytes (121), and (iii) prevents metastasis formation through inhibition of galectin-1 facilitated tumor cell–EC interactions (122).

NMR spectroscopic HSQC chemical shift mapping using uniformly ^{15}N-enriched galectin-1 was employed to identify the anginex-binding site on galectin-1, which lies on a mostly hydrophobic patch on the surface of galectin-1 near to, but not on top of, the carbohydrate-binding site (Nesmelova et al., unpublished results). Therefore, binding of anginex to galectin-1 does not block binding of lactose to galectin-1, nor does its binding block the interaction of galectin-1 to glycans on the endothelial cell surface (101). However, since anginex does block adhesion of endothelial cells to the extracellular matrix (114), it is likely that anginex functions to interfere somehow with the molecular interactions between complementary components of the matrix and the endothelial cell surface. A structure-based approach has been used to design both a partial peptide mimetic (123) and a fully nonpeptide, protein surface topomimetic (120) of anginex (124). Both of these agents are antagonists of galectin-1 function, with improved activity over anginex.

The site where anginex and its mimetics bind to galectin-1 is different from the proposed farnesyl-like isoprenoid-binding domain on galectin-1 reported by Rotblat et al. (92). The farnesyl-like binding pocket is actually modeled to lie within the interior surface of the β-sandwich of galectin-1, essentially incorporating hydrophobic residues L9, L11, L17, F30, F32, and I128. As mentioned above, the anginex-binding site on galectin-1, while primarily hydrophobic in nature as well, lies on the outer surface of galectin-1. Moreover, because the farnesyl-like binding pocket appears to be sandwiched within the galectin-1 fold, it is unlikely that an antagonist against this site could be developed as an effective therapeutic agent.

Two other peptides have been identified as potential galectin antagonists that inhibit metastasis-associated cancer cell adhesion. Using combinatorial bacteriophage display, Zou et al. identified 13 peptides, two of which (G3-A9 and G3-C12, with amino acid sequences PQNSKIPGPTFLDPH and ANTPCGPYTHDCPVKR, respectively) bound specifically to galectin-3 (and not to any other galectins or plant lectins) and with relatively high affinity (K_d of 80×10^{-9} M) (125). Both peptides recognized cell surface galectin-3 on cultured carcinoma cells and monocytes, blocked interaction between galectin-3 and TFAg (Thomsen–Friedenreich glycoantigen), and inhibited adhesion of human breast carcinoma cells to endothelial cells under flow conditions. However, the *in vivo* effectivity of these peptides is not known, as animal model studies have not been performed.

Moreover, although these peptides bind galectin-3 and inhibit its biological function *in vitro*, it remains unclear where they bind on the surface of galectin-3, or if they actually block the binding of carbohydrate ligand to galectin-3. Nevertheless, it seems likely that the peptides bind somewhere on the surface of the galectin and not directly in the carbohydrate-binding site, especially because both peptides carry a net positive charge like the actual carbohydrate-binding site.

Overall, it appears that there a number of sites on the surface of galectin-1 or galectin-3 that could be potential targets against which to design and optimize molecular antagonists of galectin function. However, it remains unknown whether any of these agents discussed above are specific for any one galectin.

CONCLUDING REMARKS

The beautiful simplicity of the highly conserved β-sheet sandwich structure common to all galectin CRDs is made more complicated by their ability to self-associate and form different types of dimers, and even higher order oligomers. While β-galactoside-directed glycan binding of their CRDs to various cell surface glycoconjugates (e.g., glycoproteins and glycolipids) appears paramount to their extracellular biological functions, galectins are now also becoming recognized to mediate various intracellular functions via interactions with nonglycosylated nuclear and cytosolic biomolecules. Given the rapidly growing importance of galectins in biology, and not only in immunology any more, numerous efforts are underway to identify effective antagonists of galectin function. While this chapter has discussed many of these drug discovery efforts, it is by no means exhaustive. Only recently have galectins been accepted as valid therapeutic targets for clinical intervention, and sometime in the near future it is likely that we will have one or more of these agents in the clinic to combat inflammatory diseases and cancer.

ACKNOWLEDGMENTS

We would like to thank our collaborates Professors Arjan Griffioen and Linda Baum, as well as Dr. Victor Thijssen, for introducing us to the world of galectins. In addition, we acknowledge the generous research support of the National Institutes of Health (National Cancer Institute grant CA-096090) that made possible work described in this chapter.

REFERENCES

1. Barondes SH, Castronovo V, Cooper DN, Cummings RD, Drickamer K, Feizi T, Gitt MA, Hirabayashi J, Hughes C, Kasai K. Galectins: a family of animal beta-galactoside-binding lectins. Cell 1994;76:597–598.
2. Leffler H. Galectins structure and function—a synopsis. Results Probl Cell Differ 2001;33:57–83.
3. Liu FT, Rabinovich GA. Galectins as modulators of tumour progression. Nat Rev Cancer 2005;5:29–41.
4. Camby I, Le Mercier M, Lefranc F, Kiss R. Galectin-1: a small protein with major functions. Glycobiology 2006;16:137R–157R.
5. Teichberg VI, Silman I, Beitsch DD, Resheff G. A beta-D-galactoside binding protein from electric organ tissue of *Electrophorus electricus*. Proc Natl Acad Sci USA 1975;72: 1383–1387.

6. Cooper DN, Barondes SH. God must love galectins; He made so many of them. Glycobiology 1999;9:979–984.
7. Cooper DN. Galectinomics: finding themes in complexity. Biochim Biophys Acta 2002;1572:209–231.
8. Hirabayashi J, Kasai K. The family of metazoan metal-independent beta-galactoside-binding lectins: structure, function and molecular evolution. Glycobiology 1993; 3:297–304.
9. Kasai K, Hirabayashi J. Galectins: a family of animal lectins that decipher glycocodes. J Biochem (Tokyo) 1996;119:1–8.
10. Lopez-Lucendo MF, Solis D, Andre S, Hirabayashi J, Kasai K, Kaltner H, Gabius HJ, Romero A. Growth-regulatory human galectin-1: crystallographic characterisation of the structural changes induced by single-site mutations and their impact on the thermodynamics of ligand binding. J Mol Biol 2004;343:957–970.
11. Lobsanov YD, Gitt MA, Leffler H, Barondes SH, Rini JM. X-ray crystal structure of the human dimeric S-Lac lectin, L-14-II, in complex with lactose at 2.9-Å resolution. J Biol Chem 1993;268:27034–27038.
12. Seetharaman J, Kanigsberg A, Slaaby R, Leffler H, Barondes SH, Rini JM. X-ray crystal structure of the human galectin-3 carbohydrate recognition domain at 2.1-Å resolution. J Biol Chem 1998;273:13047–13052.
13. Sorme P, Arnoux P, Kahl-Knutsson B, Leffler H, Rini JM, Nilsson UJ. Structural and thermodynamic studies on cation–Pi interactions in lectin–ligand complexes: high-affinity galectin-3 inhibitors through fine-tuning of an arginine–arene interaction. J Am Chem Soc 2005;127:1737–1743.
14. Tomizawa T, Kigawa T, Saito K, Koshiba S, Inoue M, Yokoyama S. Solution structure of the C-terminal gal-bind lectin domain from human galectin-4. Forthcoming.
15. Leonidas DD, Vatzaki EH, Vorum H, Celis JE, Madsen P, Acharya KR. Structural basis for the recognition of carbohydrates by human galectin-7. Biochemistry 1998;37: 13930–13940.
16. Swaminathan GJ, Leonidas DD, Savage MP, Ackerman SJ, Acharya KR. Selective recognition of mannose by the human eosinophil Charcot–Leyden crystal protein (galectin-10): a crystallographic study at 1.8 Å resolution. Biochemistry 1999;38: 13837–13843.
17. Leonidas DD, Elbert BL, Zhou Z, Leffler H, Ackerman SJ, Acharya KR. Crystal structure of human Charcot–Leyden crystal protein, an eosinophil lysophospholipase, identifies it as a new member of the carbohydrate-binding family of galectins. Structure 1995;3: 1379–1393.
18. Ackerman SJ, Liu L, Kwatia MA, Savage MP, Leonidas DD, Swaminathan GJ, Acharya KR. Charcot–Leyden crystal protein (galectin-10) is not a dual function galectin with lysophospholipase activity but binds a lysophospholipase inhibitor in a novel structural fashion. J Biol Chem 2002;277:14859–14868.
19. Bourne Y, Bolgiano B, Liao DI, Strecker G, Cantau P, Herzberg O, Feizi T, Cambillau C. Crosslinking of mammalian lectin (galectin-1) by complex biantennary saccharides. Nat Struct Biol 1994;1:863–870.
20. Liao DI, Kapadia G, Ahmed H, Vasta GR, Herzberg O. Structure of *S*-lectin, a developmentally regulated vertebrate beta-galactoside-binding protein. Proc Natl Acad Sci USA 1994;91:1428–1432.

21. Nagae M, Nishi N, Murata T, Usui T, Nakamura T, Wakatsuki S, Kato R. Crystal structure of the galectin-9 N-terminal carbohydrate recognition domain from *Mus musculus* reveals the basic mechanism of carbohydrate recognition. J Biol Chem 2006;281: 35884–35893.

22. Bianchet MA, Ahmed H, Vasta GR, Amzel LM. Soluble beta-galactosyl-binding lectin (galectin) from toad ovary: crystallographic studies of two protein–sugar complexes. Proteins 2000;40:378–388.

23. Ahmed H, Pohl J, Fink NE, Strobel F, Vasta GR. The primary structure and carbohydrate specificity of a beta-galactosyl-binding lectin from toad (*Bufo arenarum* Hensel) ovary reveal closer similarities to the mammalian galectin-1 than to the galectin from the clawed frog *Xenopus laevis*. J Biol Chem 1996;271:33083–33094.

24. Varela PF, Solis D, Diaz-Maurino T, Kaltner H, Gabius HJ, Romero A. The 2.15 Å crystal structure of CG-16, the developmentally regulated homodimeric chicken galectin. J Mol Biol 1999;294:537–549.

25. Shirai T, Mitsuyama C, Niwa Y, Matsui Y, Hotta H, Yamane T, Kamiya H, Ishii C, Ogawa T, Muramoto K. High-resolution structure of the conger eel galectin, congerin I, in lactose-liganded and ligand-free forms: emergence of a new structure class by accelerated evolution. Structure 1999;7:1223–1233.

26. Shirai T, Matsui Y, Shionyu-Mitsuyama C, Yamane T, Kamiya H, Ishii C, Ogawa T, Muramoto K. Crystal structure of a conger eel galectin (congerin II) at 1.45 Å resolution: implication for the accelerated evolution of a new ligand-binding site following gene duplication. J Mol Biol 2002;321:879–889.

27. Shionyu-Mitsuyama C, Ito Y, Konno A, Miwa Y, Ogawa T, Muramoto K, Shirai T. *In vitro* evolutionary thermostabilization of congerin II: a limited reproduction of natural protein evolution by artificial selection pressure. J Mol Biol 2005;347:385–397.

28. Walser PJ, Haebel PW, Kunzler M, Sargent D, Kues U, Aebi M, Ban N. Structure and functional analysis of the fungal galectin CGL2. Structure 2004;12:689–702.

29. Ban M, Yoon HJ, Demirkan E, Utsumi S, Mikami B, Yagi F. Structural basis of a fungal galectin from *Agrocybe cylindracea* for recognizing sialoconjugate. J Mol Biol 2005; 351:695–706.

30. Kadoya T, Horie H. Structural and functional studies of galectin-1: a novel axonal regeneration-promoting activity for oxidized galectin-1. Curr Drug Targets 2005;6: 375–383.

31. Hirabayashi J, Kasai K. Effect of amino acid substitution by site-directed mutagenesis on the carbohydrate recognition and stability of human 14-kDa beta-galactoside-binding lectin. J Biol Chem 1991;266:23648–23653.

32. Tracey BM, Feizi T, Abbott WM, Carruthers RA, Green BN, Lawson AM. Subunit molecular mass assignment of 14,654 Da to the soluble beta-galactoside-binding lectin from bovine heart muscle and demonstration of intramolecular disulfide bonding associated with oxidative inactivation. J Biol Chem 1992;267:10342–10347.

33. Cho M, Cummings RD. Galectin-1, a beta-galactoside-binding lectin in Chinese hamster ovary cells. I. Physical and chemical characterization. J Biol Chem 1995;270: 5198–5206.

34. Giudicelli V, Lutomski D, Levi-Strauss M, Bladier D, Joubert-Caron R, Caron M. Is human galectin-1 activity modulated by monomer/dimer equilibrium?Glycobiology 1997;7:viii–x.

35. Gitt MA, Wiser MF, Leffler H, Herrmann J, Xia YR, Massa SM, Cooper DN, Lusis AJ, Barondes SH. Sequence and mapping of galectin-5, a beta-galactoside-binding lectin, found in rat erythrocytes. J Biol Chem 1995;270:5032–5038.
36. Madsen P, Rasmussen HH, Flint T, Gromov P, Kruse TA, Honore B, Vorum H, Celis JE. Cloning, expression, and chromosome mapping of human galectin-7. J Biol Chem 1995;270:5823–5829.
37. Leffler H. Introduction to galectins. Trends Glycosci Glycotechnol 1997;9:9–19.
38. Morris S, Ahmad N, Andre S, Kaltner H, Gabius HJ, Brenowitz M, Brewer F. Quaternary solution structures of galectins-1, -3, and -7. Glycobiology 2004;14:293–300.
39. Ahmad N, Gabius HJ, Kaltner H, Andre S, Kuwabara I, Liu F-T, Oscarson S, Norberg T, Brewer CF. Thermodynamic binding studies of cell surface carbohydrate epitopes to galectins-1, -3, and -7: evidence for differential binding specificities. Can J Chem 2002;80:1096–1104.
40. Rini JM. Lectin structure. Annu Rev Biophys Biomol Struct 1995;24:551–577.
41. Massa SM, Cooper DN, Leffler H, Barondes SH. L-29, an endogenous lectin, binds to glycoconjugate ligands with positive cooperativity. Biochemistry 1993;32:260–267.
42. Ochieng J, Platt D, Tait L, Hogan V, Raz T, Carmi P, Raz A. Structure–function relationship of a recombinant human galactoside-binding protein. Biochemistry 1993;32:4455–4460.
43. Kuklinski S, Probstmeier R. Homophilic binding properties of galectin-3: involvement of the carbohydrate recognition domain. J Neurochem 1998;70:814–823.
44. Woo HJ, Lotz MM, Jung JU, Mercurio AM. Carbohydrate-binding protein 35 (Mac-2), a laminin-binding lectin, forms functional dimers using cysteine 186. J Biol Chem 1991;266:18419–18422.
45. Hsu DK, Zuberi RI, Liu FT. Biochemical and biophysical characterization of human recombinant IgE-binding protein, an S-type animal lectin. J Biol Chem 1992;267: 14167–14174.
46. Mehul B, Bawumia S, Hughes RC. Cross-linking of galectin 3, a galactose-binding protein of mammalian cells, by tissue-type transglutaminase. FEBS Lett 1995;360: 160–164.
47. Ahmad N, Gabius HJ, Andre S, Kaltner H, Sabesan S, Roy R, Liu B, Macaluso F, Brewer CF. Galectin-3 precipitates as a pentamer with synthetic multivalent carbohydrates and forms heterogeneous cross-linked complexes. J Biol Chem 2004; 279:10841–10847.
48. Miyanishi N, Nishi N, Abe H, Kashio Y, Shinonaga R, Nakakita S, Sumiyoshi W, Yamauchi A, Nakamura T, Hirashima M, Hirabayashi J. Carbohydrate-recognition domains of galectin-9 are involved in intermolecular interaction with galectin-9 itself and other members of the galectin family. Glycobiology 2007;17:423–432.
49. Nishi N, Itoh A, Fujiyama A, Yoshida N, Araya S, Hirashima M, Shoji H, Nakamura T. Development of highly stable galectins: truncation of the linker peptide confers protease-resistance on tandem-repeat type galectins. FEBS Lett 2005;579:2058–2064.
50. Schwarz FP, Ahmed H, Bianchet MA, Amzel LM, Vasta GR. Thermodynamics of bovine spleen galectin-1 binding to disaccharides: correlation with structure and its effect on oligomerization at the denaturation temperature. Biochemistry 1998;37: 5867–5877.

51. Hirabayashi J, Hashidate T, Arata Y, Nishi N, Nakamura T, Hirashima M, Urashima T, Oka T, Futai M, Muller WE, Yagi F, Kasai K. Oligosaccharide specificity of galectins: a search by frontal affinity chromatography. Biochim Biophys Acta 2002;1572: 232–254.
52. Sparrow CP, Leffler H, Barondes SH. Multiple soluble beta-galactoside-binding lectins from human lung. J Biol Chem 1987;262:7383–7390.
53. Ideo H, Seko A, Ohkura T, Matta KL, Yamashita K. High-affinity binding of recombinant human galectin-4 to SO(3)(−) → 3Galbeta1 → 3GalNAc pyranoside. Glycobiology 2002;12:199–208.
54. Leffler H, Barondes SH. Specificity of binding of three soluble rat lung lectins to substituted and unsubstituted mammalian beta-galactosides. J Biol Chem 1986;261: 10119–10126.
55. Sujatha MS, Sasidhar YU, Balaji PV. Insights into the role of the aromatic residue in galactose-binding sites: MP2/6-311G++** study on galactose– and glucose–aromatic residue analogue complexes. Biochemistry 2005;44:8554–8562.
56. Alonso-Plaza JM, Canales MA, Jimenez M, Roldan JL, Garcia-Herrero A, Iturrino L, Asensio JL, Canada FJ, Romero A, Siebert HC, Andre S, Solis D, Gabius HJ, Jimenez-Barbero J. NMR investigations of protein–carbohydrate interactions: insights into the topology of the bound conformation of a lactose isomer and beta-galactosyl xyloses to mistletoe lectin and galectin-1. Biochim Biophys Acta 2001;1568:225–236.
57. Zhuang T, Leffler H, Prestegard JH. Enhancement of bound-state residual dipolar couplings: conformational analysis of lactose bound to Galectin-3. Protein Sci 2006;15:1780–1790.
58. Collins PM, Hidari KI, Blanchard H. Slow diffusion of lactose out of galectin-3 crystals monitored by X-ray crystallography: possible implications for ligand-exchange protocols. Acta Crystallogr D Biol Crystallogr 2007;63:415–419.
59. Umemoto K, Leffler H, Venot A, Valafar H, Prestegard JH. Conformational differences in liganded and unliganded states of Galectin-3. Biochemistry 2003;42: 3688–3695.
60. Leppanen A, Stowell S, Blixt O, Cummings RD. Dimeric galectin-1 binds with high affinity to alpha2,3-sialylated and non-sialylated terminal *N*-acetyllactosamine units on surface-bound extended glycans. J Biol Chem 2005;280:5549–5562.
61. Yang RY, Hsu DK, Liu FT. Expression of galectin-3 modulates T-cell growth and apoptosis. Proc Natl Acad Sci USA 1996;93:6737–6742.
62. Akahani S, Nangia-Makker P, Inohara H, Kim HR, Raz A. Galectin-3: a novel antiapoptotic molecule with a functional BH1 (NWGR) domain of Bcl-2 family. Cancer Res 1997;57:5272–5276.
63. Adams L, Scott GK, Weinberg CS. Biphasic modulation of cell growth by recombinant human galectin-1. Biochim Biophys Acta 1996;1312:137–144.
64. Nishioka T, Sakumi K, Miura T, Tahara K, Horie H, Kadoya T, Nakabeppu Y. FosB gene products trigger cell proliferation and morphological alteration with an increased expression of a novel processed form of galectin-1 in the rat 3Y1 embryo cell line. J Biochem (Tokyo) 2002;131:653–661.
65. Miura T, Takahashi M, Horie H, Kurushima H, Tsuchimoto D, Sakumi K, Nakabeppu Y. Galectin-1beta, a natural monomeric form of galectin-1 lacking its six amino-terminal

residues promotes axonal regeneration but not cell death. Cell Death Differ 2004;11: 1076–1083.

66. Nieminen J, Kuno A, Hirabayashi J, Sato S. Visualization of galectin-3 oligomerization on the surface of neutrophils and endothelial cells using fluorescence resonance energy transfer. J Biol Chem 2007;282:1374–1383.
67. Karlsson A, Follin P, Leffler H, Dahlgren C. Galectin-3 activates the NADPH-oxidase in exudated but not peripheral blood neutrophils. Blood 1998;91:3430–3438.
68. Henderson NC, Mackinnon AC, Farnworth SL, Poirier F, Russo FP, Iredale JP, Haslett C, Simpson KJ, Sethi T. Galectin-3 regulates myofibroblast activation and hepatic fibrosis. Proc Natl Acad Sci USA 2006;103:5060–5065.
69. Perillo NL, Uittenbogaart CH, Nguyen JT, Baum LG. Galectin-1, an endogenous lectin produced by thymic epithelial cells, induces apoptosis of human thymocytes. J Exp Med 1997;185:1851–1858.
70. Brewer CF. Binding and cross-linking properties of galectins. Biochim Biophys Acta 2002;1572:255–262.
71. Neri D, Bicknell R. Tumour vascular targeting. Nat Rev Cancer 2005;5:436–446.
72. Perillo NL, Pace KE, Seilhamer JJ, Baum LG. Apoptosis of T cells mediated by galectin-1. Nature 1995;378:736–739.
73. Stowell SR, Karmakar S, Stowell CJ, Dias-Baruffi M, McEver RP, Cummings RD. Human galectin-1, -2, and -4 induce surface exposure of phosphatidylserine in activated human neutrophils but not in activated T cells. Blood 2007;109:219–227.
74. Ozaki K, Inoue K, Sato H, Iida A, Ohnishi Y, Sekine A, Odashiro K, Nobuyoshi M, Hori M, Nakamura Y, Tanaka T. Functional variation in LGALS2 confers risk of myocardial infarction and regulates lymphotoxin-alpha secretion *in vitro*. Nature 2004; 429:72–75.
75. Lensch M, Lohr M, Russwurm R, Vidal M, Kaltner H, Andre S, Gabius HJ. Unique sequence and expression profiles of rat galectins-5 and -9 as a result of species-specific gene divergence. Int J Biochem Cell Biol 2006;38:1741–1758.
76. Colnot C, Ripoche MA, Scaerou F, Foulis D, Poirier F. Galectins in mouse embryogenesis. Biochem Soc Trans 1996;24:141–146.
77. Bernerd F, Sarasin A, Magnaldo T. Galectin-7 overexpression is associated with the apoptotic process in UVB-induced sunburn keratinocytes. Proc Natl Acad Sci USA 1999;96:11329–11334.
78. Polyak K, Xia Y, Zweier JL, Kinzler KW, Vogelstein B. A model for p53-induced apoptosis. Nature 1997;389:300–305.
79. Cherayil BJ, Weiner SJ, Pillai S. The Mac-2 antigen is a galactose-specific lectin that binds IgE. J Exp Med 1989;170:1959–1972.
80. Gil CD, La M, Perretti M, Oliani SM. Interaction of human neutrophils with endothelial cells regulates the expression of endogenous proteins annexin 1, galectin-1 and galectin-3. Cell Biol Int 2006;30:338–344.
81. Gitt MA, Colnot C, Poirier F, Nani KJ, Barondes SH, Leffler H. Galectin-4 and galectin-6 are two closely related lectins expressed in mouse gastrointestinal tract. J Biol Chem 1998;273:2954–2960.
82. Huflejt ME, Leffler H. Galectin-4 in normal tissues and cancer. Glycoconj J 2004;20:247–255.

83. Lahm H, Andre S, Hoeflich A, Fischer JR, Sordat B, Kaltner H, Wolf E, Gabius HJ. Comprehensive galectin fingerprinting in a panel of 61 human tumor cell lines by RT-PCR and its implications for diagnostic and therapeutic procedures. J Cancer Res Clin Oncol 2001;127:375–386.

84. Gopalkrishnan RV, Roberts T, Tuli S, Kang D, Christiansen KA, Fisher PB. Molecular characterization of prostate carcinoma tumor antigen-1, PCTA-1, a human galectin-8 related gene. Oncogene 2000;19:4405–4416.

85. Miura Y, Kainuma M, Jiang H, Velasco H, Vogt PK, Hakomori S. Reversion of the Jun-induced oncogenic phenotype by enhanced synthesis of sialosyllactosylceramide (GM3 ganglioside). Proc Natl Acad Sci USA 2004;101:16204–16209.

86. Satoh M, Ito A, Nojiri H, Handa K, Numahata K, Ohyama C, Saito S, Hoshi S, Hakomori SI. Enhanced GM3 expression, associated with decreased invasiveness, is induced by brefeldin A in bladder cancer cells. Int J Oncol 2001;19:723–731.

87. Matsumoto R, Matsumoto H, Seki M, Hata M, Asano Y, Kanegasaki S, Stevens RL, Hirashima M. Human ecalectin, a variant of human galectin-9, is a novel eosinophil chemoattractant produced by T lymphocytes. J Biol Chem 1998;273: 16976–16984.

88. Lipkowitz MS, Leal-Pinto E, Cohen BE, Abramson RG. Galectin 9 is the sugar-regulated urate transporter/channel UAT. Glycoconj J 2004;19:491–498.

89. Wada J, Ota K, Kumar A, Wallner EI, Kanwar YS. Developmental regulation, expression, and apoptotic potential of galectin-9, a beta-galactoside binding lectin. J Clin Invest 1997;99:2452–2461.

90. Andre S, Kaltner H, Lensch M, Russwurm R, Siebert HC, Fallsehr C, Tajkhorshid E, Heck AJ, von Knebel Doeberitz M, Gabius HJ, Kopitz J. Determination of structural and functional overlap/divergence of five prototype galectins by analysis of the growth-regulatory interaction with ganglioside GM1 *in silico* and *in vitro* on human neuroblastoma cells. Int J Cancer 2005;114:46–57.

91. Paz A, Haklai R, Elad-Sfadia G, Ballan E, Kloog Y. Galectin-1 binds oncogenic H-Ras to mediate Ras membrane anchorage and cell transformation. Oncogene 2001;20: 7486–7493.

92. Rotblat B, Niv H, Andre S, Kaltner H, Gabius HJ, Kloog Y. Galectin-1(L11A) predicted from a computed galectin-1 farnesyl-binding pocket selectively inhibits Ras-GTP. Cancer Res 2004;64:3112–3118.

93. Hoffman GR, Nassar N, Cerione RA. Structure of the Rho family GTP-binding protein Cdc42 in complex with the multifunctional regulator RhoGDI. Cell 2000;100:345–356.

94. Ashery U, Yizhar O, Rotblat B, Elad-Sfadia G, Barkan B, Haklai R, Kloog Y. Spatiotemporal organization of Ras signaling: rasosomes and the galectin switch. Cell Mol Neurobiol 2006;26:471–495.

95. Gorfe AA, Hanzal-Bayer M, Abankwa D, Hancock JF, McCammon JA. Structure and dynamics of the full-length lipid-modified H-Ras protein in a 1,2-dimyristoylglycero-3-phosphocholine bilayer. J Med Chem 2007;50:674–684.

96. Paron I, Scaloni A, Pines A, Bachi A, Liu FT, Puppin C, Pandolfi M, Ledda L, Di Loreto C, Damante G, Tell G. Nuclear localization of Galectin-3 in transformed thyroid cells: a role in transcriptional regulation. Biochem Biophys Res Commun 2003;302:545–553.

97. Yu F, FinleyJr RL, Raz A, Kim HR. Galectin-3 translocates to the perinuclear membranes and inhibits cytochrome *c* release from the mitochondria. A role for synexin in galectin-3 translocation. J Biol Chem 2002;277:15819–15827.
98. Liu FT. Galectins: novel anti-inflammatory drug targets. Expert Opin Ther Targets 2002;6:461–468.
99. Ingrassia L, Camby I, Lefranc F, Mathieu V, Nshimyumukiza P, Darro F, Kiss R. Anti-galectin compounds as potential anti-cancer drugs. Curr Med Chem 2006;13: 3513–3527.
100. Cao Z, Said N, Amin S, Wu HK, Bruce A, Garate M, Hsu DK, Kuwabara I, Liu FT, Panjwani N. Galectins-3 and -7, but not galectin-1, play a role in re-epithelialization of wounds. J Biol Chem 2002;277:42299–42305.
101. Thijssen VL, Postel R, Brandwijk RJ, Dings RP, Nesmelova I, Satijn S, Verhofstad N, Nakabeppu Y, Baum LG, Bakkers J, Mayo KH, Poirier F, Griffioen AW. Galectin-1 is essential in tumor angiogenesis and is a target for antiangiogenesis therapy. Proc Natl Acad Sci USA 2006;103:15975–15980.
102. Salameh BA, Leffler H, Nilsson UJ. 3-(1,2,3-Triazol-1-yl)-1-thio-galactosides as small, efficient, and hydrolytically stable inhibitors of galectin-3. Bioorg Med Chem Lett 2005;15:3344–3346.
103. Tejler J, Leffler H, Nilsson UJ. Synthesis of *O*-galactosyl aldoximes as potent LacNAc-mimetic galectin-3 inhibitors. Bioorg Med Chem Lett 2005;15:2343–2345.
104. Cumpstey I, Carlsson S, Leffler H, Nilsson UJ. Synthesis of a phenyl thio-beta-D-galactopyranoside library from 1,5-difluoro-2,4-dinitrobenzene: discovery of efficient and selective monosaccharide inhibitors of galectin-7. Org Biomol Chem 2005;3: 1922–1932.
105. Salameh BA, Sundin A, Leffler H, Nilsson UJ. Thioureido *N*-acetyllactosamine derivatives as potent galectin-7 and 9N inhibitors. Bioorg Med Chem 2006;14: 1215–1220.
106. Cumpstey I, Sundin A, Leffler H, Nilsson UJ. C2-symmetrical thiodigalactoside bis-benzamido derivatives as high-affinity inhibitors of galectin-3: efficient lectin inhibition through double arginine–arene interactions. Angew Chem Int Ed Engl 2005;44: 5110–5112.
107. Fort S, Kim HS, Hindsgaul O. Screening for galectin-3 inhibitors from synthetic lacto-*N*-biose libraries using microscale affinity chromatography coupled to mass spectrometry. J Org Chem 2006;71:7146–7154.
108. Tejler J, Tullberg E, Frejd T, Leffler H, Nilsson UJ. Synthesis of multivalent lactose derivatives by 1,3-dipolar cycloadditions: selective galectin-1 inhibition. Carbohydr Res 2006;341:1353–1362.
109. Ingrassia L, Mathieu V, Mgalizzi V, Lefranc V, Darro F, Kiss R. 2007.UNBS4209: a bilactosylated steroid with anti-galectin-1 activity. Proceedings of the 98th Annual Meeting of American Association for Cancer Research.
110. Rabinovich GA, Cumashi A, Bianco GA, Ciavardelli D, Iurisci I, D'Egidio M, Piccolo E, Tinari N, Nifantiev N, Iacobelli S. Synthetic lactulose amines: novel class of anticancer agents that induce tumor-cell apoptosis and inhibit galectin-mediated homotypic cell aggregation and endothelial cell morphogenesis. Glycobiology 2006; 16:210–220.
111. Sirois S, Giguere D, Roy R. A first QSAR model for galectin-3 glycomimetic inhibitors based on 3D docked structures. Med Chem 2006;2:481–489.

112. Giguere D, Sato S, St-Pierre C, Sirois S, Roy R. Aryl *O*- and *S*-galactosides and lactosides as specific inhibitors of human galectins-1 and -3: role of electrostatic potential at O-3. Bioorg Med Chem Lett 2006;16:1668–1672.

113. Giguere D, Patnam R, Bellefleur MA, St-Pierre C, Sato S, Roy R. Carbohydrate triazoles and isoxazoles as inhibitors of galectins-1 and -3. Chem Commun (Camb) 2006; 2379–2381.

114. Griffioen AW, van der Schaft DW, Barendsz-Janson AF, Cox A, Struijker Boudier HA, Hillen HF, Mayo KH. Anginex, a designed peptide that inhibits angiogenesis. Biochem J 2001;354:233–242.

115. Rabinovich GA. Galectin-1 as a potential cancer target. Br J Cancer 2005;92: 1188–1192.

116. Dirkx AE, oude Egbrink MG, Castermans K, van der Schaft DW, Thijssen VL, Dings RP, Kwee L, Mayo KH, Wagstaff J, Bouma-ter Steege JC, Griffioen AW. Anti-angiogenesis therapy can overcome endothelial cell anergy and promote leukocyte–endothelium interactions and infiltration in tumors. FASEB J 2006;20:621–630.

117. Hellebrekers DM, Castermans K, Vire E, Dings RP, Hoebers NT, Mayo KH, oude Egbrink MG, Molema G, Fuks F, van Engeland M, Griffioen AW. Epigenetic regulation of tumor endothelial cell anergy: silencing of intercellular adhesion molecule-1 by histone modifications. Cancer Res 2006;66:10770–10777.

118. Dings RP, Yokoyama Y, Ramakrishnan S, Griffioen AW, Mayo KH. The designed angiostatic peptide anginex synergistically improves chemotherapy and antiangiogenesis therapy with angiostatin. Cancer Res 2003;63:382–385.

119. Dings RP, Williams BW, Song CW, Griffioen AW, Mayo KH, Griffin RJ. Anginex synergizes with radiation therapy to inhibit tumor growth by radiosensitizing endothelial cells. Int J Cancer 2005;115:312–319.

120. Dings RP, Chen X, Hellebrekers DM, van Eijk LI, Zhang Y, Hoye TR, Griffioen AW, Mayo KH. Design of nonpeptidic topomimetics of antiangiogenic proteins with antitumor activities. J Natl Cancer Inst 2006;98:932–936.

121. Rubinstein N, Alvarez M, Zwirner NW, Toscano MA, Ilarregui JM, Bravo A, Mordoh J, Fainboim L, Podhajcer OL, Rabinovich GA. Targeted inhibition of galectin-1 gene expression in tumor cells results in heightened T cell-mediated rejection: a potential mechanism of tumor-immune privilege. Cancer Cell 2004;5:241–251.

122. Lotan R, Matsushita Y, Ohannesian D, Carralero D, Ota DM, Cleary KR, Nicolson GL, Irimura T. Lactose-binding lectin expression in human colorectal carcinomas. Relation to tumor progression. Carbohydr Res 1991;213:47–57.

123. Mayo KH, Dings RP, Flader C, Nesmelova I, Hargittai B, van der Schaft DW, van Eijk LI, Walek D, Haseman J, Hoye TR, Griffioen AW. Design of a partial peptide mimetic of anginex with antiangiogenic and anticancer activity. J Biol Chem 2003;278: 45746–45752.

124. Dings RP, Mayo KH. A journey in structure-based drug discovery: from designed peptides to protein surface topomimetics as antibiotic and antiangiogenic agents. Acc Chem Res 2007;40:1057–1065.

125. Zou J, Glinsky VV, Landon LA, Matthews L, Deutscher SL. Peptides specific to the galectin-3 carbohydrate recognition domain inhibit metastasis-associated cancer cell adhesion. Carcinogenesis 2005;26:309–318.

4

GALECTINS AS REGULATORS OF TUMOR GROWTH AND INVASION BY TARGETING DISTINCT CELL SURFACE GLYCANS AND IMPLICATIONS FOR DRUG DESIGN

HANS-JOACHIM GABIUS

Institut fur Physiologische Chemie, Ludwig-Maximilians-Universität, Tierärztliche Fakultät, Veterinärstr. 13, D-80539 München, Germany

ALBERT M. WU

Glyco-Immunochemistry Research Laboratory, Institute of Molecular and Cellular Biology, Chang-Gung University, Kwei-san, Tao-yuan 333, Taiwan

INTRODUCTION

What has descriptively been termed glycocalyx, an outer cell surface region rich in glycan chains attached to lipids and proteins, is currently receiving a clear functional definition. This is due to the growing realization that the carbohydrate part of cellular glycoconjugates is not simply a rather inert protective coating, irrespective of structural details. In contrast, its enormous complexity (the cellular glycome) is the platform to present a large array of oligosaccharides. This variety is generated by the enzymatic machinery for template-independent synthesis and has features of a fingerprint. In fact, cell type and developmental stage are reflected in the glycomic profile. Ironically, the key attribute of carbohydrates toward structural diversity, that is, the unsurpassed ability for isomer generation, has long been an obstacle

Galectins, Edited by Anatole A. Klyosov, Zbigniew J. Witczak, and David Platt

to accomplish sequence analysis. This factor explains the discrepancy in progress between work on nucleic acids and proteins as compared to investigations on glycans.

Viewing isomers as distinct biochemical signals or code words, the cell surface can thus be considered to be decorated by sugar-encoded messages, spatially accessibly presented by protein/lipid backbones. Biological information storage is therefore not confined to proteins and nucleic acids. The respective function of oligosaccharides is the fundamental principle of the concept of the sugar or glycan code (1, 2). Necessarily, these signals will have to be read by suitable decoding devices and then translated into responses on the level of cells. This purpose is fulfilled by endogenous lectins, their level of complexity in terms of inter- and intrafamily diversification matching that of coding glycan structures (3–6).

Fittingly, the cataloged sophistication of the structures at the branch-end sections, which are the primary sites of contact by a lectin, is equaled by evolutionary emergence and ramifications of several lectin families, which home in on distinct epitopes at this strategic position (4, 5). The galectins belong to this subgroup. Their name implies galactose to be the key binding partner. Due to its abundance in cellular glycans, a fairly high level of binding to cell surfaces with low selectivity might be expected. In fact, plant lectins targeting this carbohydrate moiety such as the ricin-like AB toxins provide precedents to strengthen this assumption (7).

GALECTINS: SENSORS FOR MORE THAN THE GALACTOSE UNIT

The preparation of labeled galectins and cytofluorimetric analysis of cell staining are straightforward means to resolve this issue. Respective experiments clearly revealed lactose-inhibitable cell surface staining, with low level of positivity in comparison to the mentioned plant lectins (7, 8). Evidently, cellular reactivity toward an endogenous lectin can be more restricted than for an exogenous protein, which shares monosaccharide specificity with the tissue lectin. While lactose is a convenient affinity ligand for galectin purification in general, inhibition assays with derivatives of lactose readily spotted substantial differences among naturally substituted β-galactosides in their capacity to interact with galectins (9, 10).

Remarkably, α2,6-sialylation precludes galectin binding, unless the *N*-acetyllactosamine (LacNAc) terminus is presented in tandem repeats. In this case, for example, in poly(LacNAc) sequences in the β1,6-antenna of complex-type *N*-glycans or core 2 extensions of mucin-type *O*-glycans, ligand properties for galectin-3 and -7 but not galectin-1 are maintained (11). If α2,3-sialylation caps a sugar branch, the extended epitope remains to be a galectin ligand, unless the subterminal portion is subject to fucosylation to generate a Lewis determinant, which in turn will be a docking point for the selectins (5, 9). These examples illustrate the close relationship between structural modifications of glycans, which can be attributed to regulatory mechanisms on the level of glycosyltransferases, glycosidases, and substrate availability, among other factors, and lectin-binding properties.

A shift from sialylation to other branch-end substitutions is a molecular switch on/off signal. Moreover, other substitutions of the lactose core such as α1,3-extensions modulate affinity to galectins differentially (9). As implied above, biochemical remodeling of the branch-end structure figures as fine-tuning mechanism.

Considering the inherent differential selectivity for various oligosaccharides, it is a current challenge to relate the known sequence divergence within the galectin family to the binding properties of individual proteins. Toward this end, it is instructive to examine the architecture of carbohydrate recognition domains. Amino acid substitutions in the vicinity of strictly conserved residues are supposed to bear upon ligand binding. By visualizing the main contact area for di- and tetrasaccharides, the respective sequence alterations can then readily be traced. This is seen for the Cys/Ala case in two closely related chicken galectins presented in Fig. 4.1. The differences of their preferences to glycans are summarized in Table 4.1 (12).

Two factors are yet to be reckoned with before drawing conclusions: (a) small-angle neutron scattering has revealed that binding of lactose to human galectin-1 causes a significant decrease in the lectin's gyration radius (13), and (b) even

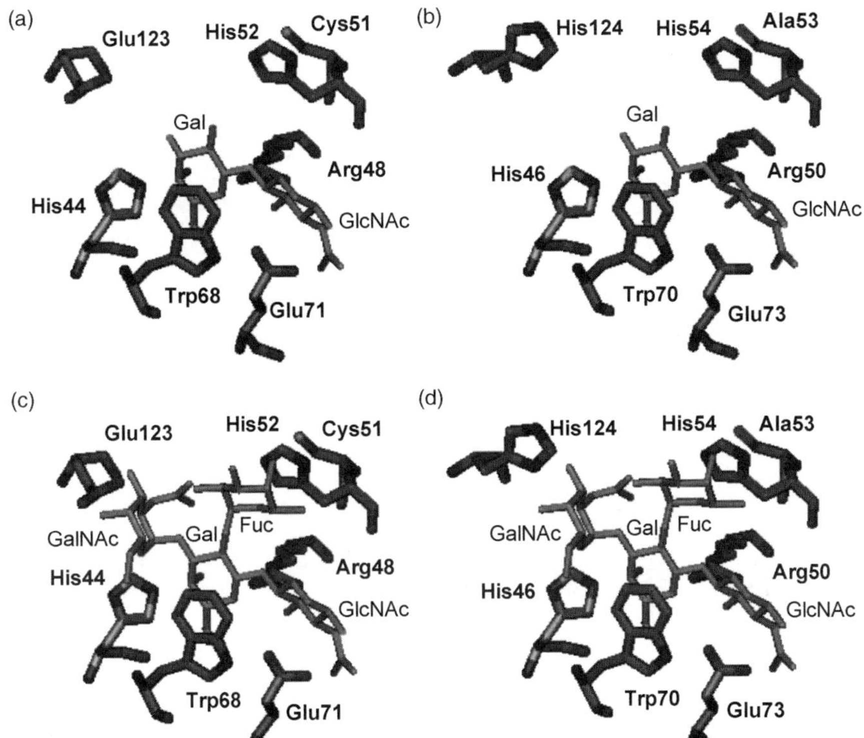

FIGURE 4.1 Graphical illustration of the carbohydrate recognition domains of CG-16 (left panel) and CG-14 (right panel). The contact sites for Galβ1-4GlcNAc (upper panel) and for histo-blood group A tetrasaccharide are given. The positioning of key contact sites was deliberately kept constant for direct comparison. Both sugars are drawn in low-energy conformations, and the comparison identifies the regions of each lectin site responsible for contact to the α1-2/3 extensions of the core galactose moiety (With permission from Reference 12). (See color insert.)

TABLE 4.1 Comparison of the Binding Properties of the Two Prototype Chicken Galectins CG-14 and CG-16

		Galectin Type	
No.	Carbohydrate Specificity	CG-14	CG-16
1	Monosaccharide specificity	β-Anomer of Gal and slightly enhanced by *N*-acetyl group in GalNAc	β-Anomer of Gal, reactivity reduced by *N*-acetyl group in GalNAc
2	(a) Ratio between α and β anomers for *p*-nitrophenyl galactoside	7.5 (α > β)	7.4 (α > β)
	(b) Ratio between α and β anomers for methyl galactoside	0.4 (β > α)	3.4 (α > β)
	(c) Hydrophobicity for Galα-anomer[a]	Yes	Yes
	(d) Hydrophobicity for Galβ-anomer[a]	No	Yes
3	Reactivity toward common branch-end epitopes expressed in decreasing order (based on nanomoles comparison)	Galβ1-4GlcNAc (**II**) > Galβ1-4Glc (**L**) > GalNAcβ1-3Gal (**P**) > GalNAcα1-3Gal (**A**) ≥ GalNAcα1-3GalNAc (**F**) ≥ Galβ1-3GalNAc (**T**) ≥ Galβ1-3GlcNAc (**I**) ≫ Galα1-3Gal (**B**) > Fucα1-2Gal (**H**) ≫ Galα1-4Gal (**E**)	Galβ1-4GlcNAc (**II**) > Galβ1-4Glc (**L**) > Galβ1-3GlcNAc (**I**) > Galα1-3Gal (**B**); Galβ1-3GalNAc(**T**) and Galα1-4Gal (**E**) were inactive
4	The most active β-galactoside	Galβ1-4GlcNAcβ1-3Galβ1-4Glc (**II**β1-3**L**) and Galβ1-4GlcNAc (**II**) and its **H**-type derivative	Galβ1-4GlcNAc (**II**) mainly, extension to **ABH** epitopes reducing activity

5	Ratio of glycotope clusters (simple multivalent form)/monomeric **II**	Triantennary glycopeptides with mostly type **II** termini and 2,4,2-branching pattern from asialofetuin was three times more active than monomeric **II**	Triantennary glycopeptides with mostly type **II** termini and 2,4,2-branching pattern from asialofetuin was three times more active than monomeric **II**
6	Substituted branch-end glycans	Histo-blood group **ABH** precursor (equivalent) gps and enhanced strongly by blood group **A**, **B** determinant sugar	Histo-blood group precursor (equivalent) gps, but hindered by **ABH** histo-blood group determinants
7	Ratio of complex polyvalent glycotopes in macromolecules/monomeric **II**	77 times more active than monomeric **II**	Only 5.5 times more active than monomeric **II**
8	The most complementary chain length	**H** active **I**β1-3**L** and Galβ1-4GlcNAcβ1-3Galβ1-4Glc (**II**β1-3**L**)	Galβ1-4GlcNAc (**II**) and Galβ1-3GlcNAcβ1-3Galβ1-4Glc (**I**β1-3**L**)

# denotes characterization number.

[a]Based on *p*-nitrophenyl to be more active than the corresponding methyl glycosides as Yes; *p*-nitrophenyl to be less active than the corresponding methyl glycosides as No; from Reference 12, with alterations.

substitutions distant to the lectin site can markedly affect the thermodynamics of ligand binding (14). These results inform us about conformational adaptation and long-range effects. To address the challenges of including flexibility at the protein level and a mutual adaptation between lectin and ligand during the molecular rendezvous, suitable modeling procedures are being devised ((12); for a movie-like illustration of the interplay between oligosaccharides and the two chicken galectins, please see this reference for details). Experimental input on the bound-state conformation of the ligand to enable modeling is provided by NMR spectroscopical methods (for further information, please see next section).

Moreover, the thorough analysis of binding data using glycoproteins as ligands revealed that the density of ligand presentation can make its mark on binding properties (15–18). Examining individual *N*-glycans, even core substitutions such as presence of a bisecting *N*-acetylglucosamine moiety affect the affinity of branch-end determinants to lectins (19–21). These results build evidence for core substitutions as modulators of ligand capacity in *N*-glycans by influencing conformational equilibria. When multivalent glycoproteins act as ligands, binding properties even become a function of binding-site occupancy.

In detail, interaction of galectins with the common test model asialofetuin with up to nine antennae in three N-glycosylation sites yielded negative cooperativity, when decreasing functional valence of the carbohydrates upon stepwise lectin binding (22). Intuitively, the mentioned factors inferred in model systems let it expect that only a limited set of cellular glycoconjugates reach the status to be physiologic and

TABLE 4.2 Binding Partners for Galectin-1 and -3

Type of Ligand	Galectin-1	Galectin-3
Glycan	CA125, CD2, CD3, CD4, CD7, CD43, CD45, CEA, fibronectin (tissue), GI mucin, hsp90-like glycoprotein, $\alpha_1/\alpha_4/\alpha_5/\alpha_7/\beta_1$- and $\alpha_4\beta_7$-integrins, L1, laminin, lamp-1, Mac-2 binding protein, thrombospondin, Thy-1, chondroitin sulfate proteoglycan, distinct neutral glycolipids, ganglioside GM_1	LI-cadherin, C4,4A (member of Ly6 family) IgE, CD7, CD11b of CD11b/CD18 (Mac-1 antigen, CR3), CD32, CD43, CD45, CD66a,b, CD71, CD95, CD98, CEA, colon cancer mucin, cubilin, haptoglobin β-subunit (after desialylation), hensin (DMBT-1), β_1-integrin (CD29), laminin, lamp-1/-2, Mac-2 binding protein, Mac-3, MAG, MP20 (tetraspanin), NG2 proteoglycan, TCR complex, tenascin, ganglioside GM_1
Protein	Gemin4, oncogenic H-Ras, OCA-B, pre-B cell receptor (human, not murine system)	AGE products, Alix/AIP-1, axin, bcl-2, β-catenin, Cys/His-rich protein, Gemin4, mSufu, nucling, oncoge-nic K-Ras, OCA-B, pCIP, PIAS1, synexin (annexin VII), TTF-1

From Reference 5, extended and modified.

functional binding partners. By compiling the literature evidence for the two most prominent and well-studied galectins, listing documented target glycoconjugates in different cell types, a rather small set is collected (Table 4.2). Case studies on galectin-1 underscore the functional significance of their interplay in distinct cell systems.

GALECTIN-1: A POTENT REGULATOR OF CELL GROWTH

This homodimeric lectin can downregulate proliferation in susceptible carcinoma cells. It appears to employ the fibronectin receptor ($\alpha_5\beta_1$-integrin) *in vitro* as contact site *en route* to increase the levels of the cyclin-dependent kinase inhibitors p21 and p27 in epithelial cancer lines, involving inhibition of the Ras-MEK-ERK pathway (23). The presence of the α_5-integrin subunit proved essential, and threonine phosphorylation of Sp1 was the molecular switch for the regulation of p27 promoter activity. The integrin is also involved in anoikis regulation by a tumor suppressor. A combined analysis using arrays for glycogenes and genes associated with pancreatic cancer, glycomic profiling by two-dimensional chromatography, and lectin staining as well as functional assays recently revealed that the tumor suppressor $p16^{INK4a}$ orchestrates a distinct protein–carbohydrate interaction to restore anoikis susceptibility.

In detail, upregulation of galectin-1 and remodeling of glycans to increase presentation of functional galectin-1 ligands led to death of cells in suspension, a chain of events initiated by the tumor suppressor (24). Looking at activated T cells as target, caspase-3, -8, and -9 were effectors of galectin-1-mediated apoptosis (25). The involvement of caspases showed characteristic galectin-type-dependent features with galectin-2 activating caspase-3 and -9 and galectin-7 caspase-1, -3, and -8 (25). At the level of the cell surface, the glycoprotein CD7 is a primary contact site eliciting the proapoptotic signaling by being cross-linked by galectin-1 (but not galectin-2), a molecular interplay manifested on the clinical level by protection of $CD4^+CD7^-$ cells against galectin-1-triggered T cell death in patients with Sézary syndrome (26, 27).

A variation of this theme with caspase-independent cell growth regulation is effective in human SK-N-MC neuroblastoma cells. In this case, the pentasaccharide of ganglioside GM_1 acts as a binding partner, whose upregulated presentation signals the transition from proliferation to differentiation due to increased cell surface ganglioside sialidase activity in culture (28–30).

The combination of NMR spectroscopical methods, especially two-dimensional transferred nuclear Overhauser enhancement and saturation transfer difference spectroscopy, with molecular modeling enabled to gain detailed insight into the conformation of the bound ligand (31). As previously determined for lactosides and digalactosides, the ligand is selected in a low-energy conformation without distortion (32, 33). Major contributions to the interaction energy (Coulomb/*van der Waals* energy terms) arise from the contact of the terminal galactose and *N*-acetylgalactosamine moieties (-8.5 and -10.0 kcal/mol, respectively). In addition, even the sialic acid residue at the branch point accounts for an energy contribution of −5.9 kcal/mol, underscoring the validity of the concept that the binding site of galectin-1 accommodates oligosaccharides beyond the terminal galactose unit (31).

What's more, the identification of the bound-state conformation of the pentasaccharide facilitated a comparison to the binding properties of cholera toxin. The result of this analysis illustrates what is meant by the third dimension of the sugar code, that is, the differential conformer selection of a ligand by lectins (1, 2, 31). In detail, the two proteins associate when bound to different conformers of the pentasaccharide: the Φ/Ψ-angle combinations for the Galα2-3Neu5Ac linkage are $70°/15°$ in complex with galectin-1 and $-172°/-26°$ when bound to the bacterial toxin (31). Obviously, these two conformers of the same oligosaccharide exhibit different shapes, and each shape has mutually exclusive ligand properties. In other words, arresting this glycan in a distinct shape will automatically restrict its range of ligand properties. This result has consequences for rational drug design to reduce *in vivo* cross-reactivities.

With the aim of a medical application in mind, blocking reagents, here for a toxin, can be tailored. The illustrated growth-regulatory activity entitles to explore the potential of minigalectins (oligopeptides with specificity to the crucial ligand to trigger the productive signaling). Target-specific minilectins with growth-regulatory capacity, here ganglioside GM_1-specific peptides, may harbor less cross-reactivity to other natural ligands such as CD7, thus cause less side reactions. Initial assays into this direction with 15mer peptides identify their enormous intramolecular flexibility as first major obstacle to be resolved along this way (34).

In the examples given above, galectin-1 operates as negative regulator of cell growth, and its expression may have therapeutic merit. Of note, the spectrum of galectin-1 activities also includes characteristics directly associated with the malignant phenotype. Tissue invasion in glioblastoma is such a property, and galectin-1 presence was correlated to shortened survival in animal models and patients (35, 36). Consequently, the design of inhibitors for galectins in situations, where a galectin is linked to tumor progression and bad prognosis, becomes an attractive challenge for medicinal chemistry. The combination of library approaches with screening of glycosclusters to first detect potent ligands and then their topologically optimal presentation on dendrimeric scaffolds is a means toward this aim.

COMPOUND LIBRARIES AND DENDRIMERS

Chemical libraries have simplified the screening for potent ligands. As proof-of-principle study for the potential of dynamic combinatorial libraries in this field, a panel of 1-thio derivatives of monosaccharides and aliphatic compounds was established and screened, coming up with the digalactosyldisulfide as new ligand for galectin-1 (37). Its conspicuous level of intramolecular dynamics in solution renders analysis of the relation between flexibility and thermodynamics of binding in terms of enthalpy and entropy an intriguing issue, currently under study. Opening the range of building blocks for compound assembly to amino acids in fully randomized on-bead libraries, a series of (glyco)peptides was found active as galectin-1 or galectin-3 inhibitors (Table 4.3) (38). When proceeding to assay selected compounds against four different human galectins, hereby covering the three subfamilies of molecular organization, and a ricin-like biohazardous agglutinin, clear evidence for differential activity was collected using the experimental solid-phase setting (Fig. 4.2) and the cytofluorimetric analysis on the level of cells (Fig. 4.3) (38). By introducing protein ligands of galectin-1 into the assay in the place of glycoproteins, for example, the oncogenic H-Ras (39) or other proteins listed in Table 4.2, the libraries are likely to become a source for candidates capable to interfere with protein–protein recognition of galectins (5).

TABLE 4.3 Sequences of (Glyco)peptides Identified by On-Bead Library Screening with Fluorescent Galectin-1 (1–17) and Galectin-3 (18–25) and Inhibitory Capacity of the Compounds in Solid-Phase Assays Using a Glycoprotein as Matrix

No.		% Inhibition
1	PFFISR	9
2	PFIChaFQ	14
3	IIAITCha	n.t.
4	MFVChaChaR	22
5	PTIFFF	n.i.
6	GVFIChaA	n.i.
7	YChaHChaYT	9
8	PChaNChaVY	67
9	IFRChaRY	70
10	T(Gal)IIQChaY	n.i.
11	ChaVIN(Gal)YQ	n.i.
12	PIFT(Lac)RR	n.i.
13	FRPRT(Lac)I	3
14	T(Lac)ChaRRFI	n.i.
15	AYRRT(Lac)I	n.i.
16	SASST(Lac)R	22
17	T(Lac)MRAT(Lac)Cha	38
	Gal/Lac	17/54
18	ChaChaRPMR	28
19	HHVYYH	n.i.
20	PFFFFF	76
21	NT(Lac)FVRI	44
22	PT(Lac)VAPR	n.i.
23	RVHYT(Lac)R	n.i.
24	MRT(Lac)RT(Lac)R	n.i.
25	T(Lac)ANYT(Lac)R	15
	Gal/Lac	n.i./44

Cha: cyclohexylalanine; galactosylated (Gal) or lactosylated (Lac) derivatives of threonine or asparagine are denoted by T(Gal/Lac) or N(Gal); n.t. = not tested; n.i. = not inhibitory at 10 mM; from Reference 38.

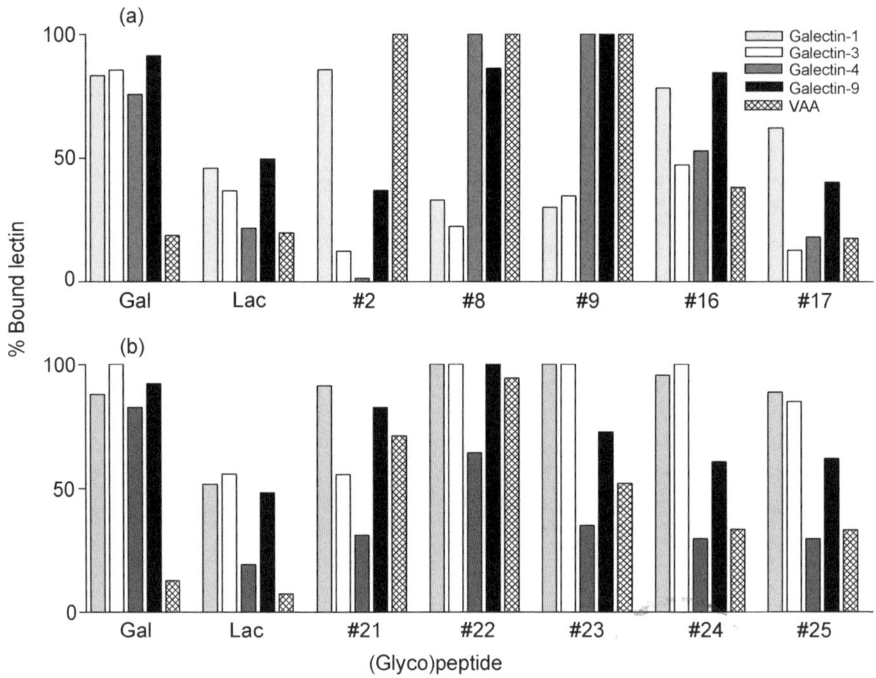

FIGURE 4.2 Representative illustration of inhibitory potency of (glyco)peptides detected by on-bead library screening with fluorescent galectin-1 (a) or galectin-3 (b) in a solid-phase assay. Extent of carbohydrate-dependent binding of four human galectins and the plant toxin from *Viscum album* L. (VAA) to a matrix of asialofetuin (0.25 μg/well for the plant lectin and 0.5 μg/well for the human lectins) was assessed spectrophotometrically in the absence and presence of 10 mM inhibitor, standard deviation in each case not exceeding 12%. For amino acid sequences of the glycopeptides and their numbering, see Table 4.3. As internal controls, galactose (Gal) and lactose (Lac) were used at the same concentration (With permission from Reference 38).

To optimize the efficacy of an inhibitor, the ligand structure counts. After all, activity also depends on topological aspects of ligand presentation, warranting to exploit dendrimeric scaffolds. Using poly(amidoamine) starburst dendrimers, the potency of lactose as inhibitor was enhanced more than 100-fold for galectin-1 and even up to 10,345-fold for a plant toxin (40). Wedge-like glycodendrimers with 3,5-di-(2-aminoethoxy)benzoic acid as branching unit were even more favorable for galectin-1 than the starburst design, yielding an up to 1667-fold increase (41). Local-density clusters appear to be more effective than decorating a sphere with uniform density of ligand, as also shown for persubstituted cyclodextrins (42). Rigidification of ligand display on triiodobenzene or pentaerythritol cores raised evidence for galectin-type-selective properties, capitalizing on the different modes of galectin aggregate formation, that is, distinguishing homodimeric galectin-1 and -7 from the pentamer of galectin-3 found in the presence of a polyvalent ligand (43, 44). Combining suitable topology with a discriminatory headgroup is a testable approach to optimize target specificity.

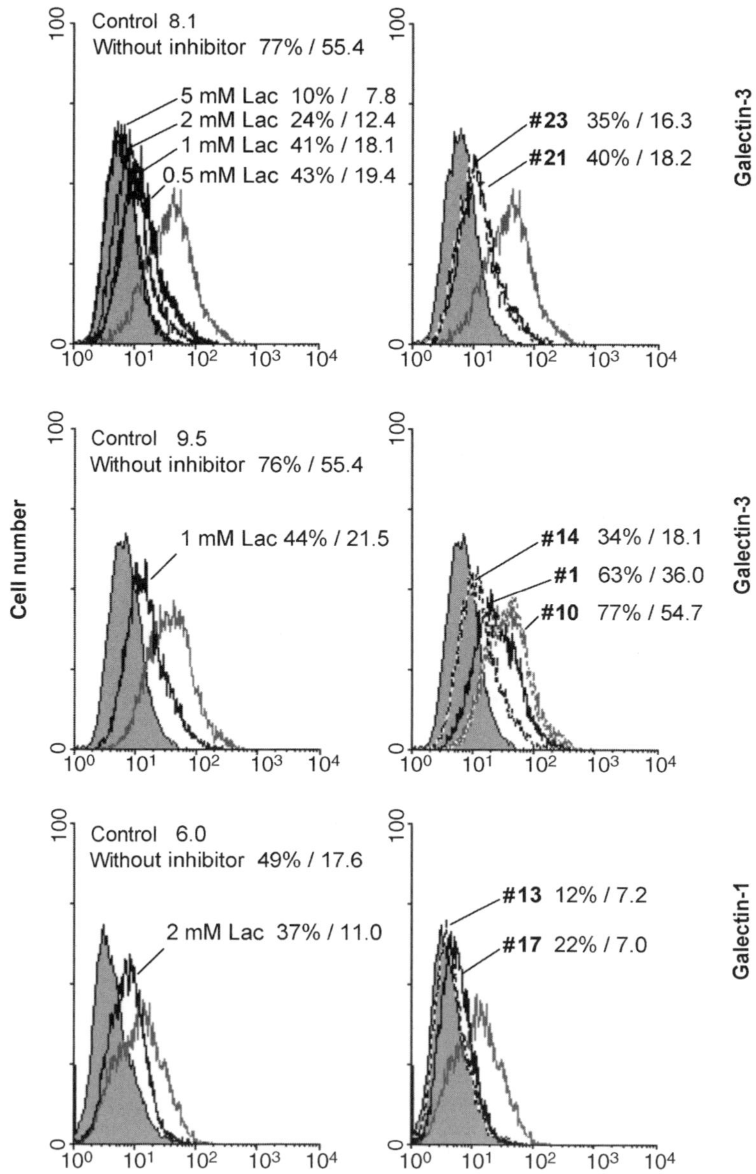

FIGURE 4.3 Semilogarithmic representation of fluorescent surface staining of cells of the human SW480 colon adenocarcinoma line (top and middle panels) and the human Capan-1 pancreatic carcinoma line reconstituted for p16 expression (bottom panel) using 10 µg/mL labeled galectin-3 (top and middle panels) and 40 µg/mL labeled galectin-1 (bottom panel). The control values representing 0% in the absence of lectin and 100% in the absence of inhibitor are given as shaded area and gray line, respectively. Inhibition was tested at concentrations of 0.5, 1, 2, and 5 mM lactose and 1 mM of the glycopeptides #21 and #23 (top panel), 1 mM lactose and 1 mM of the (glyco)peptides #1, #10, and #14 (middle panel), and 2 mM lactose and 2 mM of the glycopeptides #13 and #17 (bottom panel). For amino acid sequences of the glycopeptides, see Table 4.3. Quantitative data on percentage of positive cells (%) and mean channel fluorescence are given in each panel (With permission from Reference 38).

CONCLUSIONS

Branch-end determinants of glycan antennae are the preferential region to introduce structural variations (information storage). Spatially, they are ideal contact sites for lectins to initiate cellular responses (information transfer). Members of the galectin family teach instructive lessons on how their selectivity to distinct glycoconjugates on the cell surface is turned into growth regulation. Evidently, an intricate network of contacts beyond a central galactose unit underlies the selection of distinct low-energy conformations of certain natural glycans. The connection between cross-linking of a distinct surface ligand and growth regulation prompts development of artificial receptors with exclusive target specificity. Initial experiments encourage the design of growth-regulatory minigalectins.

As the spectrum of galectin activities also encompasses involvement in establishing certain features of the malignant phenotype, blocking of galectin binding then becomes an issue. The individual characteristics of the carbohydrate recognition domains nourish the perspective to aim at development of inhibitors with restricted range among the family members. In addition to defining optimal molecular characteristics of the headgroup, for which chemical libraries offer a great potential, the modes of its topological presentation can favorably influence inhibitory capacity, as shown by testing custom-made dendrimer scaffolds. Galectins, thus, are anticipated to enter the list of promising targets for drug design, on the grounds of their illustrated growth/invasion-regulatory activity.

REFERENCES

1. Gabius H-J. Biological information transfer beyond the genetic code: the sugar code. Naturwissenschaften 2000;87:108–121.
2. Gabius H-J, Siebert H-C, André S, Jiménez-Barbero J, Rüdiger H. Chemical biology of the sugar code. ChemBioChem 2004;5:740–764.
3. Gabius H-J. Endogenous lectins in tumors and the immune system. Cancer Invest 1987;5:39–46.
4. Gabius H-J. Animal lectins. Eur J Biochem 1997;243:543–576.
5. Gabius H-J. Cell surface glycans: the why and how of their functionality as biochemical signals in lectin-mediated information transfer. Crit Rev Immunol 2006;26:43–80.
6. Villalobo A, Nogales-Gonzalés A, Gabius H-J. A guide to signaling pathways connecting protein–glycan interaction with the emerging versatile effector functionality of mammalian lectins. Trends Glycosci Glycotechnol 2006;18:1–37.
7. Gabius H-J, Darro F, Remmelink M, André S, Kopitz J, Danguy A, Gabius S, Salmon I, Kiss R. Evidence for stimulation of tumor proliferation in cell lines and histotypic cultures by clinically relevant low doses of the galactoside-binding mistletoe lectin, a component of proprietary extracts. Cancer Invest 2001;19:114–126.
8. André S, Kojima S, Yamazaki N, Fink C, Kaltner H, Kayser K, Gabius H-J. Galectins-1 and -3 and their ligands in tumor biology. J Cancer Res Clin Oncol 1999;125:461–474.

9. Leffler H, Barondes SH. Specificity of binding of soluble rat lung lectins to substituted and unsubstituted mammalian β-galactosides. J Biol Chem 1986;261:10119–10126.
10. Gabius H-J. Influence of type of linkage and spacer on the interaction of β-galactoside-binding proteins with immobilized affinity ligands. Anal Biochem 1990;189:91–94.
11. Ahmad N, Gabius H-J, Kaltner H, André S, Kuwabara I, Liu F-T, Oscarson S, Norberg T, Brewer CF. Thermodynamic binding studies of cell surface carbohydrate epitopes to galectins-1, -3 and -7. Evidence for differential binding specificities. Can J Chem 2002;80:1096–1104.
12. Wu AM, Singh T, Liu J-H, Krzeminski M, Russwurm R, Siebert H-C, Bonvin AMJJ, André S, Gabius H-J. Activity–structure correlations in divergent lectin evolution: fine specificity of chicken galectin CG-14 and computational analysis of flexible ligand docking for CG-14 and the closely related CG-16. Glycobiology 2007;17:165–184.
13. He L, André S, Siebert H-C, Helmholz H, Niemeyer B, Gabius H-J. Detection of ligand- and solvent-induced shape alterations of cell-growth-regulatory human lectin galectin-1 in solution by small angle neutron and X-ray scattering. Biophys J 2003;85:511–524.
14. López-Lucendo MF, Solís D, André S, Hirabayashi J, Kasai K-i, Kaltner H, Gabius H-J, Romero A. Growth-regulatory human galectin-1: crystallographic characterisation of the structural changes induced by single-site mutations and their impact on the thermodynamics of ligand binding. J Mol Biol 2004;343:957–970.
15. Wu AM, Wu JH, Tsai M-S, Kaltner H, Gabius H-J. Carbohydrate specificity of a galectin from chicken liver (CG-16). Biochem J 2001;358:529–538.
16. Wu AM, Wu JH, Tsai M-S, Liu J-H, André S, Wasano K, Kaltner H, Gabius H-J. Fine specificity of domain-I of recombinant tandem-repeat-type galectin-4 from rat gastrointestinal tract (G4-N). Biochem J 2002;367:653–664.
17. Wu AM, Wu JH, Liu J-H, Singh T, André S, Kaltner H, Gabius H-J. Effects of polyvalency of glycotopes and natural modifications of human blood group ABH/Lewis sugars at the Galβ1-terminated core saccharides on the binding of domain-I of recombinant tandem-repeat-type galectin-4 from rat gastrointestinal tract (G4-N) Biochimie 2004;86:317–326.
18. Wu AM, Singh T, Wu JH, Lensch M, André S, Gabius H-J. Interaction profile of galectin-5 with free saccharides and mammalian glycoproteins: probing its fine-specificity and the effect of naturally clustered ligand presentation. Glycobiology 2006;16:524–537.
19. André S, Unverzagt C, Kojima S, Frank M, Seifert J, Fink C, Kayser K, von der Lieth C-W, Gabius H-J. Determination of modulation of ligand properties of synthetic complex-type biantennary N-glycans by introduction of bisecting GlcNAc *in silico*, *in vitro* and *in vivo*. Eur J Biochem 2004;271:118–134.
20. André S, Kojima S, Prahl I, Lensch M, Unverzagt C, Gabius H-J. Introduction of extended LEC14-type branching into core-fucosylated biantennary N-glycan. Glycoengineering for enhanced cell binding and serum clearance of the neoglycoprotein. FEBS J 2005; 272:1986–1998.
21. André S, Kozár T, Schuberth R, Unverzagt C, Kojima S, Gabius H-J. Substitutions in the N-glycan core as regulators of biorecognition: the case of core-fucose and bisecting GlcNAc moieties. Biochemistry 2007;46:6984–6995.
22. Dam TK, Gabius H-J, André S, Kaltner H, Lensch M, Brewer CF. Galectins bind to the multivalent glycoprotein asialofetuin with enhanced affinities and a gradient of decreasing binding constants. Biochemistry 2005;44:12564–12571.

23. Fischer C, Sanchez-Ruderisch H, Welzel M, Wiedenmann B, Sakai T, André S, Gabius H-J, Khachigian L, Detjen K, Rosewicz S. Galectin-1 interacts with the $\alpha_5\beta_1$ fibronectin receptor to restrict carcinoma cell growth via induction of p21 and p27. J Biol Chem 2005;280:37266–37277.

24. André S, Sanchez-Ruderisch H, Nakagawa H, Buchholz M, Kopitz J, Forberich P, Kemmner W, Böck C, Deguchi K, Detjen KM, Wiedenmann B, von Knebel Döberitz M, Gress TM, Nishimura S-I, Rosewicz S, Gabius H-J. Tumor suppressor $p16^{INK4a}$: modulator of glycomic profile and galectin-1 expression to increase susceptibility to carbohydrate-dependent induction of anoikis in pancreatic carcinoma cells. FEBS J 2007;274:3233–3256.

25. Sturm A, Lensch M, André S, Kaltner H, Wiedenmann B, Rosewicz S, Dignass AU, Gabius H-J. Human galectin-2: novel inducer of T cell apoptosis with distinct profile of caspase activation. J Immunol 2004;173:3825–3837.

26. Pace KE, Hahn HP, Pang M, Nguyen JT, Baum LG. CD7 delivers a pro-apoptotic signal during galectin-1-induced T cell death. J Immunol 2000;165:2331–2334.

27. Rappl G, Abken H, Muche JM, Sterry W, Tilgen W, André S, Kaltner H, Ugurel S, Gabius H-J, Reinhold U. $CD4^+CD7^-$ leukemic T cells from patients with Sézary syndrome are protected from galectin-1-triggered T cell death. Leukemia 2002;16:840–845.

28. Kopitz J, von Reitzenstein C, Burchert M, Cantz M, Gabius H-J. Galectin-1 is a major receptor for ganglioside GM_1, a product of the growth-controlling activity of a cell surface ganglioside sialidase, on human neuroblastoma cells in culture. J Biol Chem 1998;273:11205–11211.

29. Kopitz J, von Reitzenstein C, André S, Kaltner H, Uhl J, Ehemann V, Cantz M, Gabius H-J. Negative regulation of neuroblastoma cell growth by carbohydrate-dependent surface binding of galectin-1 and functional divergence from galectin-3. J Biol Chem 2001;276:35917–35923.

30. André S, Kaltner H, Lensch M, Russwurm R, Siebert H-C, Fallsehr C, Tajkhorshid E, Heck AJR, von Knebel-Döberitz M, Gabius H-J, Kopitz J. Determination of structural and functional overlap/divergence of five prototype galectins by analysis of the growth-regulatory interaction with ganglioside GM_1 *in silico* and *in vitro* on human neuroblastoma cells. Int J Cancer 2005;114:46–57.

31. Siebert H-C, André S, Lu S-Y, Frank M, Kaltner H, van Kuik JA, Korchagina EY, Bovin NV, Tajkhorshid E, Kaptein R, Vliegenthart JFG, von der Lieth C-W, Jiménez-Barbero J, Kopitz J, Gabius H-J. Unique conformer selection of human growth-regulatory lectin galectin-1 for ganglioside GM_1 versus bacterial toxins. Biochemistry 2003;42:14762–14773.

32. Siebert H-C, Gilleron M, Kaltner H, von der Lieth C-W, Kozár T, Bovin NV, Korchagina EY, Vliegenthart JFG, Gabius H-J. NMR-based, molecular dynamics- and random walk molecular mechanics-supported study of conformational aspects of a carbohydrate ligand (Galβ1-2Galβ1-R) for an animal galectin in the free and in the bound state. Biochem Biophys Res Commun 1996;219:205–212.

33. Asensio JL, Espinosa JF, Dietrich H, Cañada FJ, Schmidt RR, Martín-Lomas M, André S, Gabius H-J, Jiménez-Barbero J. Bovine heart galectin-1 selects a distinct (*syn*) conformation of C-lactose, a flexible lactose analogue. J Am Chem Soc 1999;121: 8995–9000.

34. Siebert H-C, Born K, André S, Frank M, Kaltner H, von der Lieth C-W, Heck AJR, Jiménez-Barbero J, Kopitz J, Gabius H-J. Carbohydrate chain of ganglioside GM_1 as a ligand: identification of the binding strategies of three 15mer peptides and their divergence

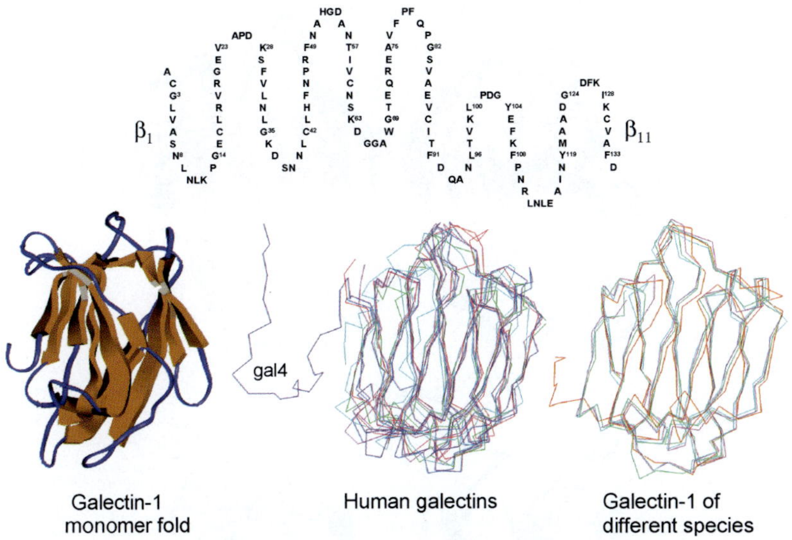

FIGURE 3.1 Superposition of CRDs from different galectin monomers. (*See text for full caption.*)

```
                                                 β1         β2         β3         β4
Gal1human  (1GZW)  .......... .......... ......ACGL VASNLNLKPG ECLRVRG... EVAPDAKSFV LNLGKDSN.. .NLCLHFNPR FNAHGDANTI
Gal2human  (1HLC)  .......... .......... ........EL EVKNMDMKPG STLKITG... SIADGTDGFV INLGQGTD.. .KLNLHFNPR FSES....TI
Gal3human  (1A3K)  .......... .......... ....LIVPYN LPLPGGVVPR MLITILG... TVKPNANRIA LDFQ...R.G NDVAFHFNPR FNENNRR.VI
Gal4human  (1X50)  GSSGSSGHQQ LNSLPTMEGP PTFNPPVPYF GRLQGGLTAR RTIIIKG... YVPPTGKSFA INFKVGSS.. GDIALHINPR MGNG....TV
Gal7human  (1BKZ)  .......... .......... ....SNVPHK SSLPEGIRPG TVLRIRG... LVPPNASRFH VNLLCGEEQG SDAALHFNPR LDTS....EV
Gal10human (1LCL)  .......... .......... ...MSLLPVP YTEAASLSTG STVTIKGRPL VCFLNEPYLQ VDFHTEMKEE SDIVFHFQVC FGR.....RV

Gal1 bovine (1SLA) .......... .......... ......ACGL VASNLNLKPG ECLRVRG... EVAADAKSFL LNLGKDDN.. .NLCLHFNPR FNAHGDVNTI
Gal9 mouse (2D6L)  .......... GSMALFSAQS PYINPIIPFT GPIQGGLQEG LQVTLQG... TTKSFAQRFV VNFQNSFN.G NDIAFHFNPR FEEGG...YV
Gal16avian (1QMJ)  .......... .......... .......QGL VVTQLDVQPG ECVKVKG... KILSDAKGFS VNVGKDSS.. .TLMLHFNPR FDCHGDVNTV
Gal1toad   (1A78)  .......... .......... .....ASAGV AVTNLNLKPG HCVEIKG... SIPPDCKGFA VNLGEDAS.. .NFLLHFNAR FDLHGDVNKI
Congerin I  (1C1F) .......... .......... ......SGGL QVKNFDFTVG KFLTVGG... FINNSPQRFS VNVGESMN.. .SLSLHLDHR FNYGADQNTI
Congerin II (1IS5) .......... .......... ......SDRA EVRNIPFKLG MYLTVGG... VVNSNATRFS INVGESTD.. .SIAMHMDHR FSYGADQNVL
ACG fungal  (1WW7) .......... .....TTSAV NIYNISAGAS VDLAAPVTTG DIVTFFSSAL NLSAGAGSPN NTALNLLSEN GAYLLHIAFR LQEN....VI
CGL2fungal  (1UL9) .......... .......... MLYHLFVNNQ VKLQNDFKPE SVAAIRS... SAFNSKGGTT VFNFLSAG.. ENILLHISIR PGEN....VI

              β5         β6             β7         β8         β9                    β10        β11
Gal1human     VCNS..KDGG AWGTEQR... EAVFPFQPGS VAEVCITFDQ ANLTVKLPDG YEFKFPNRLN .LEAINYMAA DGDFKIKCVA FD........ ...
Gal2human     VCNS..LDGS NWGQEQR... EDHLCFSPGS EVKFTVTFES DKFKVKLPDG HELTFPNRLG .HSHLSYLSV RGGFNMSSFK LKE....... ...
Gal3human     VCNT..KLDN NWGREER... QSVFPFESGK PFKIQVLVEP DHFKVAVNDA HLLQYNHRVK KLNEISKLGI SGDIDLTSAS YTMI...... ...
Gal4human     VRNS..LLNG SWGSEEKK.. ITHNPFGPGQ FFDLSIRCGL DRFKVYANGQ HLFDFAHRLS AFQRVDTLEI QGDVTLSYVQ ISGPSSG... ...
Gal7human     VFNS..KEQG SWGREER... GPGVPFQRGQ PFEVLIIASD DGFKAVVGDA QYHHFRHRLP .LARVRLVEV GGDVQLDSVR IF........ ...
Gal10human    VMNS..REYG AWKQQVE... SKNMPFQDGQ EFELSISVLP DKYQVMVNGQ SSYTFDHRIK .PEAVKMVQV WRDISLTKFN VSYLKR.... ...

Gal1bovine    VCNS..KDAG AWGAEQR... ESAFPFQPGS VVEVCISFNQ TDLTIKLPDG YEFKFPNRLN .LEAINYLSA GGDFKIKCVA FE........ ...
Gal9mouse     VCNT..KQNG QWGPEER... KMQMPFQKGM PFELCFLVQR SEFKVMVNKK FFVQYQHRVP .YHLVDTIAV SGCLKLSFIT FQTQNFRPAH QA.
Gal16avian    VCNS..KEDG TWGEEDR... KADFPFQQGD KVEICISFDA AEVKVKVP.E VEFEFPNRLG .MEKIQYLAV EGDFKVKAIK FS........ ...
Gal1toad      VCNS..KEAD AWGSEQR... EEVFPFQQGA EVMVCFEYQT QKIIIKFSSG DQFSFPVRKV .LPSIPFLSL EG.LAFKSIT TE........ ...
Congerin I    VMNSTLKGDN GWETEQR... STNFTLSAGQ YFEITLSYDI NKFYIDILDG PNLEFPNRYS .KEFLPFLSL AGDARLTLVK LE........ ...
Congerin II   VLN.SLVHNV GWQQEER... SKKFPFTKGD HFQTTITFDT HTFYIQLSNG ETVEFPNRNK .DAAFNLIYL AGDARLTFVR LE........ ...
ACGfungal     VFNSR.QPNA PWLVEQRVSN VANQFIGSGG KAMVTVFDHG DKYQVVINEK TVIQYTKQIS GTTSSLSYNS TEGTSIFSTV VEAVTYTGLA ...
CGL2fungal    VFNSR.LKNG AWGPEERI.. PYAEKFRPPN PSITVIDHG. DRFQIRFDYG TSIYYNKRIK ENAAAIAYNA ENSLFSSPVT VDVHGLLPPL PPA
```

FIGURE 3.2 Amino acid sequences of galectins. The amino acid sequences of galectins, whose three-dimensional folded structures have been reported, are shown aligned. References for these sequences are provided in Table 3.1. The 11 β-strands common to these galectins are indicated at the top by hatched bars and labeled β1 through β11. Conserved residues that form the lactose-binding pocket are identified in red.

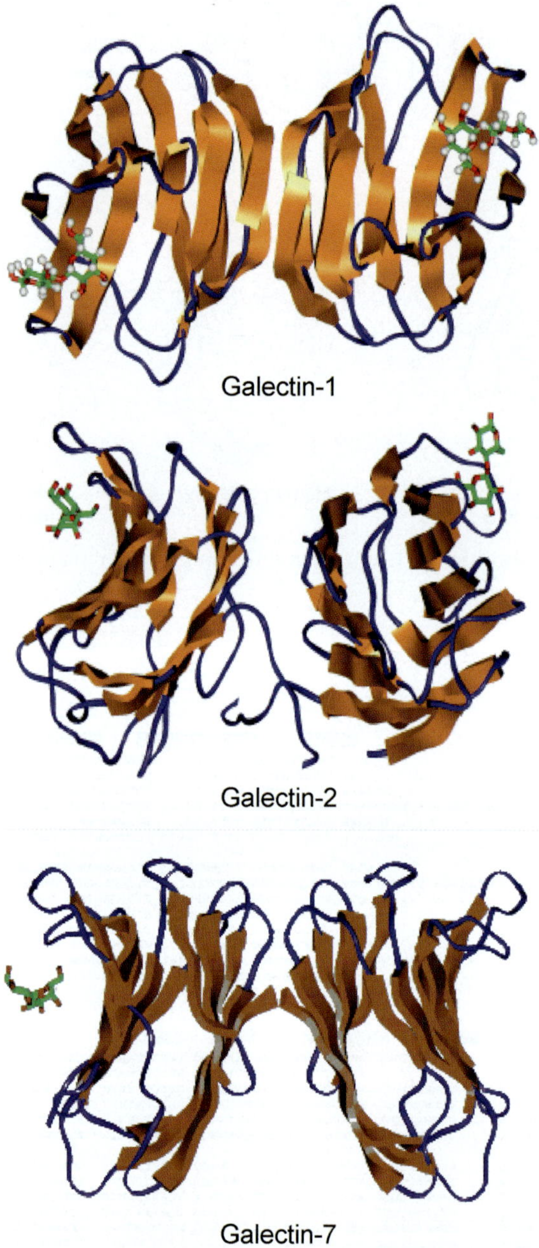

FIGURE 3.3 Galectin dimers. The different types of galectin dimers that can form are shown here. The CRD of the galectin-1 "terminal" dimer (PDB access code lgzw) (10) is shown at the top of the figure, followed by the "nonsymmetric sandwich" dimer of fungal galectin-2 (1ul9) (28) in the middle, and then the "symmetric sandwich" dimer of human galectin-7 (1bkz) (15) at the bottom. The carbohydrate-binding sites in each dimer are indicated by the lactose molecules shown as ball-and-stick structures.

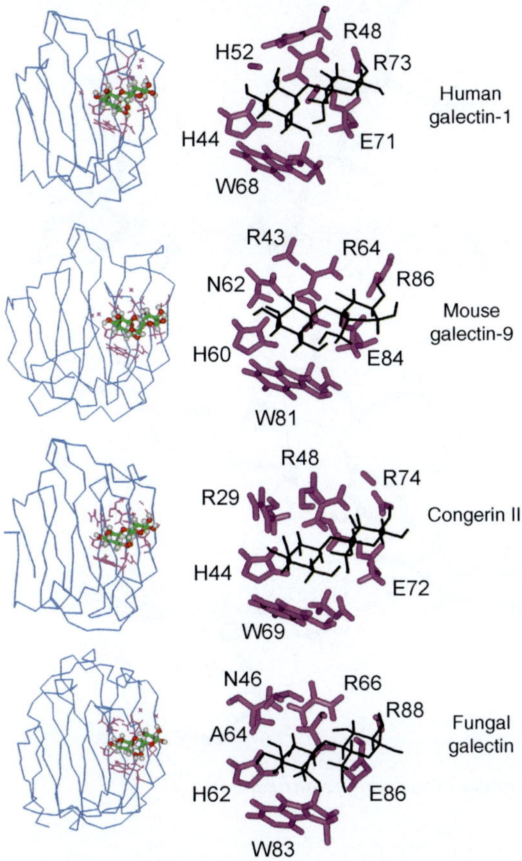

FIGURE 3.4 Lactose recognition site is similar in all galectins. (*See text for full caption.*)

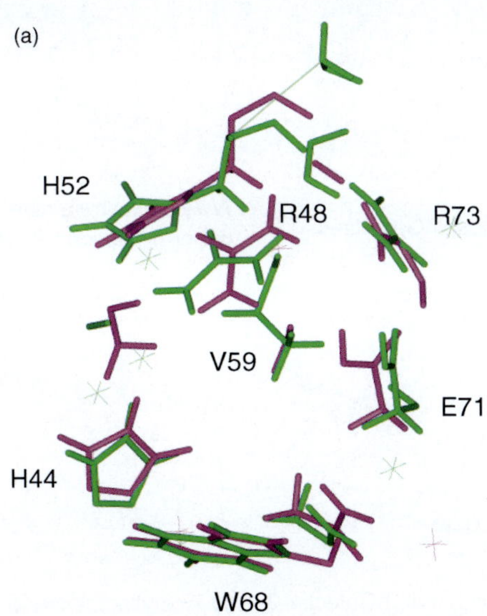

FIGURE 3.5 Carbohydrate-binding site of galectin-1. (*See text for full caption.*)

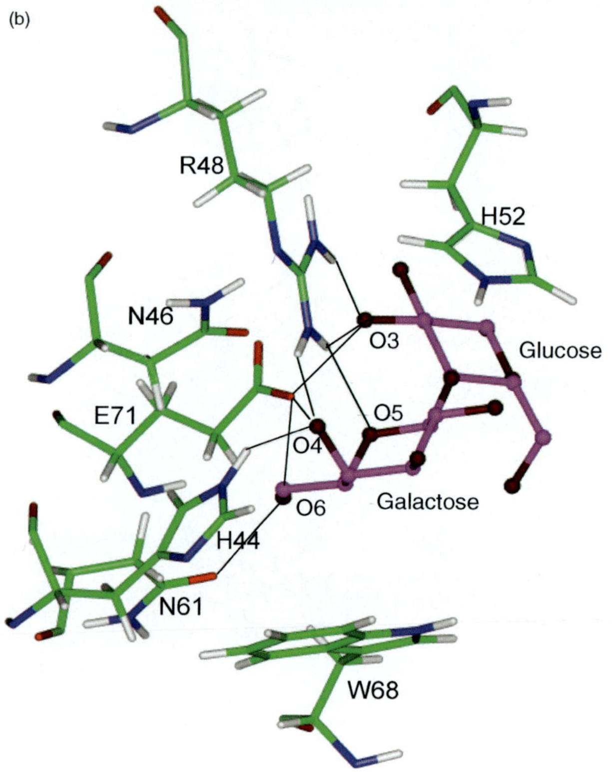

FIGURE 3.5 (*Continued*) (*See text for full caption.*)

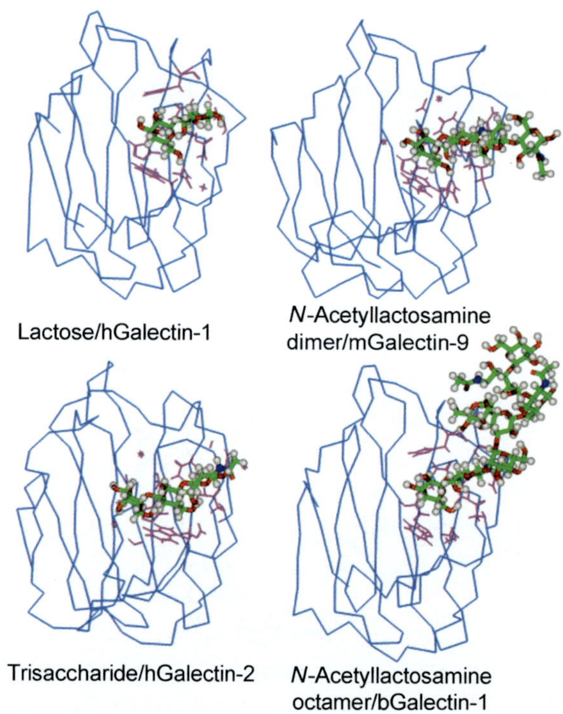

FIGURE 3.6 Carbohydrate-binding sites of some galectins. (*See text for full caption.*)

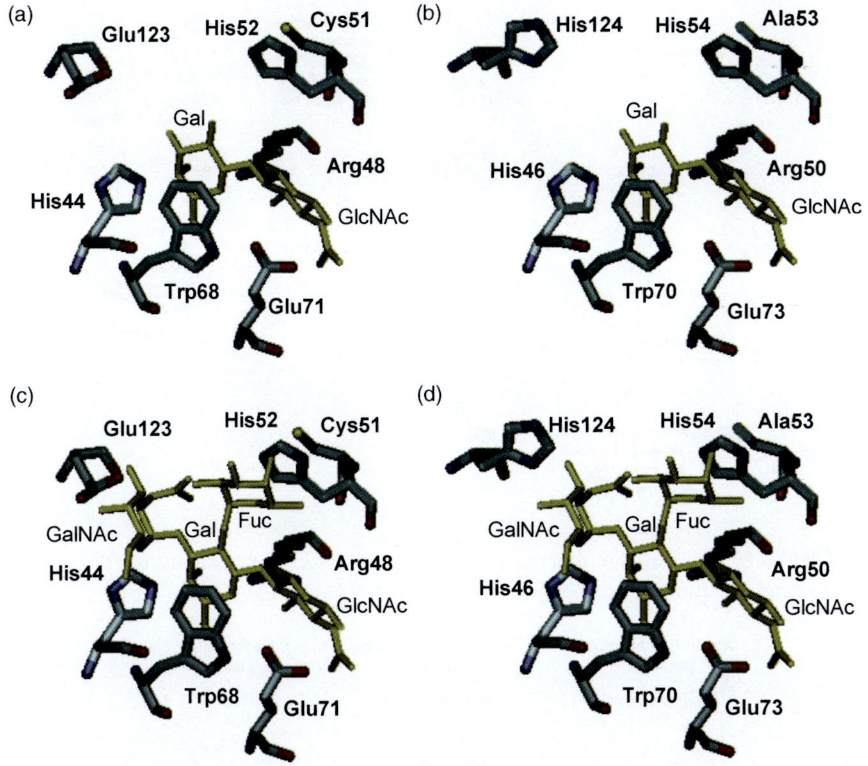

FIGURE 4.1 Graphical illustration of the carbohydrate recognition domains of CG-16 (left panel) and CG-14 (right panel). (*See text for full caption.*)

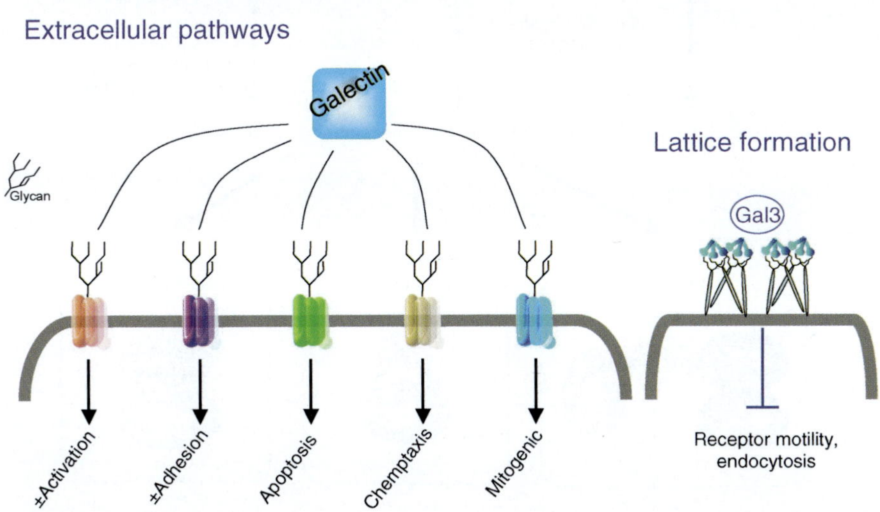

FIGURE 6.1 Extracellular functions of galectins related to immune and inflammatory responses. (*See text for full caption.*)

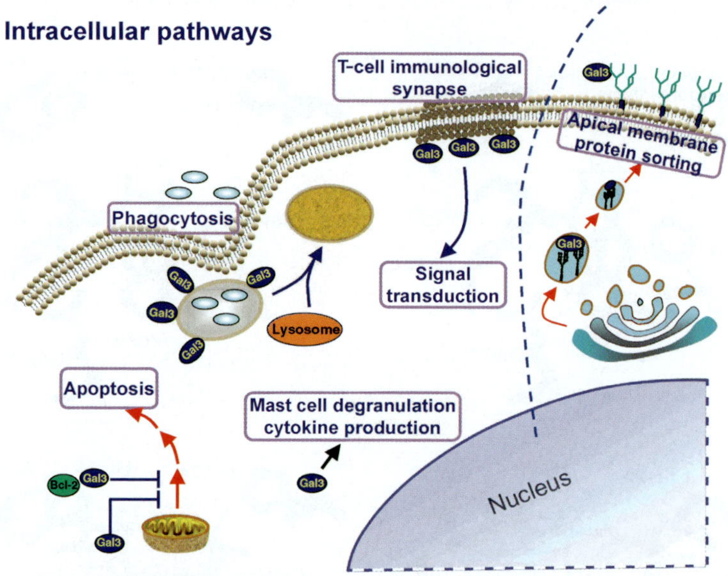

FIGURE 6.2 Intracellular functions of galectin-3 associated with its regulation of the immune and inflammatory responses. (*See text for full caption.*)

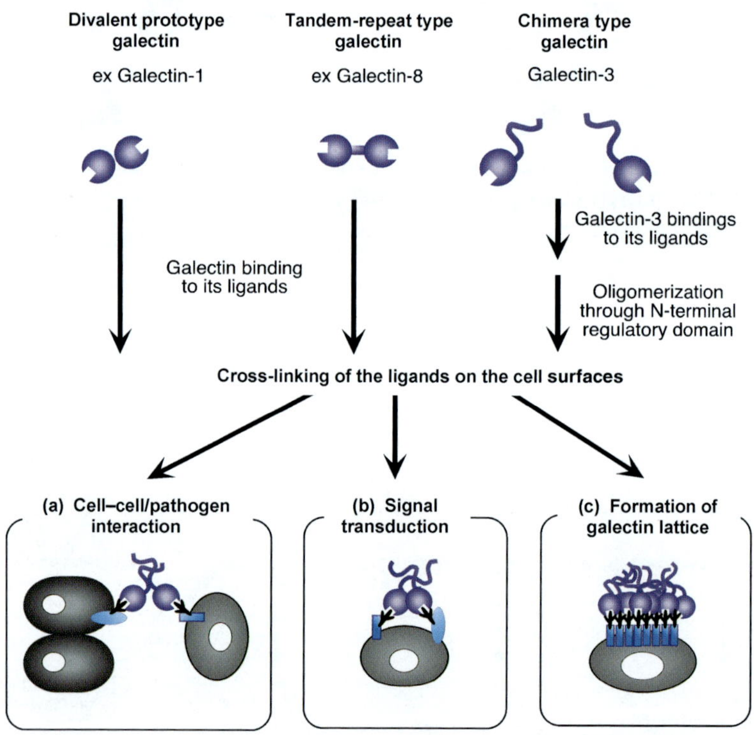

FIGURE 7.2 Cross-linking by galectins. Galectins cross-link the ligands on the cell surface, leading to mediating cell–cell/pathogen interaction (a), receptor clustering (b), and formation of galectin–glycoprotein lattice (c).

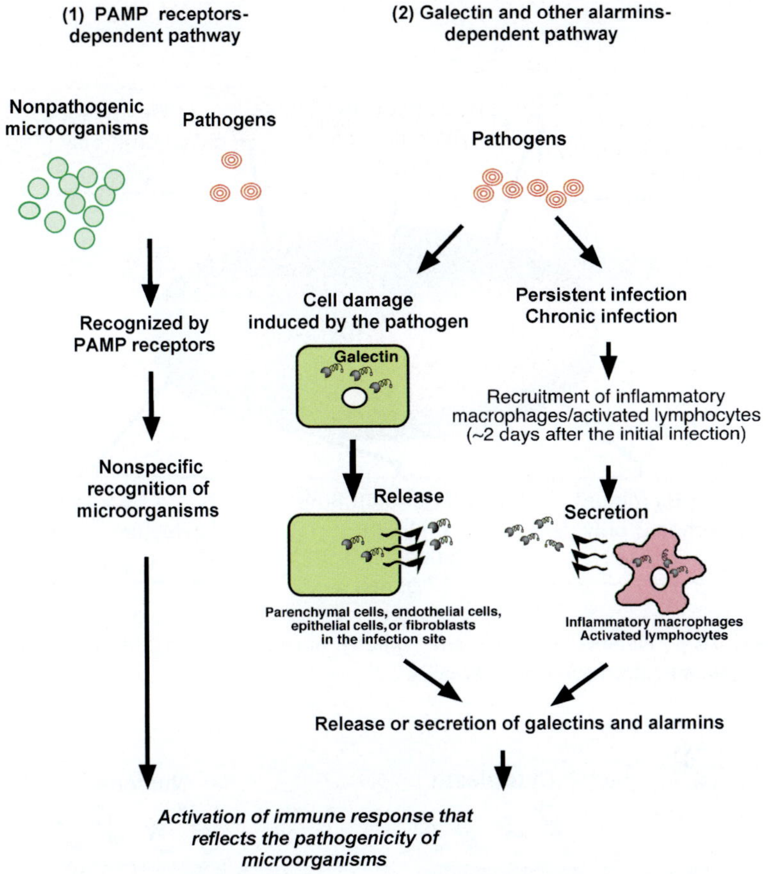

FIGURE 7.3 Innate immune recognition of pathogen by DAMPs (damage-associated molecular patterns).

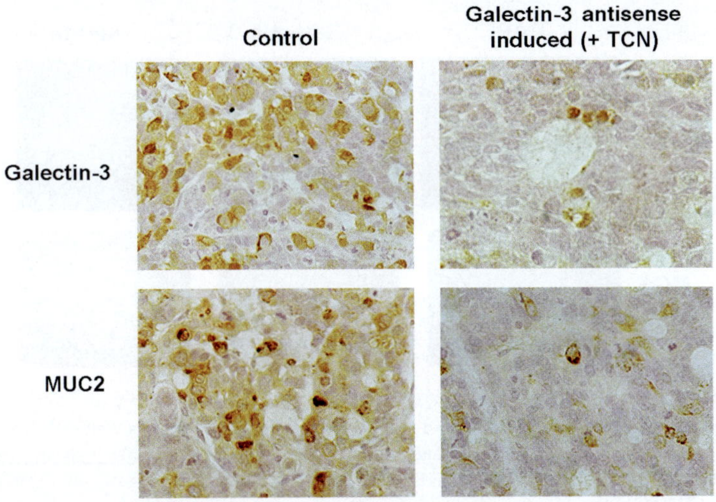

FIGURE 10.6 Effect of induction of galectin-3 antisense on galectin-3 and MUC2 expression *in vivo*. (*See text for full caption.*)

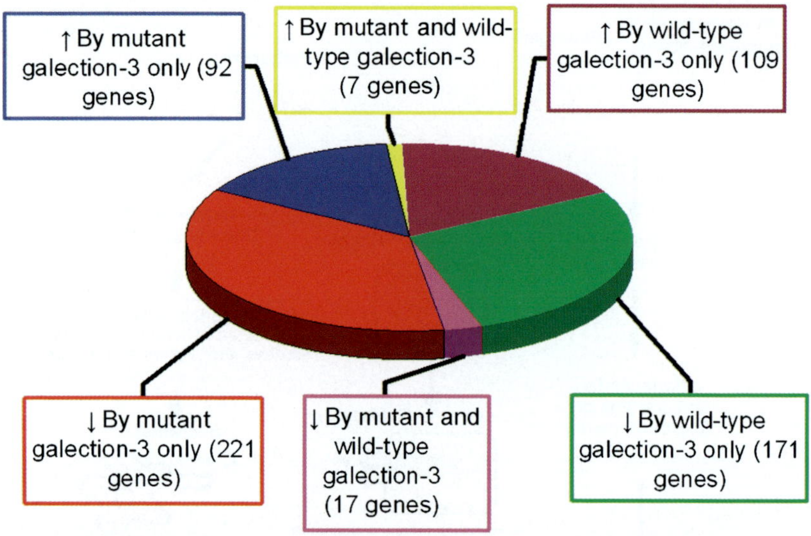

FIGURE 10.9 Number of genes differentially modulated by wild-type galectin-3 and phospho-mutant galectin-3 in BT-549 cells.

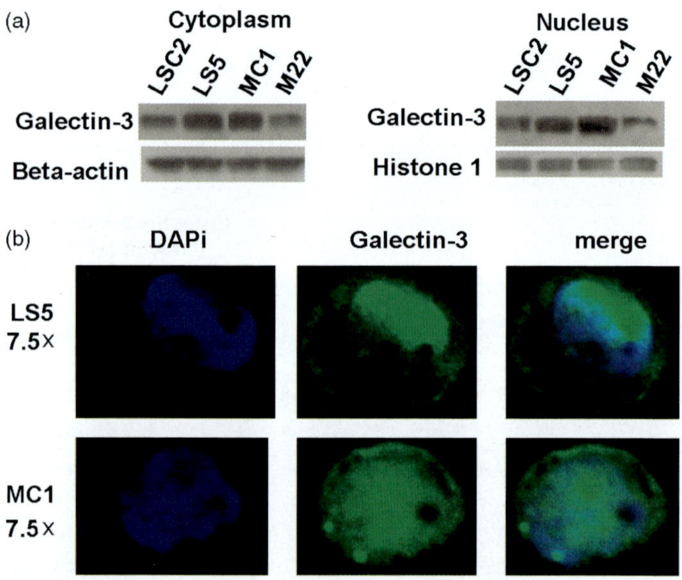

FIGURE 10.10 Subcellular location of galectin-3 in galectin-3 sense and antisense clones of colon cancer cells. (a) Western blots of cytoplasmic and nuclear fractions with antigalectin-3, beta-actin for cytoplasm loading control, and histone 1 for nucleus loading control. (b) Indirect immunofluorescence performed in LS5 and MC1 cells using antigalectin-3 antibody (TIB166, green), followed by DAPI nuclear counterstaining (blue). The merge of galectin-3 (green) with DAPI (blue) also is shown. Reproduced from Reference 7 with permission.

from the binding modes of growth-regulatory galectin-1 and cholera toxin. Chem Eur J 2006;12:388–402.

35. Rorive S, Belot N, Decaestecker C, Lefranc F, Gordower L, Micik S, Maurage C-A, Kaltner H, Ruchoux M-M, Danguy A, Gabius H-J, Salmon I, Kiss R, Camby I. Galectin-1 is highly expressed in human gliomas with relevance for modulation of invasion of tumor astrocytes into the brain parenchyma. Glia 2001;33:241–255.
36. Camby I, Belot N, Lefranc F, Sadeghi N, de Launoit Y, Kaltner H, Musette S, Darro F, Danguy A, Salmon I, Gabius H-J, Kiss R. Galectin-1 modulates human glioblastoma cell migration into the brain through modifications to the actin cytoskeleton and levels of expression of small GTPases. J Neuropathol Exp Neurol 2002;61:585–596.
37. André S, Pei Z, Siebert H-C, Ramstrm O, Gabius H-J. Glycosyldisulfides from dynamic combinatorial libraries as O-glycoside mimetics for plant and mammalian lectins: their reactivities in solid-phase and cell assays and conformational analysis by molecular dynamics simulations. Bioorg Med Chem 2006;14:6314–6323.
38. André S, Maljaars CEP, Halkes KM, Gabius H-J, Kamerling JP. Discovery of galectin ligands in fully randomized combinatorial one-bead-one-compound (glyco)peptide libraries. Bioorg Med Chem Lett 2007;17:793–798.
39. Rotblat B, Niv H, André S, Kaltner H, Gabius H-J, Kloog Y. Galectin-1(L11A) predicted from a computed galectin-1 farnesyl-binding pocket selectively inhibits Ras-GTP. Cancer Res 2004;64:3112–3118.
40. André S, Cejas Ortega PJ, Alamino Perez M, Roy R, Gabius H-J. Lactose-containing starburst dendrimers: influence of dendrimer generation and binding-site orientation of receptors (plant/animal lectins and immunoglobulins) on binding properties. Glycobiology 1999;9:1253–1261.
41. André S, Pieters RJ, Vrasidas I, Kaltner H, Kuwabara I, Liu F-T, Liskamp RMJ, Gabius H-J. Wedgelike glycodendrimers as inhibitors of binding of mammalian galectins to various glycoproteins, lactose maxiclusters and cell surface glycoconjugates. ChemBioChem 2001;2:822–830.
42. André S, Kaltner H, Furuike T, Nishimura S-I, Gabius H-J. Persubstituted cyclodextrin-based glycoclusters as inhibitors of protein–carbohydrate recognition using purified plant and mammalian lectins and wild-type and lectin-gene-transfected tumor cells as targets. Bioconjug Chem 2004;15:87–98.
43. André S, Liu B, Gabius H-J, Roy R. First demonstration of differential inhibition of lectin binding by synthetic tri- and tetravalent glycoclusters from cross-coupling of rigidified 2-propynyl lactoside. Org Biomol Chem 2003;1:3909–3916.
44. Ahmad N, Gabius H-J, André S, Kaltner H, Sabesan S, Roy R, Liu B, Macaluso F, Brewer CF. Galectin-3 precipitates as a pentamer with synthetic multivalent carbohydrates and forms heterogeneous cross-linked complexes. J Biol Chem 2004;279:10841–10847.

5

NUCLEAR AND CYTOPLASMIC LOCALIZATION OF GALECTIN-1 AND GALECTIN-3 AND THEIR ROLES IN PRE-mRNA SPLICING

JOHN L. WANG, KEVIN C. HAUDEK AND PATRICIA G. VOSS

Department of Biochemistry and Molecular Biology, Michigan State University, East Lansing, MI 48824, USA

RONALD J. PATTERSON

Department of Microbiology and Molecular Genetics, Michigan State University, East Lansing, MI 48824, USA

INTRODUCTION

Galectin-1 (Gal1) and galectin-3 (Gal3) were initially isolated as galactose-specific carbohydrate-binding proteins, which provided the basis for coining the name for this family of proteins (1). To date, this family contains 15 members and they have been classified, according to the number and organization of carbohydrate recognition domains (CRDs), into the prototype, the tandem repeat, and the chimera subgroups (2–5). Each CRD, ~130 amino acids, has highly conserved amino acid residues that interact with carbohydrate, as revealed by X-ray crystallography (6). From equilibrium dialysis experiments using radioactive lactose (Lac) as the saccharide ligand, the K_D values ranged from 2×10^{-5} to 5×10^{-5} M (7, 8).

In addition to binding to galactose-containing glycoconjugates, many members of the galectin family share another property in terms of their cell biology. They exhibit dual localization (9), being found in both the intracellular (cytoplasm and nucleus) and

Galectins, Edited by Anatole A. Klyosov, Zbigniew J. Witczak, and David Platt

the extracellular (cell surface and medium) compartments. The mechanism of externalization appears to be unusual because there does not seem to be a typical signal sequence for sequestration into the endomembrane pathway for secretion (10). For the most part, the literature on the galectins has been dominated by studies focused on their activity on the extracellular side, based on their binding to cell surface carbohydrates and adhesive glycoproteins of the extracellular matrix (11–13).

However, many members of the galectin family are predominantly intracellular proteins, and observations of nuclear localization have been reported for 11 of the 15 known galectins (3, 14–16). Do the localization data offer insights on the intracellular activities of the galectins? In the context of the present proceedings of a special symposium dedicated to the galectins, it is the purpose of this chapter to review the information on the nuclear localization of Gal1 and Gal3, particularly in terms of their role in pre-mRNA splicing.

NUCLEAR LOCALIZATION OF GAL1 AND GAL3

Mouse 3T3 fibroblasts, fixed with paraformaldehyde and permeabilized with Triton X-100, yielded nuclear and cytoplasmic staining with rabbit anti-Gal3 (17, 18). The nuclear staining is usually diffuse, covering the entire nucleus with the exception of a few circles. These small circles, devoid of fluorescence, correspond to nucleoli. A similar diffuse nucleoplasmic and cytoplasmic staining pattern is also obtained with human HeLa cells for both Gal1 and Gal3 (19).

The nuclear versus cytoplasmic distribution of Gal3 was found to be dependent on the proliferation state of the cells under analysis. In quiescent cultures of 3T3 fibroblasts (e.g., serum-deprived), Gal3 was predominantly cytoplasmic; proliferating cultures of the same cells (e.g., serum-stimulated), however, showed intense nuclear staining (20). These conclusions were substantiated by quantitative immunoblotting of subcellular fractions derived from quiescent and proliferating cultures. Moreover, this correlation between a predominantly cytoplasmic localization in quiescent cells and a nuclear localization in proliferating cells has also been observed in human fibroblasts as they senesce during *in vitro* culture. Proliferating young (low cumulative population doubling) cells exhibit nuclear localization of Gal3, whereas the protein appears to be excluded from the nuclei of senescent cells (high cumulative population doubling) that have lost replicative competence (21). Nuclear exclusion of Gal3 has also been observed in cancer cells of the colon. While Gal3 is concentrated in the nuclei of the differentiated cells of normal colonic epithelia, it is present only in the cytoplasm of adenoma and carcinoma cells (22).

A more informative labeling pattern was obtained when cells were permeabilized with Triton X-100 without prior fixation, subjected to extraction with 0.25 M ammonium sulfate, followed by fixation and staining. This procedure extracts the majority of the nonhistone nuclear proteins, leaving chromatin, nuclear matrix, and associated RNAs (23, 24). A distinct speckled fluorescence pattern was observed in 3T3 cells for Gal3 (23, 24), as well as in HeLa cells for both Gal1 and Gal3 (19, 25). Speckles have been observed for a number of nuclear proteins involved in the splicing of pre-mRNA: (a) the Sm polypeptides of the small nuclear ribonucleoprotein particles (snRNPs); and (b) the Ser-, Arg-rich (SR) family of non-snRNP splicing factors (26). Indeed, double immunofluorescence experiments have documented that both of the galectins can be colocalized in speckled structures revealed by anti-Sm and anti-SC 35 (a member of the SR family) (19, 25).

ASSOCIATION OF GAL3 WITH RIBONUCLEOPROTEIN COMPLEXES

The association of the speckled structures with ribonucleoprotein components of the nuclear matrix is also consistent with the observation that digestion of RNA released Gal3 from the nucleus, as documented in Fig. 5.1. Permeabilized, but unfixed, 3T3 cells were treated with ribonuclease A; this resulted in the loss of nuclear staining for Gal3. Such ribonuclease digestion did not perturb the DNA content as revealed by the staining with the DNA-specific dye, DAPI. However, deoxyribonuclease I digestion completely removed the DAPI fluorescence while Gal3 staining was retained. This pattern of nuclease sensitivity of Gal3 matched that observed with the Sm antigens, which are known to be associated with RNA particles (Fig. 5.1). These experiments, documented here at the level of light microscopy (23), have been confirmed at the ultrastructural level (24).

This notion of Gal3 association with RNA particles is also consistent with its position of sedimentation when nucleoplasm is fractionated over a cesium sulfate

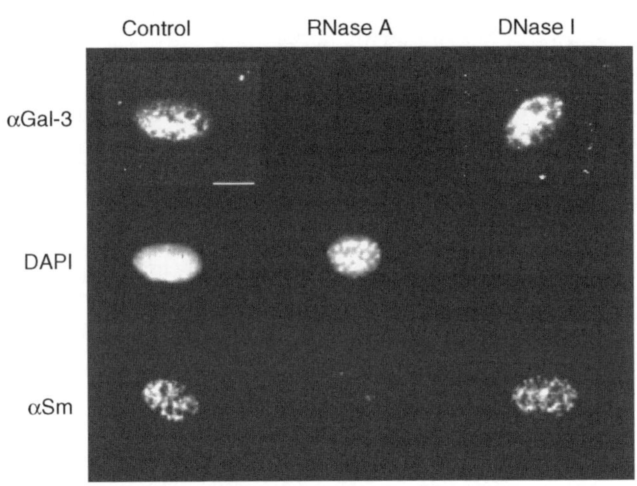

FIGURE 5.1 Effect of enzyme treatments on the intranuclear staining pattern for galectin-3 and the Sm antigens of snRNPs. Mouse 3T3 fibroblasts were permeabilized with Triton X-100 (0.5%) (column under control). These permeabilized cells were treated with ribonuclease A (100 μg/mL) for 1 h at room temperature (column under RNase A). Parallel samples of the permeabilized cells were also treated with deoxyribonucleae I (100 U/mL) for 20 min at room temperature (column under DNase I). All cells were then fixed with paraformaldehyde (4%) and stained with rabbit anti-galectin-3 (αGal-3) or with human autoimmune anti-Sm (αSm) and counterstained with the DNA-specific dye, 4′,6-diamidino-2-phenylindole-2-HCl (DAPI; 1 μg/mL).

gradient ranging in density from 1.25 g/mL at the top to 1.7 g/mL at the bottom. When the individual fractions of the gradient were subjected to immunoblotting for Gal3, it was found in fractions with densities 1.3–1.35 g/mL (23). These densities correspond to those reported for heterogeneous nuclear ribonucleoprotein complexes (hnRNP) and the snRNPs (27). Because both of these RNPs are involved in the processing of pre-mRNAs in the nucleus (28, 29), the intriguing possibility was raised that Gal3 might also be involved in the splicing reaction.

DEPLETION OF GAL1 AND GAL3 FROM NUCLEAR EXTRACTS (NEs) AND RECONSTITUTION WITH RECOMBINANT PROTEINS

The most persuasive evidence for an involvement of Gal1 and Gal3 in pre-mRNA splicing comes from depletion–reconstitution experiments on nuclear extracts capable of carrying out the splicing reaction in a cell-free assay (25, 30). The depletion can be accomplished with either lactose–agarose (LAC-A) beads, to which both Gal1 and Gal3 adsorb on the basis of their saccharide-binding activity, or with beads covalently coupled with antibodies directed against Gal1 and Gal3. The NE is incubated with the beads in Hepes buffer (pH 7.9) containing high salt (0.5 M NaCl), which dissociates the protein complexes. The bound fraction can quantitatively account for the galectins present in the original NE and the unbound fraction (UB) is essentially devoid of the two proteins (25, 30). The depleted extract is then dialyzed against Hepes buffer (pH 7.9) containing low salt (60 mM KCl) in the presence and absence of proteins to be tested for splicing activity.

As documented in Fig. 5.2 (left panel), the original NE exhibited good splicing activity, capable of converting >40% of the pre-mRNA substrate into the mature RNA product (Fig. 5.2, left, lane 1). The unbound fraction of the double antibody (anti-Gal1 and anti-Gal3) depletion has lost this splicing activity (Fig. 5.2, left, lane 2). This depleted extract can be reconstituted with either recombinant Gal3 (Fig. 5.2, left, lane 3) or with recombinant Gal1 (Fig. 5.2, left, lane 5), so the two proteins appear to be redundant. Each of the reconstituted splicing activities is sensitive to inhibition by thiodigalactoside (TDG), a high affinity saccharide ligand of the galectins (Fig. 5.2, left, lanes 4 and 6). This recapitulates the observation that the splicing activity of nondepleted NEs is sensitive to inhibition by saccharide ligands (e.g., Lac or TDG) of the galectins (30).

In the above analysis (Fig. 5.2, left panel), the splicing reaction was allowed to proceed for 45 min and gel electrophoresis was carried out under denaturing conditions so that various RNA species (starting pre-mRNA substrate, mature mRNA product, and intermediates) can be resolved. If the splicing reaction was analyzed early (15–20 min) by gel electrophoresis under nondenaturing conditions, however, the various steps of spliceosome assembly can be discerned. Initially, the pre-mRNA is complexed with hnRNP to form the H-complex. Addition of the uracil-rich snRNPs results in higher order complexes, A- and B-complexes, as well as the active spliceosome in the C-complex.

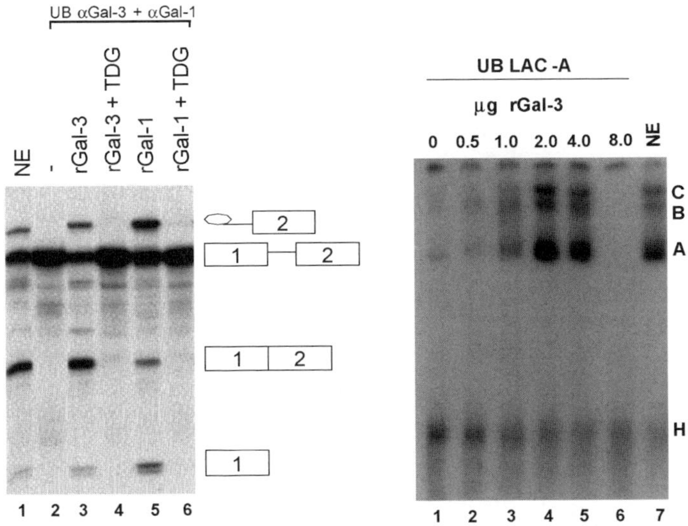

FIGURE 5.2 Left: Comparison of the splicing activities of various nuclear extracts. Lane 1, complete nuclear extract (NE). Lane 2, NE depleted of the galectins (UB fraction of αGal3 + αGal1). Lane 3, depleted extract reconstituted with recombinant Gal3. Lane 4, depleted extract with Gal3 and the saccharide ligand thiodigalactoside (TDG). Lane 5, depleted extract reconstituted with recombinant Gal1. Lane 6, depleted extract with Gal1 and TDG. The positions of migration of the pre-mRNA, the mRNA product, and intermediates of the splicing reaction are indicated on the right. In these schematic structures, the pre-mRNA contains two exons (rectangles labeled 1 and 2) connected by an intron (solid line between rectangles). The mature mRNA product has the two exons joined together. Free exon1 and intron lariat-exon2 are intermediates of the reaction. Right: Comparison of the various complexes formed during spliceosome assembly. Lane 1, NE depleted of the galectins (UB fraction of lactose agarose (LAC-A) column). Lane 2–6, depleted extract reconstituted with different doses of recombinant Gal3. Lane 7, complete NE. The positions of migration of the various spliceosomal complexes (H, A, B, and C) are indicated on the right.

As documented in Fig. 5.2 (right panel), the original, nondepleted NE yields the A-, B-, and C-complexes (Fig. 5.2, right, lane 7). The depleted extract used in this experiment was derived from the unbound fraction of the lactose–agarose column; there was no splicing activity in this depleted extract and the pre-mRNA appears to be stuck at a position corresponding to the H-complex (Fig. 5.2, right, lane 1). Upon reconstitution by the addition of recombinant Gal3, the higher order spliceosomal complexes were restored (Fig. 5.2, right, lanes 2–5). Thus, it appears that the presence of the galectins is required for the progression from the H-complex to the higher order complexes. Curiously, addition of very high doses (e.g., 8 μg to the splicing reaction mixture; ~26 μM) resulted in loss of spliceosome assembly (Fig. 5.2, right, lane 6) and no splicing activity.

PERTURBATION OF SPLICING BY FRAGMENTS OF GAL3 AND BY ANTIBODIES AGAINST GAL3

Two additional lines of evidence have now been accumulated to implicate the H-complex as the locus at which Gal1 or Gal3 is required for spliceosome assembly and splicing activity. First, we noted that like Gal1, the carboxyl-terminal domain of Gal3 is a CRD and it can reconstitute splicing activity in a galectin-depleted extract. However, at equimolar concentrations, full-length Gal3 was more potent in reconstituting splicing than either its carboxyl-terminal CRD or Gal1 (25). This suggested the possibility that Gal3 uses its amino-terminal domain (ND) to interact with components of the splicing machinery, enhancing the affinity of the full-length polypeptide relative to CRD alone. Fragments of Gal3 containing the ND might be expected, then, to compete for these interactions and inhibit the activity of Gal3 in a fashion mimicking the action of a dominant-negative mutant. Indeed, addition of Gal3 ND to a splicing competent NE resulted in a dose-dependent inhibition while parallel additions of either full-length Gal3 or the carboxyl-terminal CRD failed to yield the same effect (31). In the ND-inhibited reaction, spliceosome assembly was arrested at the H-complex.

Second, the epitope of the monoclonal antibody NCL-GAL3 has been mapped to the ND of Gal3. When added to a splicing competent NE, this antibody also inhibited the splicing reaction. In the antibody-inhibited reaction, spliceosome assembly was again arrested at the H-complex (32). Together, the depletion and the perturbation experiments all suggest that Gal3 enters the splicing cycle at the earliest complex. Consistent with this notion, we have obtained preliminary evidence to indicate that Gal3 is loaded onto the pre-mRNA scaffold during spliceosome assembly via a Gal3–U1 snRNP complex. Indeed, the binding of U1 snRNP to the 5′-splice site of the pre-mRNA is one of the hallmarks of early complex formation committed to the spliceosome assembly pathway (33).

CHALLENGING QUESTIONS AND PERSPECTIVES

The observation of Gal3 in the nucleus and its association with ribonucleoprotein complexes provided the initial hints that Gal1 and Gal3 may be involved in pre-mRNA splicing. Depletion and reconstitution experiments have established that these proteins are, indeed, required factors in splicing, as assayed in a cell-free system. Although there is recent evidence that Gal3 enters the splicing pathway early during spliceosome assembly, much remains to be elucidated in terms of molecular details. First, with how many components of the splicing machinery does Gal3 (or Gal1) interact directly and what is the identity of the interacting partner(s)? Does Gal3 play a static scaffolding role in the assembly of the splicing complex, which has been reported to contain more than 300 proteins and at least six RNA species (33–37)? Or, does Gal3 play a more dynamic role, particularly in terms of the chemistry of enzymatic catalysis itself? Finally, the original purification of Gal3 (and Gal1) was based on its carbohydrate-binding activity but is saccharide binding, *per se*, required for the splicing activity? The pursuit of these issues promises an exciting time in the years to come.

ACKNOWLEDGMENTS

The work carried out in the authors' laboratories has been supported by grants MCB-0092929 from the National Science Foundation (RJP), GM-38740 from the National Institutes of Health (JLW), and an intramural 06-IRGP-858 grant from Michigan State University (JLW).

REFERENCES

1. Barondes SH, Castronovo V, Cooper DW, Cummings RD, Drickamer K, Feizi T, Gitt MA, Hirabayashi J, Hughes C, Kasai K, Leffler H, Liu FT, Lotan R, Mercurio AM, Monsigny M, Pillai S, Poirier F, Raz A, Rigby PWJ, Rini JM, Wang JL. Galectins: a family of animal β-galactoside-binding lectins. Cell 1994;76:597–598.
2. Cooper DN, Barondes SH. God must love galectins; He made so many of them. Glycobiology 1999;9:979–984.
3. Wang JL, Gray RM, Haudek KC, Patterson RJ. Nucleocytoplasmic lectins. Biochim Biophys Acta 2004;1673:75–93.
4. Houzelstein D, Goncalves IR, Fadden AJ, Sidhu SS, Cooper DNW, Drickamer K, Leffler H, Poirier F. Phylogenetic analysis of the vertebrate galectin family. Mol Biol Evol 2004;21:1177–1187.
5. Kasai K, Hirabayashi J. Galectins: a family of animal lectins that decipher glycocodes. J Biochem 1996;119:1–8.
6. Rini JM, Lobsanov YD. New animal lectin structures. Curr Opin Struct Biol 1999;9:578–584.
7. Hsu DK, Zuberi RI, Liu F-T. Biochemical and biophysical characterization of human recombinant IgE-binding protein, an S-type animal lectin. J Biol Chem 1992;267: 14167–14174.
8. Knibbs RN, Agrwal N, Wang JL, Goldstein IJ. Carbohydrate binding protein 35. II. Analysis of the interaction of the recombinant polypeptide with saccharides. J Biol Chem 1993;268:14940–14947.
9. Arnoys EJ, Wang JL. Dual localization: proteins in extracellular and intracellular compartments. Acta Histochem 2007;109:89–110.
10. Hughes RC. Secretion of the galectin family of mammalian carbohydrate-binding proteins. Biochim Biophys Acta 1999;1473:172–185.
11. Hughes RC. Galectins as modulators of cell adhesion. Biochimie 2001;83:667–676.
12. Leffler H. Galectins: structure and function—a synopsis. Results Probl Cell Differ 2001;33:57–83.
13. Liu F-T, Rabinovich GA. Galectins as modulators of tumour progression. Nat Rev 2005;5:29–41.
14. Saal I, Nagy N, Lensch M, Lohr M, Manning JC, Decaestecker C, Andre S, Kiss R, Salmon I, Gabius HJ. Human galectin-2: expression profiling RT-PCR/immunohistochemistry and

its introduction as a histochemical tool for ligand localization. Histol Histopathol 2005;20:1191–1208.

15. Danguy A, Rorive S, Decaestecker C, Bronckart Y, Kaltner H, Hadari YR, Goren R, Zick Y, Petein M, Salmon I, Gabius HJ, Kiss R. Immunohistochemical profile of galectin-8 expression in benign and malignant tumors of epithelial, mesenchymatous and adipous origins, and of the nervous system. Histol Histopathol 2001;16:861–868.
16. Huflejt ME, Leffler H. Galectin-4 in normal tissues and cancer. Glycoconj J 2004;20:247–255.
17. Moutsatsos IK, Davis JM, Wang JL. Endogenous lectins from cultured cells: subcellular localization of carbohydrate-binding protein 35 in 3T3 fibroblasts. J Cell Biol 1986;102:477–483.
18. Patterson RJ, Wang W, Wang JL. Understanding the biochemical activities of galectin-1 and galectin-3 in the nucleus. Glycoconj J 2004;19:499–506.
19. Vyakarnam A, Lenneman AJ, Lakkides KM, Patterson RJ, Wang JL. A comparative nuclear localization study of galectin-1 with other splicing components. Exp Cell Res 1998;242:419–428.
20. Moutsatsos IK, Wade M, Schindler M, Wang JL. Endogenous lectins from cultured cells: nuclear localization of carbohydrate-binding protein 35 in proliferating 3T3 fibroblasts. Proc Natl Acad Sci USA 1987;84:6452–6456.
21. Openo KP, Kadrofske MM, Patterson RJ, Wang JL. Galectin-3 expression and subcellular localization in senescent human fibroblasts. Exp Cell Res 2000;255:278–290.
22. Lotz MM, Andrews CW, Korzelius CA, Lee EC, Steele GD, Clarke A, Mercurio AM. Decreased expression of Mac-2 (carbohydrate binding protein 35) and loss of its nuclear localization are associated with the neoplastic progression of colon carcinoma. Proc Natl Acad Sci USA 1993;90:3466–3470.
23. Laing JG, Wang JL. Identification of carbohydrate binding protein 35 in heterogeneous nuclear ribonucleoprotein complex. Biochemistry 1988;27:5329–5334.
24. Hubert M, Wang S-Y, Wang JL, Seve A-P, Hubert J. Intranuclear distribution of galectin-3 in mouse 3T3 fibroblasts: comparative analyses by immunofluorescence and immunoelectron microscopy. Exp Cell Res 1995;220:397–406.
25. Vyakarnam A, Dagher SF, Wang JL, Patterson RJ. Evidence for a role for galectin-1 in pre-mRNA splicing. Mol Cell Biol 1997;17:4730–4737.
26. Spector DL, Fu XD, Maniatis T. Associations between distinct pre-mRNA splicing components and the cell nucleus. EMBO J 1991;10:3467–3481.
27. Mayrand S, Pederson T. Nuclear ribonucleoprotein particles probed in living cells. Proc Natl Acad Sci USA 1981;78:2208–2212.
28. Padget RA, Mount SM, Steitz JA, Sharp PA. Splicing of messenger RNA precursors is inhibited by antisera to small nuclear ribonucleoprotein. Cell 1983;35:101–107.
29. Choi YD, Grabowski P, Sharp PA, Dreyfuss D. Heterogeneous nuclear ribonucleoproteins: role in RNA splicing. Science 1986;231:1534–1539.
30. Dagher SF, Wang JL, Patterson RJ. Identification of galectin-3 as a factor in pre-mRNA splicing. Proc Natl Acad Sci USA 1995;92:1213–1217.
31. Park JW, Voss PG, Grabski S, Wang JL, Patterson RJ. Association of galectin-1 and galectin-3 with Gemin4 in complexes containing the SMN protein. Nucleic Acids Res 2001;29:3595–3602.

32. Gray RM, Davis MJ, Ruby KM, Voss PG, Patterson RJ, Wang JL. Distinct effects on splicing of two monoclonal antibodies directed against the amino-terminal domain of galectin-3. Arch. Biochem. Biophys. 2008; submitted for publication.
33. Brow DA. Allosteric cascade of spliceosome activation. Annu Rev Genet 2002; 36:333–360.
34. Jurica MS, Licklider LJ, Gygi SR, Grigorieff N, Moore MJ. Purification and characterization of native spliceosomes suitable for three-dimensional structural analysis. RNA 2002;8:426–439.
35. Hartmuth K, Urlaub H, Vornlocher HP, Will CL, Gentzel M, Wilm M, Luhrmann R. Protein composition of human prespliceosomes isolated by a tobramycin affinity-selection method. Proc Natl Acad Sci USA 2002;99:16719–16724.
36. Zhou Z, Licklider LJ, Gygi SP, Reed R. Comprehensive proteomic analysis of the human spliceosome. Nature 2002;419:182–185.
37. Rappsilber J, Ryder U, Lamond AI, Mann M. Large-scale proteomic analysis of the human spliceosome. Genome Res 2002;12:1231–1245.

6

GALECTINS IN REGULATION OF INFLAMMATION AND IMMUNITY

Fu-Tong Liu, Daniel K. Hsu, Ri-Yao Yang, Huan-Yuan Chen, and Jun Saegusa

Department of Dermatology, University of California, Davis, School of Medicine, 3301 C Street, Suite 1400, Sacramento, CA 95816, USA

INTRODUCTION

Since 1994 when the name galectins was given to a family of animal lectins containing just four identified members, a great deal of progress has been made in our understanding of this family. Fifteen members have so far been identified in mammals and their homologues have been found in various animal species as well as insects and plants. The protein sequences of all the recognized members are known and the 3D structures of some of them have been resolved. Expression of the members in various cells and tissues in normal and disease conditions has been reported. The functions of many members have been studied, some rather extensively, and their roles in a variety of physiological and pathological processes have been described. This chapter focuses on the roles of selected galectins in the immune and inflammatory responses. Because a number of reviews have been published on this subject, we will provide only an overview, by citing some selected studies, and refer the readers to published reviews.

Extracellular Functions and Oligovalency

Most functions of galectins were demonstrated by adding recombinant proteins to cells or tissues, primarily *in vitro*, and a few were described *in vivo*. These functions

Galectins, Edited by Anatole A. Klyosov, Zbigniew J. Witczak, and David Platt

are related to galectins' intrinsic bivalency or tendency to dimerize or oligomerize. Galectins with one carbohydrate recognition domain (CRD) often exist as dimers, and those with two inherent CRDs may have two carbohydrate-binding sites and are thus at least bivalent. Galectin-3 forms pentamers in the presence of multivalent carbohydrate ligands (reviewed in Reference 1). Most activities of galectins are demonstrated by the use of proteins at high submicromolar to low micromolar concentrations. However, recent studies with galectin-1 suggest that these proteins may be more effective when presented by extracellular matrices. In addition, an increasing number of studies have revealed extracellular functions of endogenous galectins by demonstrating that the stated functions are suppressed by specific inhibitors added to the extracellular space.

Formation of Complexes with Cell Surface Glycans

The current view is that secreted galectins can function in an autocrine or paracrine manner by binding to and cross-linking selected glycoproteins or glycolipids present on the cell surfaces. Unlike cytokines or growth factors, galectins do not appear to have specific individual receptors. Instead, each is able to bind to a number of different cell surface glycoproteins (and glycolipids) that carry suitable galactose-containing oligosaccharides. For example, galectin-3 binds to high-affinity IgE receptor (FcεRI) among a limited number of other glycoproteins on mast cells (as reflected by the number of protein bands visualized by SDS-PAGE with samples affinity purified by galectin-3 conjugated to Sepharose-4B) (2). It also binds to a number of glycoproteins on neutrophils, including CD66a, CD66b, and lysosome-associated membrane glycoprotein-1 and -2 (3, 4).

Affinity purification of membrane glycoproteins on T cells that bind to galectin-1 identified five major bands and four minor bands by SDS-PAGE (5). Specific antibodies to various CD molecules allowed the identification of six CD molecules as ligands of galectin-1. A similar approach in conjunction with mass spectrometry showed that galectin-3 binds to over 16 glycoproteins on the surface of T cells (6). These two galectins not only bind to some common glycoproteins, but also recognize different species and thus each binds to a different set of glycoproteins.

When added to neutrophils, galectin-3 was found to cluster at a site (capping) on the cell surface (7). Pace et al. (5) showed that galectin-1 caused aggregation of various CD molecules on the surface of T cells. CD3 and CD45 coclustered and CD7 and CD43 coclustered, but these assemblies were segregated from each other. Stillman et al. (6) showed that galectin-3 caused aggregation of CD45 and CD71.

The work by James Dennis' group demonstrated the formation of lattices between galectin-3 and cell surface glycoproteins, in particular, glycoproteins modified by the *N*-acetylglucosaminyltransferase, Mgat5. One example is T-cell receptor (TCR), and the authors provided evidence that such lattice formation results in restricted TCR motility and thus a downregulation of the T-cell response (8). This group subsequently showed that glycoproteins, such as epidermal growth factor receptor and transforming growth factor-β receptor, also form complexes with galectin-3. This results in a delay in receptor endocytosis (9). Similarly, Ohtsubo et al. (10) showed that glucose transporter forms aggregates with galectin-9, resulting in a downregulation of endocytosis of the transporter. In the absence of appropriate glycosylation, these receptors form weaker complexes with galectins and are more readily internalized.

Intracellular Functions

All galectins lack a classical signal peptide and transmembrane domains required for secretion through the classical secretory pathway and for display on the cell surface, respectively (reviewed in Reference 11). A number of intracellular functions have been reported and, for some of these, intracellular proteins with which galectins interact have been identified, leading to the suggestion that the binding partners mediate these functions. Notably, galectins bind to them through protein–protein interactions and not lectin–carbohydrate interactions. Because glycans do exist inside the cells, intracellular galectins may have the opportunity to bind to some of them. A very interesting and unexpected association between galectin-3 and glycoproteins contained in post-Golgi vesicles was reported by Delacour et al. (12). These glycoproteins are destined to be exported to the apical side of the cell. Knocking down of galectin-3 with siRNA resulted in the apical glycoproteins being misdirected to the basolateral cell surfaces. Their findings implicate galectin-3 with a role in controlling the trafficking of cell surface glycoproteins.

GALECTIN-1

The most extensively studied function of galectin-1 is regulation of apoptosis in immune cells. This subject has been reviewed extensively and only some of the highlights will be mentioned here, followed by a brief review of other functions related to the immune and inflammatory responses.

Regulation of Apoptosis

Recombinant galectin-1 induces apoptosis in activated human T cells, as well as T leukemia cell lines (reviewed in References 13 and 14). Most recently, Toscano et al. (15) showed that galectin-1 binds to Th1, Th2, and Th17 cells differentially and induces apoptosis in Th1 and Th17 cells, but not Th2 cells. As mentioned above, galectin-1 binds to a restricted set of T-cell surface glycoproteins, including CD45, CD43, and CD7. CD7 is essential for galectin-1-induced cell death, while CD45 regulates the process in a fashion that is dependent of its glycosylation state (reviewed in Reference 16).

Whether endogenous galectin-1 is capable of inducing apoptosis has been addressed. Galectin-1 secreted and presented by thymic epithelial cells is effective in inducing apoptosis in T-cell lines cocultured with the former cells (17). Likewise, Perone (18) found that dendritic cells overexpressing galectin-1-induced apoptosis in cocultured activated T cells and they provided evidence that this is due to galectin-1 released by transfected dendritic cells and binding to T cells.

In addition, by using an antisense strategy, it has been shown that galectin-1 secreted by mouse melanoma cells can induce apoptosis of activated T cells targeting the tumor cells (19). Furthermore, splenocytes from galectin-1-deficient (gal1$^{-/-}$) mice produce higher IFN-γ (a Th1 cytokine) and IL-17 (15), which is consistent with the higher sensitivity of Th1 and Th17 cells to apoptosis induced by galectin-1.

Caspases are key effector molecules in the apoptosis pathway. One study showed that galectin-1 activated caspases (20). However, another study documented that galectin-1 triggered other apoptosis-related events, but not caspase activation (21). Yet another group reported that galectin-1 induced the exposure of phosphatidylserine on the cell surface of T-cell lines, an early event of apoptosis, but not DNA fragmentation, a hallmark of apoptosis (22). Therefore, it appears that variations in experimental conditions can induce different endpoints in cells exposed to galectin-1 along apoptotic pathways, which may not be apoptosis.

Other Functions

Galectin-1 has been shown to inhibit TCR-mediated responses in T cells, including IL-2 production, when the cells are mixed with antigen-presenting cells (23). Galectin-1 has also been shown to inhibit the allogeneic T-cell response (20) and induce IL-10 production in T cells (24). Galectin-1 is expressed by regulatory T cells and inhibition of its activity significantly reduces the suppressive effects of these cells (25). In addition, regulatory T cells obtained from gal1$^{-/-}$ mice exhibit lower suppressive activity. Thus, galectin-1 appears to contribute to the suppressive function of regulatory T cells.

Galectin-1 has been shown to inhibit T-cell adhesion to extracellular matrix glycoproteins, such as fibronectin and laminin (26). As T cells migrate through endothelial monolayers, galectin-1 expressed on the surface of the latter cells can retard the process (27). In total, galectin-1 is apparently immunosuppressive with regard to the T-cell response, through induction of apoptosis, suppression of the T-cell response, or mediation of the T regulatory cell activities.

With regard to other immune cell types, galectin-1 inhibits transendothelial migration and chemotaxis of neutrophils (28). Galectin-1 also induces exposure of phosphatidylserine on the surface of this cell type and results in conferring greater sensitivity to phagocytosis by macrophages (22). Galectin-1 was found to induce FcγRI expression on monocytes, remarkably at a relatively low concentration of 0.4 μg/mL, where dimerization is not expected to occur. This effect was associated with an increase in FcγRI-mediated phagocytic activity of these cells. In contrast, galectin-1 inhibits the IFN-γ-induced expression of FcγRI. This occurs at micromolar concentrations. Similarly, galectin-1 inhibits both constitutive and IFN-γ-induced expression of MHC-II, also at micromolar concentrations (29). The ability of endogenous galectin-1 to inhibit expression of MHC-II and phagocytosis has been confirmed by studying cells from gal1$^{-/-}$ mice.

However, galectin-1 induces maturation of dendritic cells, as evidenced by the expression of cell surface maturation markers, production of certain proinflammatory cytokines, and enhanced ability to migrate through the extracellular matrix (30).

Finally, in the studies mentioned above using dendritic cells overexpressing galectin-1, the authors found that transfectants had a more mature phenotype, as evidenced by increased expression of proinflammatory cytokines and enhanced ability to stimulate naive T cells. These results suggest that galectin-1 can also promote activation of dendritic cells.

Functions of Galectin-1 Demonstrated *In Vivo*

A large number of studies have demonstrated anti-inflammatory properties of galectin-1 *in vivo* in several models of acute and chronic inflammation and autoimmunity. While most studies employed recombinant galectin-1, some involved engraftment of cells transfected with galectin-1 cDNA (reviewed in Reference 13).

Some studies revealed that administration of galectin-1 can result in skewing the T-cell response toward a Th2-polarized direction, which is consistent with the ability of the lectin to induce apoptosis in Th1 cells and not Th2 cells. Other studies demonstrated that treatment with recombinant galectin-1 in mice resulted in the promotion of a Th2 response and potentiation of T regulatory cell activity, as evidenced by the resulting cytokine profile (32). In addition, regulatory T cells obtained from galectin-1-treated mice demonstrated stronger suppressive activity on the immune response, when adoptively transferred to naïve recipients. The role of endogenous galectin-1 in tumor-induced immunosuppression *in vivo* has also been demonstrated by comparing control melanoma cells to those with galectin-1 expression suppressed by antisense oligonucleotides (19). Specifically, down-regulation of galectin-1 in cancer cells resulted in heightened T-cell-mediated tumor rejection associated with increased survival of IFN-γ-producing Th1 cells. Finally, $gal1^{-/-}$ mice were found to be highly susceptible to development of experimental autoimmune encephalitis, a Th1-mediated disease, which is consistent with galectin-1's inhibition of the Th1 response (15).

GALECTIN-3

The functions of galectin-3 in the immune and inflammatory responses have been extensively reviewed (31, 33). In comparison with galectin-1, there are more studies of intracellular functions (34).

Homeostatic Regulation of Immune Cells

The roles of galectin-3 in differentiation and growth of various immune cells have been reviewed in Reference (33). Like galectin-1, galectin-3 induces apoptosis in T cells. One study concluded that CD7 and CD29 (β1 integrin) mediate galectin-3-induced apoptosis (35), while another study showed that CD45 and CD71 are involved (6). Galectin-3 causes cytochrome *c* release and caspase-3 activation, but not caspase-8 activation (35). The lectin has also been shown to induce apoptosis in neutrophils (36).

By using the gene transfection and the antisense approaches, galectin-3 has been found to have antiapoptotic activity in a number of cell types against a diverse array of apoptotic stimuli, including immune and inflammatory cells (reviewed in Reference 34). This was first demonstrated by studying transfectants of the human T-cell line Jurkat overexpressing the protein treated with apoptotic stimuli (37) and subsequently with another human T-cell line CEM treated with other apoptotic stimuli, galectin-1 (21) and C(2)-ceramide (35). The antiapoptotic activity of galectin-3 has also been documented in human B lymphoma (38). Studies of cells from galectin-3-deficient (gal3$^{-/-}$) mice confirmed the antiapoptotic activity of galectin-3, in particular macrophages (39).

Mechanisms by which intracellular galectin-3 confers resistance to apoptosis remain to be fully elucidated. Existing information suggests that this may involve interaction with other regulators of apoptosis operating in the mitochondria (reviewed in Reference 40). First, galectin-3 translocates to the mitochondria in cells undergoing apoptosis. Second, expression of galectin-3 suppresses the loss in the mitochondrial potential associated with apoptosis. Third, galectin-3 interacts with other apoptosis regulators known to function in the mitochondria, such as Bcl-2. The antiapoptotic activity of galectin-3 has been shown to be dependent on phosphorylation of serine at position 6. In addition, the Asn-Trp-Gly-Arg (NWGR) motif in the C-terminal domain of the protein, known to be at the carbohydrate-binding site, is indispensable. This motif is contained in well-characterized members of the Bcl-2 family of apoptosis regulators and is critical in their functions.

Thus, endogenous galectin-3 in T cells is antiapoptotic, but extracellular galectin-3 secreted by other cells may kill T cells. This may be generalized to other functions and other galectins in that activities demonstrated by using recombinant proteins added to cells may not demonstrate the functions of endogenous galectins in that particular cell type.

Functional Regulation of Immune Cells

In a carbohydrate-dependent manner, galectin-3 can bind to extracellular matrix proteins and influence cell adhesion to extracellular matrices. It can also interact with integrins $\alpha 1\beta 1$ and $\alpha M\beta 1$ (CD11b/18) and potentiate cell adhesion through these integrins. The functions of galectin-3 in the immune and inflammatory processes by modulating cell adhesion in various cell types have been reviewed in Reference 41.

Recombinant galectin-3 was found to promote adhesion of human neutrophils to laminin (7) and an endothelial cell line (42). It has also been shown to activate a number of lymphoid and myeloid cells, including mast cells, neutrophils, monocytes, and T cells. These result in different responses from mediator release, to superoxide anion production, to cytokine production (reviewed in Reference 33). The cell surface receptors for each of these have not been definitively identified. It is likely that different receptors are involved in different cell types. Recombinant galectin-3 has also been shown to serve like a chemokine in inducing migration of human monocytes and macrophages (43). Similar to that of chemokines, the activity is mediated through a pertussis toxin (PTX)-sensitive (G-protein-coupled) pathway and associated with a Ca^{2+} influx. Thus, specific chemokine receptors may be involved.

Galectin-3 can also exert a suppressive effect on myeloid cells, as exemplified by inhibition of IL-5 production in human eosinophils (44). This activity is not surprising, as galectin-3 should not be expected to bind only to and engage those receptors promoting activation, but may engage inhibitory receptors containing ITIM motifs. Indeed, galectin-3 appears to downregulate IL-5 production by engaging FcγRII (45), an inhibitory receptor.

Galectin-3 was described to be associated with the TCR complex, in a fashion that is dependent on glycosylation of TCR controlled by a glycosyltransferase, β1,6 *N*-acetylglucosaminyltransferase V (Mgat5) (8), resulting in the formation of lattices. This can cause a restricted lateral motility of TCR and a lower TCR-mediated cell response. $Mgat5^{-/-}$ T cells in which TCR forms weaker complexes with galectin-3 exhibited higher TCR-mediated responses compared to $Mgat5^{+/+}$ cells. In addition, inhibition of galectin-3 on T-cell surfaces by lactose increased TCR clustering in response to an agonist and enhanced the T-cell response. This suggests that galectin-3 can serve as a negative regulator of TCR-initiated signal transduction through binding to the receptor.

Most of the above studies employed recombinant protein, but there are a few studies addressing the activity of the endogenous protein. For example, Swarte et al. (46) showed that adhesion of T cells to dendritic cells or macrophages was inhibited by a known galectin-3 sugar ligand and antigalectin-3 antibody. The above-mentioned studies of $Mgat5^{-/-}$ T cells also demonstrated the suppressive effect of endogenous galectin-3. Finally, the activity of endogenous galectin-3 has been addressed by using cells from galectin-3-deficient ($gal3^{-/-}$) mice, as summarized below.

Functions Suggested by Studying $Gal3^{-/-}$ Cells

The function of galectin-3 in phagocytosis has been demonstrated by comparing macrophages from $gal3^{-/-}$ mice and wild-type mice (47): $gal3^{-/-}$ macrophages were defective in phagocytosis mediated by Fcγ receptor. Likewise, the role of galectin-3 in mast cell functions has been revealed by studying mast cells from $gal3^{-/-}$ mice: $gal3^{-/-}$ mast cells produced lower levels of granular mediators and cytokines, when activated by cross-linkage of cell surface IgE receptor, compared to wild-type cells (48).

We have compared the cytokine responses of $gal3^{-/-}$ and $gal3^{+/+}$ $CD4^+$ T cells upon activation by TCR engagement and found that $gal3^{-/-}$ $CD4^+$ T cells secreted more IFN-γ compared to $gal3^{+/+}$ cells (unpublished observation). Also, galectin-3 appears to suppress secretion of IL-12 by dendritic cells as $gal3^{-/-}$ cells produced higher levels of IL-12 than wild-type cells (49).

The mechanisms by which galectin-3 regulates these activities still remain to be elucidated. In all these cases, the response in wild-type cells are not reduced by lactose, suggesting that the endogenous protein does not function by being released by the cells and acting extracellularly through lectin action by binding to the cell surfaces. In addition, in wild-type macrophages undergoing phagocytosis, galectin-3 was detected in phagocytic cups and concentrated in phagosomes, suggesting that the protein may function at these locations. Similarly, galectin-3 is localized in the immunological synapse in T cells exposed to antigen-presenting cells (unpublished observation).

Gal3$^{-/-}$ mast cells contain lower levels of c-jun-N-terminal kinase-1 (JNK1) mRNA and protein, suggesting that the lectin regulates JNK1 transcription (48). Because JNK1 plays a central role in the signaling pathways leading to production of selected cytokines, this could explain how galectin-3 controls cytokine production in mast cells. How galectin-3 regulates JNK and only one of the two isoforms remains to be elucidated. The intriguing possibility is that it may do so by regulating pre-mRNA splicing, which is one of the demonstrated activities of galectin-3 *in vitro*.

Biological Effects Demonstrated *In Vivo*

In vitro studies suggest that galectin-3 can promote inflammatory responses through its functions on cell activation, cell migration, and inhibition of apoptosis (thus prolonging the survival of inflammatory cells). However, as described above, other *in vitro* functions suggest that galectin-3 may suppress the immune and inflammatory responses. Thus far, *in vivo* studies using gal3$^{-/-}$ mice support the notion that galectin-3 promotes inflammatory responses. Reduced IgE-mediated responses of mast cells *in vivo* were observed in gal3$^{-/-}$ mice.

In addition, gal3$^{-/-}$ mice exhibited attenuated infiltration of leukocytes relative to wild-type mice in a model of peritoneal inflammation (39, 50). Gal3$^{-/-}$ mice developed lower lung eosinophilia compared to similarly treated wild-type mice following airway antigen challenge (51) in a mouse model of atopic asthma. These studies also revealed the regulation of Th1/Th2 polarization by galectin-3: compared to wild-type mice, gal3$^{-/-}$ mice showed (i) decreased IL-4 and IgE levels (markers of the Th2 response) and (ii) elevated IFN-γ levels and IgG2a/IgG1 ratios (an index for the Th1 response), in bronchoalveolar fluid and serum.

It should be mentioned that other investigators observed reduced eosinophil infiltration following airway antigen challenge in rats and mice treated by intranasal delivery of cDNA encoding galectin-3 (52, 53). These contrasting results may be explained by a potentiating role for endogenous galectin-3 in the airway inflammatory response, but a suppressive effect of pharmacological concentrations of galectin-3 applied to the airways. This is consistent with the statement made above that the activities demonstrated with recombinant proteins added to cells may not represent the functions of the endogenous protein. However, findings made with exogenously added recombinant proteins or DNA are certainly relevant in terms of applications of galectins in therapeutics.

We have recently studied the role of galectin-3 in the development of atopic dermatitis, which is a chronic inflammatory skin disease characterized by spongiotic skin lesions attributable to a Th2-mediated inflammatory response. We used a mouse model of allergic skin inflammation with a Th2-mediated immune response, induced by repeated epicutaneous sensitization with a protein antigen. Gal3$^{-/-}$ mice exhibited less marked epidermal thickening, lower eosinophil infiltrations in the dermis, and lower total and antigen-specific IgE levels, but a higher antigen-specific IgG2a/IgG1 ratio in the sera, compared to similarly treated gal3$^{+/+}$ mice. In addition, the former had lower IL-4 (Th2) mRNA expression, but higher IFN-γ (Th1) mRNA expression than the latter at the antigen-treated skin sites (unpublished results). Thus, galectin-3 is critical for the development of the Th2 inflammatory response to epicutaneously introduced protein antigens.

The role of galectin-3 in innate immunity against microorganisms has also been revealed by studying gal3$^{-/-}$ mice. Consistent with the role of galectin-3 in promoting the inflammatory response, gal3$^{-/-}$ mice infected with the parasites *Toxoplasma* (*T.*) *gondii* exhibited lower inflammatory scores in gut, liver, and brain (but not the lungs) compared to similarly infected wild-type mice (49). Also consistent with the information mentioned above, gal3$^{-/-}$ mice infected with the parasites exhibited a Th1-polarized response relative to wild-type mice, in that serum levels of IL-12 and IFN-γ were highly elevated in the former (49). In a mouse model of *Mycobacterium tuberculosis* infection, Beatty et al. (54) observed that gal3$^{-/-}$ mice had impaired capacities in clearing the late infection compared to wild-type mice, suggesting involvement of galectin-3 in innate defense against the microorganism.

OTHER GALECTINS

Galectin-2

Galectin-2 can also induce T-cell apoptosis (55). Its effect is associated with the activation of caspase-3 and -9, cytochrome *c* release, disruption of the mitochondrial membrane potential, and DNA fragmentation.

Galectin-4

Galectin-4 has been shown to play a role in the pathogenesis of inflammatory bowel disease (56), an immune-mediated intestinal inflammatory condition. Galectin-4 was found to induce IL-6 production, an inflammatory cytokine known to contribute to the progression of the disease, in CD4^{+} T cells from mice that developed colitis. This is relevant to the pathogenesis of colitis, as (i) the response was correlated with the severity of colitis in the host from which the CD4^{+} cells were isolated and (ii) administration of a neutralizing galectin-4 antibody into mice suppressed the development of colitis.

Galectin-9

Human galectin-9 was initially identified as an eosinophil chemoattractant produced by T lymphocytes (57). Subsequent studies suggest that this lectin plays a role in adhesion of eosinophils to other cell types (58).

Galectin-9 can also induce T-cell apoptosis (59, 60). Both $CD4^+$ and $CD8^+$ cells are susceptible to apoptosis induced by this lectin, and the cells are more sensitive to galectin-9 if they are first activated with anti-CD3 antibody. Galectin-9 induces apoptosis via the caspase-1 pathway and caspase-8, -9, and -10 are not involved. Galectin-9 also induces apoptosis in cell lines of the B cell, monocytic cell, and promyelocytic cell lineage (60).

Galectin-9 was recently identified as a ligand for Tim-3, a T helper 1 (Th1)-specific cell surface molecule, and shown to induce apoptosis through binding to this protein (61). Consistent with this, the lectin preferentially induces killing in Th1 cells over Th2 cells. The ability of galectin-9 to kill Th1 cells *in vivo* was demonstrated by the fact that recombinant galectin-9 suppressed the development of experimental autoimmune encephalitis in mice, which is a Th1-mediated disease model.

Galectin-9 has been shown to directly induce apoptosis in eosinophils from healthy subjects. However, it suppresses apoptosis in this cell type from patients with eosinophilia (62). Furthermore, it either enhances or suppresses apoptosis in eosinophils from both eosinophilic patients and healthy subjects induced by apoptotic stimuli, dependent on the stimuli used. The reason for these differential effects is not known.

Similar to galectin-1, most *in vitro* data suggest that galectin-9 suppresses the immune response. *In vivo* evidence for this activity is emerging. Administration of galectin-9 to mice reduced Th2-associated airway inflammation (63).

CONCLUSIONS

Galectins have been shown to exert a variety of functions related to the immune and inflammatory responses. The field of galectins is faced with the challenge that no unified theme has been developed with respect to the functions of the family members and, to the contrary, an array of different activities have been demonstrated for each of the different member. The added challenge is that both extracellular and intracellular functions need to be considered.

Through extracellular mode of action, galectins can activate or inhibit cellular responses, modulate cell adhesion, influence cell migration, induce apoptosis, and affect cell growth (Fig. 6.1). These are demonstrated mostly by the use of recombinant proteins added to cells. Whether they represent those of endogenous proteins remains to be proven in many cases. In general, relatively high concentrations of galectins are required to demonstrate the functions. However, under certain conditions, some galectins have been shown to be active in relatively low concentrations. In addition, an increasing number of studies have demonstrated functions associated with the endogenous proteins that are secreted and function outside the cell. Moreover, *in vivo* data are accumulating that are consistent with the extracellular functions of galectins demonstrated *in vitro*.

The cell surface receptors responsible for galectins' action also remain to be clarified. Both galectin-1 and -3 have been shown to bind to a large number of different glycoproteins. It appears that each galectin invariably binds to a number of

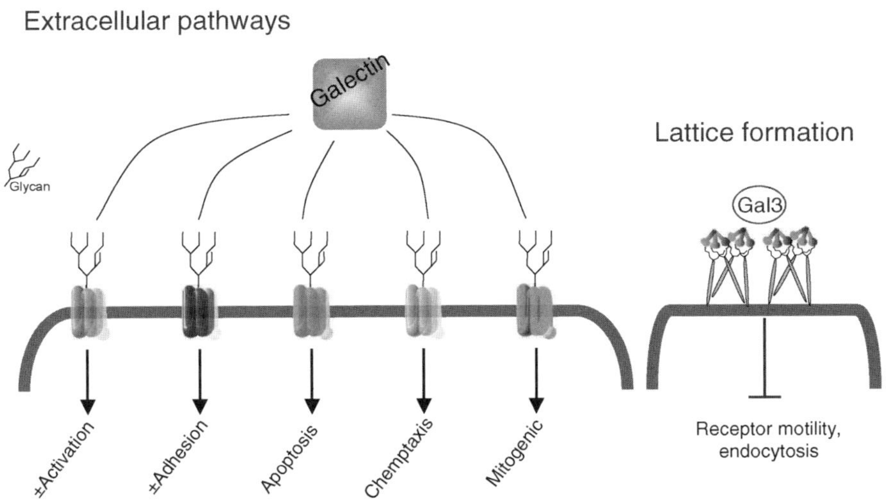

FIGURE 6.1 Extracellular functions of galectins related to immune and inflammatory responses. Various galectins can either potentiate or suppress responses of immune and inflammatory cells, modulate their cell adhesion, influence their migration, induce apoptosis, and modulate their growth. These functions are believed to result from the galectins binding to and aggregating cell surface glycoproteins. Some galectins have been suggested to form lattices with cell surface glycoproteins, thereby causing their delay in endocytosis or restricting their motility. This results in modulation of the cellular response. (See color insert.)

different glycoproteins on the cell surfaces and thus do not have specific individual receptors. It also appears that the lectin binds to different cell surface glycoproteins on different cell types. It will continue to be a challenge to establish which glycoprotein(s) recognized by a given lectin is (are) responsible for its function exerted on a specific cell type.

Through intracellular actions, galectins can regulate cell growth, cell differentiation, apoptosis, cytokine production, and phagocytosis, as well as some fundamental process, such as pre-mRNA splicing. Those associated with galectin-3 that are relevant to inflammation and immunity are illustrated in Fig. 6.2. These intracellular functions are not expected for a protein with lectin properties, but are consistent with the protein's intracellular localization. Many intracellular functions of galectin-3 were revealed from studies involving gene transfection and antisense nucleotides to influence expression, and others from correlation with intracellular localization of the proteins. More recently, some of them have been confirmed by the use of cells from mice deficient in a given galectin.

The intracellular action is supported by the fact that the activities in question are not affected by lactose added to the culture medium, which would inhibit galectins' carbohydrate-dependent extracellular actions. In addition, intracellular proteins with which galectins interact with have been identified. It is to be noted that most of these involved protein–protein interactions and not lectin–carbohydrate interactions. However, as described earlier in this review, galectin-3 has been identified in

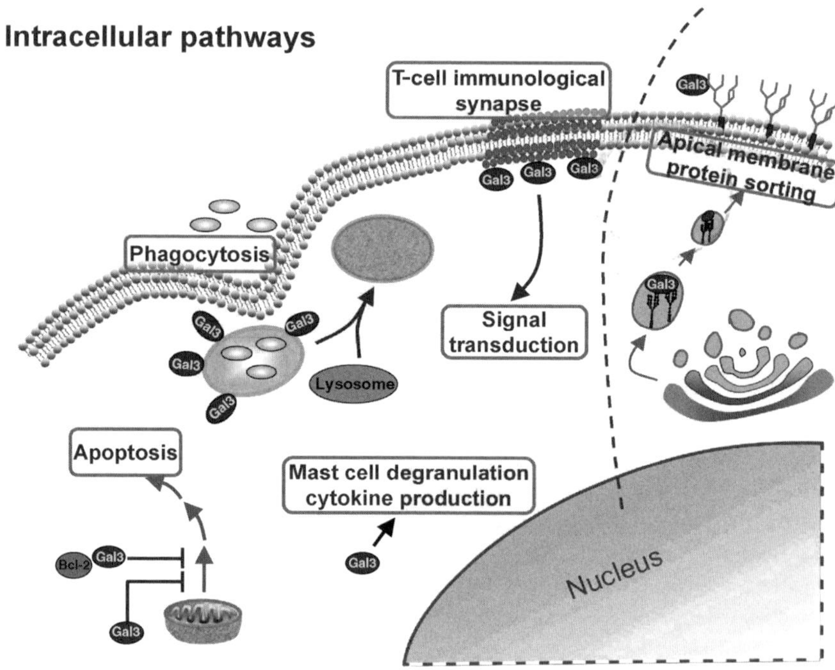

FIGURE 6.2 Intracellular functions of galectin-3 associated with its regulation of the immune and inflammatory responses. Galectin-3 is antiapoptotic in T and B cells, as well as macrophages. It contributes to phagocytosis by macrophages and mediator release/cytokine production by mast cells. It is localized at the immunological synapse in T cells and regulates the T-cell response. Recent data suggest that galectin-3 is localized in the post-Golgi vesicles carrying glycoproteins destined for expression at the apical surface of cells. Thus, the lectin may be responsible for sorting of glycoproteins into these vesicles. Whether this activity is related to other demonstrated intracellular functions of galectin-3 is unknown. (See color insert.)

post-Golgi intracellular vesicles transporting glycoproteins destined for apical export (12) and may be responsible for sorting these glycoproteins to these vesicles (Fig. 6.2). Whether this function is related to some of the other intracellular activities demonstrated for galectin-3 and whether other galectins have a similar role remain to be determined.

Despite the complexity, a picture that has emerged is that galectins regulate the immune and inflammatory responses through regulating apoptosis of immune cells, in particular T cells. Galectin-1, -2, -3, and -9 all induce T-cell apoptosis through binding to T-cell surface glycoproteins. Galectin-3 also inhibits apoptosis in T cells by functioning inside the cells. Through differential apoptosis-inducing or -suppressing activities on different subsets of T cells, galectins can regulate T-cell polarization to Th1, Th2, or Th17. A number of galectins have also been shown to affect various inflammatory cells and thus can regulate the inflammatory responses.

Each of the relatively well-studied members (i.e., galectin-1, -3, and -9) has activities *in vitro* that would correspond to opposite effects (i.e., either potentiation or suppression) on inflammatory responses, depending on the cell types involved and the experimental conditions used. For example, galectin-9 not only is a chemoattractant for eosinophils, but also induces apoptosis in these cells under certain conditions. Thus, it can not only attract eosinophils to the inflammatory sites, but also induce apoptosis in these cells resulting in a downregulation of the inflammatory response. Nevertheless, *in vivo* studies have shown that galectin-1 consistently suppresses the response in a variety of animal models, when administered exogenously, and endogenous galectin-3 is proven to promote the response in a number of mouse models. Overall, current information suggests that inhibitors of some galectins, such as galectin-3, may be useful for treatment of inflammatory diseases, while other galectins, such as galectin-1 and galectin-9, may themselves be used to suppress these diseases.

ACKNOWLEDGMENTS

The work has been supported by grants from the National Institutes of Health (RO1AI20958, RO1AI39620, and PO1 AI50498).

REFERENCES

1. Brewer CF, Miceli MC, Baum LG. Clusters, bundles, arrays and lattices: novel mechanisms for lectin–saccharide-mediated cellular interactions. Curr Opin Struct Biol 2002;12:616–623.
2. Frigeri LG, Zuberi RI, Liu FT. eBP, a β-galactoside-binding animal lectin, recognizes IgE receptor (FceRI) and activates mast cells. Biochemistry 1993;32:7644–7649.
3. Feuk-Lagerstedt E, Jordan ET, Leffler H, Dahlgren C, Karlsson A. Identification of CD66a and CD66b as the major galectin-3 receptor candidates in human neutrophils. J Immunol 1999;163:5592–5598.
4. Sarafian V, Jadot M, Foidart JM, Letesson JJ, Van den Bržle F, Castronovo V, Wattiaux R, Wattiaux-De Coninck S. Expression of Lamp-1 and Lamp-2 and their interactions with galectin-3 in human tumor cells. Int J Cancer 1998;75:105–111.
5. Pace KE, Lee C, Stewart PL, Baum LG. Restricted receptor segregation into membrane microdomains occurs on human T cells during apoptosis induced by galectin-1. J Immunol 1999;163:3801–3811.
6. Stillman BN, Hsu DK, Pang M, Brewer CF, Johnson P, Liu FT, Baum LG. Galectin-3 and galectin-1 bind distinct cell surface glycoprotein receptors to induce T cell death. J Immunol 2006;176:778–789.
7. Kuwabara I, Liu FT. Galectin-3 promotes adhesion of human neutrophils to laminin. J Immunol 1996;156:3939–3944.
8. Demetriou M, Granovsky M, Quaggin S, Dennis JW. Negative regulation of T-cell activation and autoimmunity by Mgat5 N-glycosylation. Nature 2001;409:733–779.

9. Partridge EA, Le Roy C, Di Guglielmo GM, Pawling J, Cheung P, Granovsky M, Nabi IR, Wrana JL, Dennis JW. Regulation of cytokine receptors by Golgi *N*-glycan processing and endocytosis. Science 2004;306:120–124.
10. Ohtsubo K, Takamatsu S, Minowa MT, Yoshida A, Takeuchi M, Marth JD. Dietary and genetic control of glucose transporter 2 glycosylation promotes insulin secretion in suppressing diabetes. Cell 2005;123:1307–1321.
11. Hughes RC. Secretion of the galectin family of mammalian carbohydrate-binding proteins. Biochim Biophys Acta 1999;1473:172–185.
12. Delacour D, Cramm-Behrens CI, Drobecq H, Le Bivic A, Naim HY, Jacob R. Requirement for galectin-3 in apical protein sorting. Curr Biol 2006;16:408–414.
13. Rabinovich GA, Baum LG, Tinari N, Paganelli R, Natoli C, Liu FT, Iacobelli S. Galectins and their ligands: amplifiers, silencers or tuners of the inflammatory response? Trends Immunol 2002;23:313–320.
14. Rabinovich GA, Toscano MA, Ilarregui JM, Rubinstein N. Shedding light on the immunomodulatory properties of galectins: novel regulators of innate and adaptive immune responses. Glycoconj J 2004;19:565–573.
15. Toscano MA, Bianco GA, Ilarregui JM, Croci DO, Correale J, Hernandez JD, Zwirner NW, Poirier F, Riley EM, Baum LG, Rabinovich GA. Differential glycosylation of T(H)1, T(H)2 and T(H)-17 effector cells selectively regulates susceptibility to cell death. Nat Immunol 2007;8:825–834.
16. Hernandez JD, Baum LG. Ah, sweet mystery of death! Galectins and control of cell fate. Glycobiology 2002;12:127R–136R.
17. He J, Baum LG. Presentation of galectin-1 by extracellular matrix triggers T cell death. J Biol Chem 2004;279:4705–4712.
18. Perone MJ, Bertera S, Tawadrous ZS, Shufesky WJ, Piganelli JD, Baum LG, Trucco M, Morelli AE. Dendritic cells expressing transgenic galectin-1 delay onset of autoimmune diabetes in mice. J Immunol 2006;177:5278–5289.
19. Rubinstein N, Alvarez M, Zwirner NW, Toscano MA, Ilarregui JM, Bravo A, Mordoh J, Fainboim L, Podhajcer OL, Rabinovich GA. Targeted inhibition of galectin-1 gene expression in tumor cells results in heightened T cell-mediated rejection: a potential mechanism of tumor-immune privilege. Cancer Cell 2004;5:241–251.
20. Rabinovich GA, Ramhorst RE, Rubinstein N, Corigliano A, Daroqui MC, Kier-Joffe EB, Fainboim L. Induction of allogenic T-cell hyporesponsiveness by galectin-1-mediated apoptotic and non-apoptotic mechanisms. Cell Death Differ 2002;9:661–670.
21. Hahn HP, Pang M, He J, Hernandez JD, Yang RY, Li LY, Wang X, Liu FT, Baum LG. Galectin-1 induces nuclear translocation of endonuclease G in caspase- and cytochrome *c*-independent T cell death. Cell Death Differ 2004;11:1277–1286.
22. Dias-Baruffi M, Zhu H, Cho M, Karmakar S, McEver RP, Cummings RD. Dimeric galectin-1 induces surface exposure of phosphatidylserine and phagocytic recognition of leukocytes without inducing apoptosis. J Biol Chem 2003;278:41282–41293.
23. Vespa GNR, Lewis LA, Kozak KR, Moran M, Nguyen JT, Baum LG, Miceli MC. Galectin-1 specifically modulates TCR signals to enhance TCR apoptosis but inhibit IL-2 production and proliferation. J Immunol 1999;162:799–806.
24. van der Leij J, van den Berg A, Harms G, Eschbach H, Vos H, Zwiers P, van Weeghel R, Groen H, Poppema S, Visser L. Strongly enhanced IL-10 production using stable galectin-1 homodimers. Mol Immunol 2007;44:506–513.

25. Garin MI, Chu CC, Golshayan D, Cernuda-Morollon E, Wait R, Lechler RI. Galectin-1: a key effector of regulation mediated by $CD4^{+}CD25^{+}$ T cells. Blood 2007;109:2058–2065.
26. Rabinovich GA, Ariel A, Hershkovitz R, Hirabayashi J, Kasai KI, Lider O. Specific inhibition of T-cell adhesion to extracellular matrix and proinflammatory cytokine secretion by human recombinant galectin-1. Immunology 1999;97:100–106.
27. He J, Baum LG. Endothelial cell expression of galectin-1 induced by prostate cancer cells inhibits T-cell transendothelial migration. Lab Invest 2006;86:578–590.
28. La M, Cao TV, Cerchiaro G, Chilton K, Hirabayashi J, Kasai K, Oliani SM, Chernajovsky Y, Perretti M. A novel biological activity for galectin-1: inhibition of leukocyte–endothelial cell interactions in experimental inflammation. Am J Pathol 2003;163: 1505–1515.
29. Barrionuevo P, Beigier-Bompadre M, Ilarregui JM, Toscano MA, Bianco GA, Isturiz MA, Rabinovich GA. A novel function for galectin-1 at the crossroad of innate and adaptive immunity: galectin-1 regulates monocyte/macrophage physiology through a nonapoptotic ERK-dependent pathway. J Immunol 2007;178:436–445.
30. Fulcher JA, Hashimi ST, Levroney EL, Pang M, Gurney KB, Baum LG, Lee B. Galectin-1-matured human monocyte-derived dendritic cells have enhanced migration through extracellular matrix. J Immunol 2006;177:216–226.
31. Rabinovich GA, Liu FT, Hirashima M, Anderson A. An emerging role for galectins in tuning the immune response: lessons from experimental models of inflammatory disease, autoimmunity and cancer. Scand J Immunol 2007;66:143–158.
32. Toscano MA, Commodaro AG, Ilarregui JM, Bianco GA, Liberman A, Serra HM, Hirabayashi J, Rizzo LV, Rabinovich GA. Galectin-1 suppresses autoimmune retinal disease by promoting concomitant Th2- and T regulatory-mediated anti-inflammatory responses. J Immunol 2006;176:6323–6332.
33. Liu FT. Regulatory roles of galectins in the immune response. Int Arch Allergy Immunol 2005;136:385–400.
34. Liu FT, Patterson RJ, Wang JL. Intracellular functions of galectins. Biochim Biophys Acta 2002;1572:263–273.
35. Fukumori T, Takenaka Y, Yoshii T, Kim HR, Hogan V, Inohara H, Kagawa S, Raz A. CD29 and CD7 mediate galectin-3-induced type II T-cell apoptosis. Cancer Res 2003;63: 8302–8311.
36. Fernandez GC, Ilarregui JM, Rubel CJ, Toscano MA, Gomez SA, Beigier Bompadre M, Isturiz MA, Rabinovich GA, Palermo MS. Galectin-3 and soluble fibrinogen act in concert to modulate neutrophil activation and survival: involvement of alternative MAPK pathways. Glycobiology 2005;15:519–527.
37. Yang RY, Hsu DK, Liu FT. Expression of galectin-3 modulates T cell growth and apoptosis. Proc Natl Acad Sci USA 1996;93:6737–6742.
38. Hoyer KK, Pang M, Gui D, Shintaku IP, Kuwabara I, Liu FT, Said JW, Baum LG, Teitell MA. An anti-apoptotic role for galectin-3 in diffuse large B-cell lymphomas. Am J Pathol 2004;164:893–902.
39. Hsu DK, Yang RY, Yu L, Pan Z, Salomon DR, Fung-Leung WP, Liu FT. Targeted disruption of the galectin-3 gene results in attenuated peritoneal inflammatory responses. Am J Pathol 2000;156:1073–1083.
40. Hsu DK, Yang RY, Liu FT. Galectins in apoptosis. Methods Enzymol 2006;417:256–273.
41. Hughes RC. Galectins as modulators of cell adhesion. Biochimie 2001;83:667–676.

42. Sato S, Ouellet N, Pelletier I, Simard M, Rancourt A, Bergeron MG. Role of galectin-3 as an adhesion molecule for neutrophil extravasation during streptococcal pneumonia. J Immunol 2002;168:1813–1822.
43. Sano H, Hsu DK, Yu L, Apgar JR, Kuwabara I, Yamanaka T, Hirashima M, Liu FT. Human galectin-3 is a novel chemoattractant for monocytes and macrophages. J Immunol 2000;165:2156–2164.
44. Cortegano I, del Pozo V, Cardaba B, de Andres B, Gallardo S, del Amo A, Arrieta I, Jurado A, Palomino P, Liu FT, Lahoz C. Galectin-3 down-regulates IL-5 gene expression on different cell types. J Immunol 1998;161:385–389.
45. Cortegano I, Del Pozo V, Cardaba B, Arrieta I, Gallardo S, Rojo M, Aceituno E, Takai T, Verbeek S, Palomino P, Liu FT, Lahoz C. Interaction between galectin-3 and FcgammaRII induces down-regulation of IL-5 gene: implication of the promoter sequence IL-5REIII. Glycobiology 2000;10:237–242.
46. Swarte VV, Mebius RE, Joziasse DH, Van den Eijnden DH, Kraal G. Lymphocyte triggering via L-selectin leads to enhanced galectin-3-mediated binding to dendritic cells. Eur J Immunol 1998;28:2864–2871.
47. Sano H, Hsu DK, Apgar JR, Yu L, Sharma BB, Kuwabara I, Izui S, Liu FT. Critical role of galectin-3 in phagocytosis by macrophages. J Clin Invest 2003;112:389–397.
48. Chen HY, Sharma BB, Yu L, Zuberi R, Weng IC, Kawakami Y, Kawakami T, Hsu DK, Liu FT. Role of galectin-3 in mast cell functions: galectin-3-deficient mast cells exhibit impaired mediator release and defective JNK expression. J Immunol 2006;177:4991–4997.
49. Bernardes ES, Silva NM, Ruas LP, Mineo JR, Loyola AM, Hsu DK, Liu FT, Chammas R, Roque-Barreira MC. *Toxoplasma gondii* infection reveals a novel regulatory role for galectin-3 in the interface of innate and adaptive immunity. Am J Pathol 2006;168: 1910–1920.
50. Colnot C, Ripoche MA, Milon G, Montagutelli X, Crocker PR, Poirier F. Maintenance of granulocyte numbers during acute peritonitis is defective in galectin-3-null mutant mice. Immunology 1998;94:290–296.
51. Zuberi RI, Hsu DK, Kalayci O, Chen HY, Sheldon HK, Yu L, Apgar JR, Kawakami T, Lilly CM, Liu FT. Critical role for galectin-3 in airway inflammation and bronchial hyperresponsiveness in a murine model of asthma. Am J Pathol 2004;165:2045–2053.
52. del Pozo V, Rojo M, Rubio ML, Cortegano I, Cardaba B, Gallardo S, Ortega M, Civantos E, Lopez E, Martin-Mosquero C, Peces-Barba G, Palomino P, Gonzalez-Mangado N, Lahoz C. Gene therapy with galectin-3 inhibits bronchial obstruction and inflammation in antigen-challenged rats through interleukin-5 gene downregulation. Am J Respir Crit Care Med 2002;166:732–737.
53. Lopez E, Del Pozo V, Miguel T, Sastre B, Seoane C, Civantos E, Llanes E, Baeza ML, Palomino P, Cardaba B, Gallardo S, Manzarbeitia F, Zubeldia JM, Lahoz C. Inhibition of chronic airway inflammation and remodeling by galectin-3 gene therapy in a murine model. J Immunol 2006;176:1943–1950.
54. Beatty WL, Rhoades ER, Hsu DK, Liu FT, Russell DG. Association of a macrophage galactoside-binding protein with *Mycobacterium*-containing phagosomes. Cell Microbiol 2002;4:167–176.
55. Sturm A, Lensch M, Andre S, Kaltner H, Wiedenmann B, Rosewicz S, Dignass AU, Gabius HJ. Human galectin-2: novel inducer of T cell apoptosis with distinct profile of caspase activation. J Immunol 2004;173:3825–3837.

56. Hokama A, Mizoguchi E, Sugimoto K, Shimomura Y, Tanaka Y, Yoshida M, Rietdijk ST, de Jong YP, Snapper SB, Terhorst C, Blumberg RS, Mizoguchi A. Induced reactivity of intestinal $CD4^+$ T cells with an epithelial cell lectin, galectin-4, contributes to exacerbation of intestinal inflammation. Immunity 2004;20:681–693.

57. Matsumoto R, Matsumoto H, Seki M, Hata M, Asano Y, Kanegasaki S, Stevens RL, Hirashima M. Human ecalectin, a variant of human galectin-9, is a novel eosinophil chemoattractant produced by T lymphocytes. J Biol Chem 1998;273:16976–16984.

58. Asakura H, Kashio Y, Nakamura K, Seki M, Dai S, Shirato Y, Abedin MJ, Yoshida N, Nishi N, Imaizumi T, Saita N, Toyama Y, Takashima H, Nakamura T, Ohkawa M, Hirashima M. Selective eosinophil adhesion to fibroblast via IFN-gamma-induced galectin-9. J Immunol 2002;169:5912–5918.

59. Wada J, Ota K, Kumar A, Wallner EI, Kanwar YS. Developmental regulation, expression, and apoptotic potential of galectin-9, a β-galactoside binding lectin. J Clin Invest 1997;99:2452–2461.

60. Kashio Y, Nakamura K, Abedin MJ, Seki M, Nishi N, Yoshida N, Nakamura T, Hirashima M. Galectin-9 induces apoptosis through the calcium-calpain-caspase-1 pathway. J Immunol 2003;170:3631–3636.

61. Zhu C, Anderson AC, Schubart A, Xiong H, Imitola J, Khoury SJ, Zheng XX, Strom TB, Kuchroo VK. The Tim-3 ligand galectin-9 negatively regulates T helper type 1 immunity. Nat Immunol 2005;6:1245–1252.

62. Saita N, Goto E, Yamamoto T, Cho I, Tsumori K, Kohrogi H, Maruo K, Ono T, Takeya M, Kashio Y, Nakamura K, Hirashima M. Association of galectin-9 with eosinophil apoptosis. Int Arch Allergy Immunol 2002;128:42–50.

63. Katoh S, Ishii N, Nobumoto A, Takeshita K, Dai SY, Shinonaga R, Niki T, Nishi N, Tominaga A, Yamauchi A, Hirashima M. Galectin-9 inhibits CD44–hyaluronan interaction and suppresses a murine model of allergic asthma. Am J Respir Crit Care Med 2007;176:27–35.

7

GALECTINS AS DANGER SIGNALS IN HOST–PATHOGEN AND HOST–TUMOR INTERACTIONS: NEW MEMBERS OF THE GROWING GROUP OF "ALARMINS"?

SACHIKO SATO

Glycobiology Laboratory, Research Centre for Infectious Diseases, Faculty of Medicine, Laval University, Québec, Canada

GABRIEL A. RABINOVICH

Laboratory of Immunopathology, Institute of Biology and Experimental Medicine, IBYME/CONICET, Argentina

INTRODUCTION

Nearly 30 years passed since several glycobiology laboratories independently found that various cell extracts contained soluble β-galactoside-binding proteins (1–19). As all mammalian lectins are either released through the classical secretory pathway or synthesized as membrane proteins, it was expected that these soluble β-galactoside-binding lectins also followed this biosynthetic pathway for their externalization to the extracellular space, where the majority of β-galactoside-containing glycoconjugates are found. Independently, immunologists also identified a protein synthesized by immune cells, following activation with inflammatory stimuli, that was first called Mac-2 or IgE-binding protein and has since then been renamed "galectin-3" (20, 21). Having being found as a well-defined surface marker for activated macrophages and

Galectins, Edited by Anatole A. Klyosov, Zbigniew J. Witczak, and David Platt

having the ability to bind to IgE, galectin-3 was also assumed to be secreted through the normal secretory pathway.

Successful cloning of those β-galactoside-binding proteins in the late 1980s (17, 21–27), together with genetic analysis, revealed that all these proteins belong to a lectin family (28), now officially named the galectin family (29, 30). While all galectins contain either one or two evolutionarily well-conserved carbohydrate-binding domain(s), their sequences do not contain any evident signal peptide, which is essential to access the classical secretory pathway. Indeed, immunohistochemistry studies confirmed that galectins are predominantly cytosolic, but are also found to be associated with the cell surface. Although it is now known that under certain conditions, cytosolic galectins are actively secreted, one could wonder why only the galectin family, among all lectin families, has been chosen to be localized in the cytoplasm, isolated from their natural glycan ligands. One could also wonder that their carbohydrate-binding properties might not be responsible for their critical biological activities, but rather a vestige of protein evolution. As a matter of fact, there is a significant number of evidences indicating that galectins have intracellular functions in cellular homeostasis through protein–protein or protein–RNA interactions (31, 32). At the same time, research progress in the last decade established a firm ground for the extracellular roles of galectins, most of which depend on their carbohydrate-binding activity; ~40% of publications in the last 5 years state that the biological roles of galectins are related to immunity. In the 1990s, it has become well established that galectins can be released through the "leaderless" secretory pathway without compromising membrane integrity, especially when certain cells are differentiated or activated by inflammatory stimuli (27, 33–38).

The list of immune cells that can release galectins is growing, which strongly supports their immunomodulatory activities and justifies the hypothesis that this intracellular lectin family plays a role in immunity, even though many galectin-related activities require exogenous addition of galectins when reproduced *in vitro*. Thus, we may argue that galectins differentially regulate cell physiology either inside or outside the cells. Since all reported lectin families, except galectins, are synthesized and released through the classical secretory pathway, an important question remains to be addressed: What is the biological significance of the cytosolic localization of the galectin family?

INNATE IMMUNITY AND DISRUPTION OF HOST INTEGRITY

In the past 20 years, there has been a quiet revolution in our understanding of the recognition mechanisms used by our immune defense system to maintain individual integrity. Immunity that has two pillars, the innate and the adaptive immune systems, protects the host from invasion by microorganisms and from aberrant growth and dissemination of transformed cancerous cells. In most cases, when a foreign entity invades the host, the innate immune response is triggered first, featured by potent recruitment of neutrophils to the affected site (39) and by complement activation (40). If this first wave of innate immune response fails to control the invasion, cells that are more closely associated with adaptive immunity will be recruited. However, recruitment of lymphocytes often takes at least one day after the initiation of invasion, implying that in the early stage of infection, innate immunity is required to recognize and clear the foreign entity without any support of the adaptive immunity.

In contrast to the adaptive immunity, which recognizes defined antigenic epitopes to initiate its defense reaction, the molecules recognized by innate immunity have been ambiguous and have often been stated as nonspecific. The term nonspecific was used since the defined antigenic epitope (i.e., structures of the epitopes) for innate recognition could not be easily identified at the molecular level. Thus, innate immunity was initially referred to as a "nonself" recognizing response. Then, in 1989, Janeway (41–43) proposed that innate immunity recognizes molecules called "pathogen-associated molecular patterns (PAMPs)." Receptors for PAMPs include Toll-like receptors (TLRs) and C-type lectins (e.g., mannose receptors, dectins, and DC-SIGN). These receptors on the surfaces of innate immune cells (neutrophils, macrophages, and dendritic cells) bind/recognize epitopes mostly unique to nonhost microorganisms, thereby triggering the innate defense system in a "nonself"/microorganism-specific manner. Yet, it should be noticed that the host harbors many nonvirulent/nonpathogenic and commensal microorganisms. These nonpathogenic microorganisms also express PAMPs. They do not compromise host integrity, and the innate immune system does not appear to be strongly stimulated by their existence.

Thus, while PAMPs recognition is an essential part of the innate detection system, one could argue that there should be another mechanism to distinguish the tiny fraction of pathogens from the immense population of nonpathogenic microorganisms. In 1994, Matzinger proposed a "danger model," arguing that innate immunity responds after detecting leakage or release of cytosolic molecules, which is a result of tissue damage caused by the foreign entity (44–46). In other words, when microorganisms compromise the integrity of cells/tissues, damaged or necrotic cells release their cytosolic contents, a part of which may have the potential to initiate or modulate immune responses.

Indeed, during the previous 10 years, several lines of evidence had indicated that some cytosolic proteins can induce proinflammatory reactions, even though the immunological significance of such activities was puzzling, until this model was proposed. These cytosolic "danger" proteins include high mobility group box 1 (HMGB1, which binds to nucleosomes and promotes DNA bending) (47, 48), S100 proteins (dominant cytosolic protein in neutrophils) (49, 50), heat shock proteins (which play an essential role as chaperone for protein folding) (51, 52), and annexins (lipocortins, which suppress phospholipase A) (53). Thus, it has become apparent that innate immunity recognizes two types of molecules to initiate appropriate response toward pathogenic invasion: PAMPs and the leakage of some cytosolic proteins.

Last year, an EMBO workshop entitled "Innate Danger Signals" was held in Italy. The objective of this workshop was to shed light on cytosolic proteins that can regulate the innate immune response according to the "danger model." During this meeting, the new term "alarmin" was proposed to identify endogenous cytosolic molecules that signal tissue and cell damages. Together, alarmins and PAMPs constitute the large family of "damage-associated molecular pattern (DAMPs)."

The following requirements were proposed for a molecule to be categorized as an alarmin: (1) rapid release following nonprogrammed cell death, but no release by apoptotic cells; (2) cells of the immune system can also be induced to produce and release alarmins without dying by using the "leaderless" secretory pathway; (3) these molecules may recruit and activate innate immune cells, including dendritic cells, and may thus regulate adaptive immunity; and (4) alarmins should also restore homeostasis by promoting the reconstruction of tissue that was destroyed either by direct insult or by the secondary effects of the initial immune reaction (54). Many of the proposed features of alarmins intriguingly overlap with the main properties of galectins. However, likely due to the complexity of the galectin family, it is expected that some reported features might not fit completely with the requirements for being considered alarmins. Nevertheless, in the following sections, we will introduce several immunomodulatory effects of galectins using the conceptual framework of "alarmins."

As reviewed in the other chapters, galectins regulate cytokine synthesis, phagocytosis, and homeostasis of immune cells. Some of these functions are independent of the extracellular properties of this protein family (55–59). Thus, interpretation of the roles of galectins in immunity becomes complicated, as both cytoplasmic and extracellular galectins independently modulate immune responses. As this chapter's goal is to discuss the potential role of galectins as alarmins, we will not focus our attention on the role of intracellular (cytoplasmic/nuclear) galectins in the regulation of immune responses. Thus, we request readers to refer to other recent reviews as well as other related chapters in this book for a comprehensive overview of galectins' intracellular functions.

BIOCHEMISTRY AND CELL BIOLOGY ASPECTS OF GALECTIN

Structure

The galectin family is defined by conserved peptide sequence elements in the carbohydrate recognition domain (CRD), consisting of ∼130 amino acids (Fig. 7.1) (30).

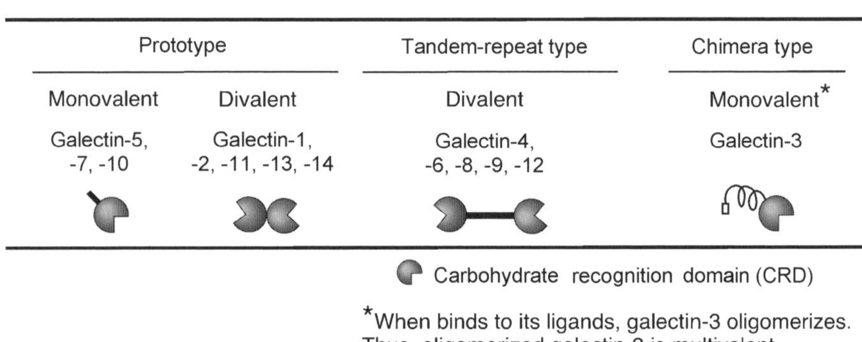

FIGURE 7.1 Structures of galectins.

Up to 15 galectins (galectin-1 to -15) have been found in mammals so far, as well as in many other phyla including birds, amphibians, fish, nematodes, drosophila, sponges, and fungi (60). While all galectins share a core sequence in their CRD, galectins exhibit interesting structural differences in the presentation of the CRD (Fig. 7.1) (29). Some galectins contain one CRD (prototype) and exist as monomers (galectin-5, -7, and -10) or dimers (galectin-1, -2, -11, -13, and -14), while other galectins, such as galectin-4, -6, -8, -9, and -12, contain two CRD connected by a short linker region (tandem repeat) (29). In contrast, galectin-3 uniquely occurs as a chimeric protein with one CRD and an additional non-CRD domain, which is involved in the oligomerization of galectin-3. Upon binding to its glycan ligands at the cell surface, the conformation of galectin-3 appears to be altered, and galectin-3 oligomerizes through the self-assembly of the N-terminal regulatory domain of the galectin-3 molecules. This oligomerization results in the formation of galectin-3 molecules with multivalent CRDs (61, 62).

Expression

Galectin-1, -3, -9, and -12 are found in cells involved in immune responses (63, 64). Galectin-1 is expressed in thymic epithelial cells, endothelial cells, trophoblasts, activated T cells, Mϕ, activated B cells, follicular DCs, and $CD4^+CD25^+$ regulatory T cells (38, 65–71). Galectin-3 and -9 are preferentially expressed in myeloid cells, such as macrophages, as well as in stromal cells of the lymph nodes, fibroblasts, endothelial, and epithelial cells (63, 72–74). Galectin-3 is expressed in human neutrophils but not in mouse neutrophils (75). Expression of galectin-1, -3, and -9 is increased during the activation of these cells (63, 72–74, 76) and human inflamed lymph nodes contain as much as 10 μM galectins (69, 77). In addition, galectin-3 is known as an established activated microglia marker (78–80).

Galectin expression is also increased during immune responses (64), including inflammation and infections, such as pneumonia with *Streptococcus pneumoniae*, infection with *Trypanosoma cruzi*, *Toxiplasma gondii*, human immunodeficiency virus-1 (HIV-1), human T-lymphotropic virus-1 (HTLV-1), *Helicobactoer pylori*, *Porphylomonas gingivalis*, and helminthes (55, 81–89). In addition, the expression of galectins has been well documented in cancer cells and cancer-associated stromal cells of different tumor types including gliomas, melanoma and breast, ovary, thyroid, colon, and prostate carcinomas as well as in hematological malignancies (90–93). Interestingly, in most cases this expression correlates with the aggressiveness of these tumors and the acquisition of metastatic phenotype (93).

Glycan-Binding Specificity

The minimal saccharide unit recognized by galectins is the galactose residue, linked to adjacent monosaccharide in the β configuration (called β-galactoside), such as lactosamine residues (Galβ1-4GlcNAc, where Gal is galactose and GlcNAc is *N*-acetylglucosamine) (72, 73, 94–96). This saccharide structure is often found in N-linked "complex type" glycans and O-linked glycans on host glycoproteins.

However, the affinity of each galectin member to its potential ligands varies depending on substitutions in this core β-galactoside, which results in subtle yet significant differences in the CRD. For example, the α-*N*-acetylgalactosamine modification of Gal residues significantly decreases the affinity of galectin-1 for this residue, but increases the affinity of galectin-3 (96). Alternatively, modification with α2,6-linked sialic acid, but not α2,3-linked sialic acid, reduces the affinity mainly of galectin-1, as well as of galectin-3 (94–98). Recently, this modification has been shown to play a crucial role in galectin-1-dependent skewing of T cells toward a Th2 cytokine profile (in both mice and humans) (99).

While Th1 and Th17 cells express the repertoire of cell-surface glycans (poly-*N*-acetyllactosamine units $(\text{Gal}\beta1\text{-}4\text{GlcNAc})_n$), which are critical for galectin-1 binding and cell death, Th2 cells are protected from galectin-1 binding through α2, 6-sialylation of cell-surface glycoproteins (99). In general, it is assumed that galectins exhibit higher affinity for glycans that contain a higher number of lactosamine residues (like polylactosamine or tri/tetraantennary complex type). In fact, this polylactosamine effect has been reported for galectin-3 and -9 but not for galectin-1, -7, and -8 (94–96, 98). In addition to differences in the detailed structure of their CRD pockets, which have a significant impact on ligand selectivity, the presentation of the CRD in the molecules should have an impact on the selectivity for their ligands (for more detail the reader may refer to recent reviews on different ligands of galectin-1 and -3 (90, 100)).

Multivalency and Avidity

The affinity of galectins' CRD toward their glycan ligands is often lower ($\sim10^{-6}$ M) than those observed in typical protein–protein interactions ($\sim10^{-8}$ M) (96). Despite a weak affinity of their CRD, galectins achieve a stable interaction with their ligands through their multivalency, as binding to multiple ligands leads to increases in their avidity (61, 101). For example, at 4°C, both galectin-3 and truncated galectin-3, which lacks the N-terminal domain necessary for oligomerization, bind tightly to the glycocalyx of an endothelial cell layer. However, at 37°C, where membrane fluidity is increased, truncated galectin-3 that does not oligomerize cannot associate with the cell surface (62). In contrast, full-length galectin-3 molecules bind tightly to the cell surface through oligomerization (62, 75). Soluble monomeric galectin-1 also binds weakly to lactosamine residues compared to dimeric galectin-1 (102). Importantly, the affinity for glycan ligands is similar and independent of the number of CRDs in the molecule when galectins are immobilized to a solid surface (96). In conclusion, high avidity of soluble galectins is achieved by oligomerization of galectin molecules and is essential for soluble galectins to induce their activities (101).

Cross-Linking

Since the majority of the members of the galectin family are multivalent in glycan binding, they can cross-link specific ligands to induce various activities such as cell–cell or cell–pathogen interactions (Fig. 7.2a), signal transduction through

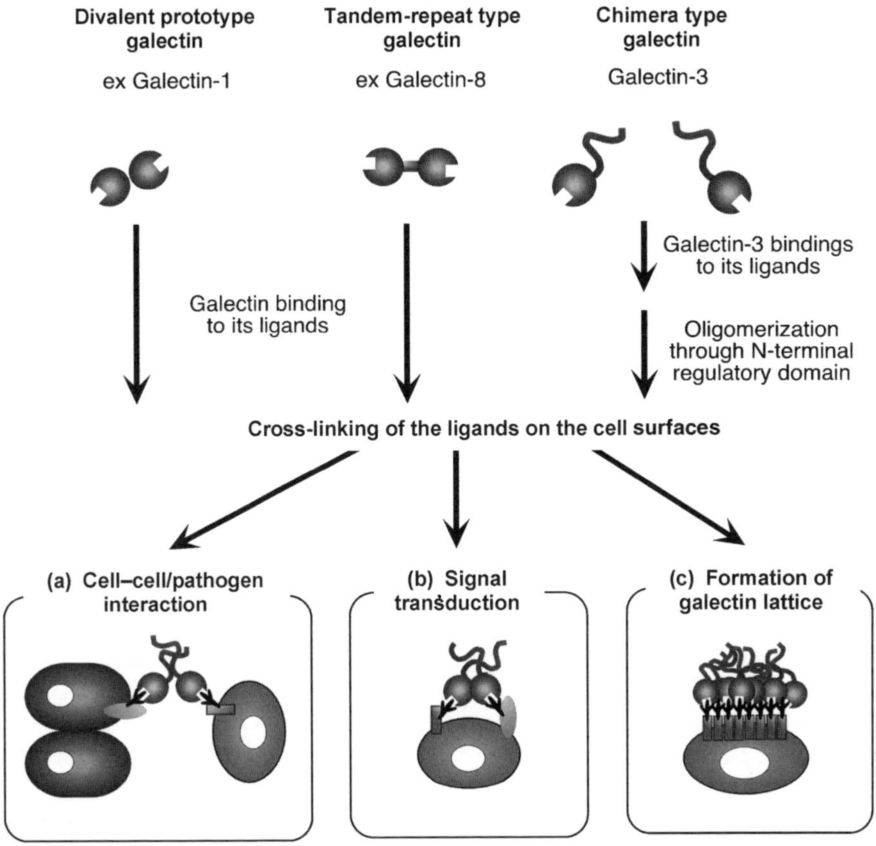

FIGURE 7.2 Cross-linking by galectins. Galectins cross-link the ligands on the cell surface, leading to mediating cell–cell/pathogen interaction (a), receptor clustering (b), and formation of galectin–glycoprotein lattice (c). (See color insert.)

receptor clustering (Fig. 7.2b), or the formation of a multivalent galectin-3–glycoprotein lattice on the cell surface (Fig. 7.2c) (62, 72, 73). We recently suggested that depending on the immune status and the microenvironment surrounding neutrophils, galectin-3 oligomerization occurs through three different modes (62, 75). The cross-linking results in the induction of cell–cell adhesion, small clustering of ligands, or the formation of lattices, which are stable on the surface of naïve neutrophils (up to 15 min) and endothelial cell layers (up to 60 min) (75, 103). Interestingly, several recent reports indicate a biological significance for galectin–carbohydrate cell surface lattices in immune regulation as well as in the retention of receptors/transporters on the cell surface. Dennis and colleagues (104, 105) clearly demonstrated that galectin-3 lattices can retain cytokine receptors, such as transforming growth factor (TGF)-β receptor, on the surface of invasive tumor cells, by interfering with its endocytosis. On the surface of T lymphocytes, galectin-3 lattices are suggested to restrict T-cell receptor recruitment to the site of antigen presentation, thereby increasing the threshold for T-lymphocyte activation (106).

Further, Ohtsubo et al. (107) recently suggested that galectin-9–glycan lattices are involved in the metabolism of glucose through their ability to retain glucose receptor (GLUT2) on the cell surface. In fact, mice deficient in GlcNAcT-IVa glycosyltransferase, which do not express galectin-9 ligands, develop type 2 diabetes as a result of a lower expression of glucose transporters on their cell surfaces (107). Thus, galectins' cross-linking is involved in a wide range of galectin functions, such as cell activation/repression and cell adhesion through three different modes of action: receptor clustering, lattice formation, and cell–cell interactions (Fig. 7.2). Different types of galectin–carbohydrate lattices might provide rational explanatory mechanisms for the different, and in many cases paradoxical, functions exerted by these carbohydrate-binding proteins.

EXPORT OF GALECTINS FROM THE CYTOSOL TO THE EXTRACELLULAR SPACE—CONNECTION TO ALARMINS?

As mentioned previously, innate immune cells must distinguish among a vast range of microorganisms and determine their level of pathogenicity, which ranges from nonvirulent to highly pathogenic. This task may be complicated by the fact that the host immune condition and the frequency of infection (recurrent versus acute) could affect the innate immune response to a given microorganism. Thus, a "weakly pathogenic" microorganism could be considered as "highly pathogenic" in different settings, as shown by the possible pathogenesis profiles for an infection with *Pseudomonas aeruginosa*. A microorganism's pathogenicity notwithstanding the innate immune system has been tailored, through evolutionary selection, to protect individual integrity by eliminating microorganisms that cause damage to the host (45, 46). Thus, the "danger model" and the definition of "alarmins" as key molecules in this process have emerged to provide a link between the pathogenicity of different microbes and the immune response evoked in the host (45, 46, 54). This original definition proposes that alarmins are proteins that have their own cytosolic functions and are stored inside cells, such as cytoplasm and nucleus. Upon invasion of a pathogenic entity that causes an insult on the tissue, these cytosolic alarmins are exported and modulate immune responses. Thus, the timing of export of alarmins to the extracellular space apparently holds the key to measure the level of "damage" or pathogenicity. Thus, two criteria have been proposed to define alarmins: (1) passive release from damaged cells, but not from apoptotic cells; and (2) upregulation of synthesis and active secretion through the "leaderless" secretory pathway (Fig. 7.3).

During a pathogenic infection, expression of galectins is often upregulated, as mentioned above (55, 64, 81, 83–89). No reports clearly examined whether galectins are passively released from damaged cells, but not from apoptotic cells; thus, the first criterion for export still remains unresolved. Yet, one could speculate that cytosolic stored galectins could be passively released from damaged cells at the affected site, as galectins are expressed in many cells directly involved in innate immunity or parenchymal cells aligned at the entry sites of pathogenic

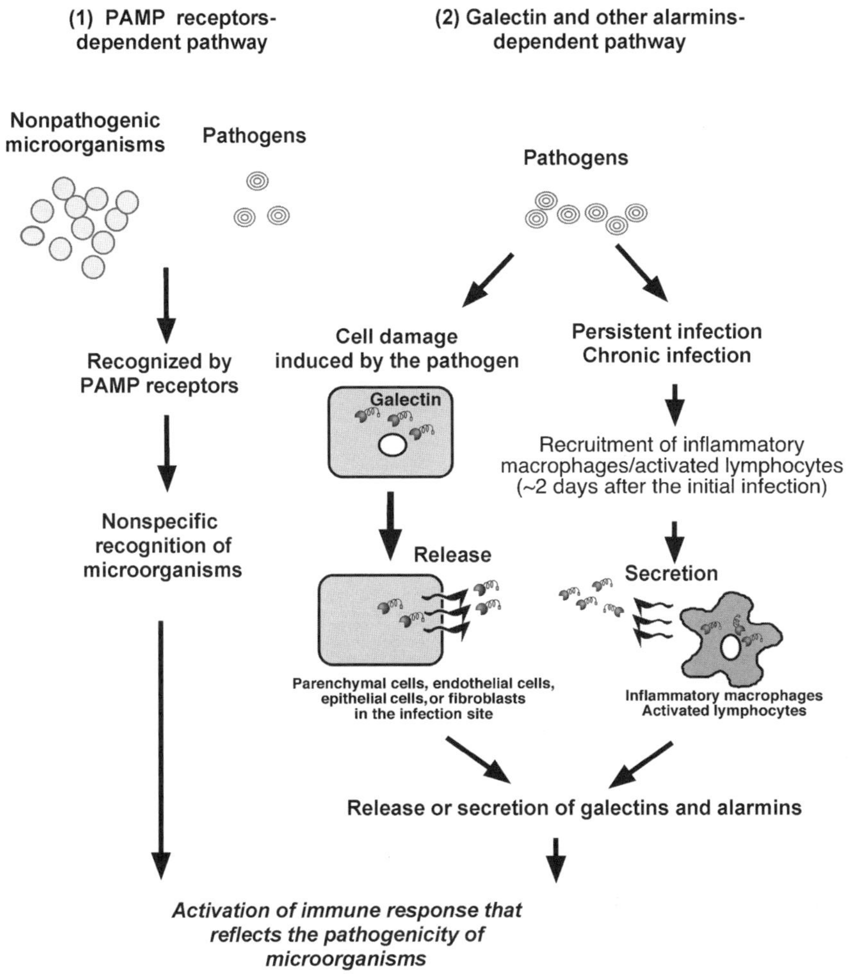

FIGURE 7.3 Innate immune recognition of pathogen by DAMPs (damage-associated molecular patterns). (See color insert.)

microorganisms. In addition, export of galectin-3 to alveoli is reported to be observed only when mice are infected with pathogenic *S. pneumoniae*, but not with less pathogenic *E. coli* (73, 81). These data imply that the release of galectins (e.g., galectin-3 from the lungs) may correlate with the level of virulence of a given invading microorganism (Fig. 7.3).

As mentioned above, the nascent peptide chains of galectins do not contain any significant peptide stretch, such as a signal peptide sequence or a hydrophobic peptide cluster required for entry into the classical secretory pathway. In fact, galectins are synthesized and accumulated in the cytoplasms of cells, segregated from their glycan ligands. While many types of cells involved in immunity store galectins in their cytosol without exporting them, some of those cells start secreting galectins (without compromising membrane integrity) when differentiated or activated by receptor engagement or cytokines. This secretion occurs through a "leaderless" secretory pathway, which is also used by fibroblast growth factors, and IL-1 (33, 34, 37, 54, 108–113). Galectin-1 is secreted by activated B cells, activated but not quiescent T cells, $CD4^{+}CD25^{+}$ regulatory T cells, activated Mϕ, and certain epithelial cells (36, 38, 66, 70, 82, 113). Activated lymphocytes also secrete galectin-9 (114). Gastric epithelial cells rapidly secrete galectin-3 when infected with *H. pylori* (85). Neutrophil migration across endothelial cells induces accumulation of galectin-3 on the endothelial cell surface, implying that secretion of galectin-3 occurs during the transmigration process (115). In the case of myelomonocytic cells, inflammatory macrophages, but not peritoneal resident or immature macrophages, actively secrete galectin-3(27, 34, 35).

Although it is proposed that alarmins are able to be secreted from live cells, at first glance, active secretion of galectins from inflammatory macrophages (galectin-1 and -3) and activated lymphocytes (galectin-1 and -9), without any cellular damage, may not seem to fit very well with the "danger model." However, when we take into consideration the timing of the recruitment of those leukocytes, this notable exception can be considered as part of the "danger model." In general, neutrophils are the first type of leukocytes to be recruited to the infection site. If the infection is persistent and exudated neutrophils have not controlled the infection, then inflammatory macrophages and activated lymphocytes are recruited to these sites (116, 117). Thus, recruitment of these cells represents a critical situation by which the primary defense response involves the transmission of "danger signals" to the immune system for further protection (Fig. 7.3). As discussed before, galectin-1 and -3 are actively secreted from inflammatory macrophages but not from resident macrophages. Galectin-1 and -9 are also secreted from activated lymphocytes. Thus, such strict control of galectin secretion from these inflammatory macrophages and activated lymphocytes could also be a clear indication that galectins are "inducible" molecules evoking "danger signals" for innate immunity (72, 73).

GALECTINS IN HOST–MICROBES INTERACTIONS

Once exported to the extracellular space, galectins bind to their host ligands containing β-galactosides, such as *N*-acetyllactosamine-containing glycoconjugates. Enveloped viruses acquire host-type glycans on their surface as viruses utilize the host protein synthesis machinery, including glycosylation. Indeed, galectin-1 binds to some enveloped viruses, such as HIV-1 and Nipah virus (77, 118, 119). For Nipah virus, galectin-1 inhibits its infection, while HIV-1 appears to exploit galectin-1 to stabilize virus attachment to $CD4^{+}$ T lymphocytes and macrophages, thus promoting viral infection in these target cells (77, 118, 119). Many bacteria also carry "lactosamine" residues in lipooligosaccharides or capsular polysaccharides of bacteria, such as *H. pylori*, *Neisseria meningitides*, *N. gonorrhoeae*, *Klebsiella pneumoniae*, and *S. pneumoniae*, which are recognized by galectins (120, 121).

In addition to polylactosamine-dependent binding, other types of β-galactosides expressed on parasite surfaces (*Leishmania major* and helminthes) are reported to be recognized by galectins. *Leishmania major* is a species that belongs to a genus of trypanosome protozoa, *Leishmania*, which is responsible for the disease leishmaniasis (affecting 12 million people). The surface of this parasite is covered by glycoconjugates, called phosphoglycans (PG). Galectin-3 and -9, but not galectin-1, bind to the poly-β-galactoside (Galβ1-3)$_n$ of PG (122, 123). Parasitic worms, helminthes, express unique β-galactosides, GalNAcβ1-4GlcNAc (LacdiNAc, where GalNAc is *N*-acetylgalactosamine). Galectin-3 but not galectin-1 recognizes this LacdiNAc (84, 124).

The recognition of parasites by galectins appears to lead to a promotion of internalization/binding of these parasites by macrophages: galectin-9 promotes internalization of *L. major* and galectin-3 increases phagocytosis of LacdiNAc-coated beads by macrophages (84, 122). In the case of galectin-3, its binding to *L. major*-specific (Galβ1-3)$_n$ of major surface glycoconjugate of phosphoglycans leads to cleavage of galectin-3 by the parasite surface protease gp63. Since truncated galectin-3 cannot oligomerize, the high avidity necessary for a stable interaction with the membrane is lost, therefore deactivating galectin-3 (123).

Galectin-3 also binds to *Trypanosoma cruzi*, promoting adhesion of this parasite to coronary artery smooth muscle cells (125). Inhibition of this lectin in muscle cells reduces *T. cruzi* adhesion, suggesting that endogenous galectin-3 is implicated in host–parasite interactions. More detailed studies *in vivo* are necessary to investigate the immunological significance of the PAMPs-receptor-like activity of galectins.

Host-specific recognition of pathogenic fungi by immune cells has recently attracted attention as macrophage transmembrane lectin, dectin-1, initiates an antifungal response by binding to β-glucan, one of the major components of fungi. Since β-glucan moieties are only exposed when yeasts are dividing, dectin-1-mediated detection provides the host with a mechanism to restrict the innate immune response to potentially invasive fungal forms (126–128). As β-glucans are expressed in both pathogenic *Candida albicans* and nonpathogenic *Saccharomyces cerevisiae*, Jouault et al. (129) searched other mechanism to detect the *C. albicans*-specific glycan, β-mannoside (Manβ1–2)$_n$. This initial study rediscovered galectin-3, a β-galactoside-binding lectin, as a β-mannoside-binding lectin. More recently, this group reported an immunomodulatory role (induction of TNF-α production) by galectin-3 through cross talk with Toll-like receptor TLR-2 (130). Interestingly, galectin-3 does not appear to be involved in phagocytosis of fungi.

Another elegant study showed that binding of galectin-3 to (Manβ1-2)$_n$ of *C. albicans* directly kills fungi, implying a microbicidal activity of this lectin (131). Interestingly, other established members of alarmins, HMGB1, and S100 protein family also possess antimicrobial activities (132). In addition, a classical host bactericidal cationic peptide family, defensins, has been originally suggested by Oppenheimer et al. as alarmins (133, 134). Thus, the activity of galectin-3 as an antifungal protein provides novel insights into our understanding of galectins in host–microbe interactions.

MODULATION OF ACTIVATION AND RECRUITMENT OF INNATE IMMUNE CELLS

In the initial stages of infection, a number of cells are recruited to initiate the innate immune response. These include neutrophils, tissue resident macrophages, dendritic cells, mast cells, and eosinophils. In this regard, another criterion required for being members of the alarmin family is the ability of these molecules to recruit and activate some of these innate immune cells, including dendritic cells, which can transfer pathogen information and orchestrate the subsequent adaptive immune response.

Neutrophils

Both galectin-1 and -3 activate neutrophils, leading to the secretion of superoxide and IL-8 (a chemokine for neutrophils) (75, 135–137). Galectin-3 facilitates neutrophil adhesion and migration into alveoli infected with pathogenic *S. pneumoniae* but not nonpathogenic *E. coli* (62, 81, 103, 138). This effect appears to depend on the extracellular presence of galectin-3, thereby implying that galectin-3-induced migration reflects the pathogenecity of the invading microorganism (103). Late but not acute neutrophil migration in the peritoneal cavity is impaired in mice deficient in galectin-3, also suggesting its role in efficient migration of neutrophils during chronic inflammation, which is often associated with sustained infection (139, 140). In contrast to galectin-3, galectin-1 suppresses both *in vivo* acute neutrophil migration and *in vitro* transmigration of neutrophils across endothelial cells (141, 142).

Eosinophils/Mast Cells

Galectin-9 is a potent chemoattractant for eosinophils (114). It also promotes adhesion of eosinophils (but not neutrophils) to interferon (IFN)-γ-activated fibroblasts (143). As the expression of galectin-9 on the surface is also increased in IFN-γ-treated endothelial cells (144), it is possible that galectin-9 facilitates eosinophil migration during inflammation and/or infection. In addition, galectin-3 has been reported to induce mast cell degranulation (145).

Macrophages

Galectins also regulate various functions of macrophages (64). Galectin-3 is a chemoattractant for macrophage (146), potentiates LPS-induced IL-1 productions (147), and induces Ca^{2+} influx in monocytes (146).

Dendritic Cells

Galectin-1 induces the secretion of IL-6 and TNF-α, but not IL-12, by human monocyte-derived dendritic cells. In addition, this glycan-binding protein positively influences the migration of these cells (148, 149). Galectin-3 mediates adhesion of L-selectin-activated lymphocytes on dendritic cells. Recent work by

Trottien et al. (150) suggest that galectin-3 controls the magnitude of T-cell priming by dendritic cells, although it remains unknown whether this activity is related to the extracellular activity of this protein (150). Galectin-9 also induces the maturation of human monocyte-derived dendritic cells, secreting high levels of IL-12, but not IL-10. This extracellular activity appears to be independent of the carbohydrate-binding domain (151).

REGULATORY ROLES OF GALECTINS IN IMMUNE RESPONSES

Although a potent innate immune response is essential for maintaining tissue homeostasis and protection against infection, it also contributes to tissue injury that is recognized to be involved in many diseases, including atherosclerosis, cancer, asthma, chronic obstructive pulmonary disease, and arthritis. In addition, tissue reconstitution by fibrosis in some organs, such as lungs, displaces functional cells, compromising their functions. Thus, termination and resolution of the initial response is required to be appropriately regulated. As featured in a special supplement of *Nature Immunology* in 2005 entitled "Dampening inflammation" (152, 153), this restoring process is not merely the passive termination of inflammation but rather an actively regulated program (152). Natural resolution of innate immune responses is believed to be driven mainly by three types of processes: (1) removal of the initial stimulus; (2) decrease in proinflammatory mediators; and (3) removal of the innate immune cells and remaining cell debris to allow final tissue repair to occur. During the termination of innate immunity, tissue repair is initiated. In fact, one of the proposed criteria for being classified as an alarmin is the ability of the molecule to restore homeostasis by promoting the reconstruction of tissue that was destroyed because of either the direct insult or the secondary effect of the initial immune reaction (54).

Cummings and colleagues (154–156) reported that galectin-1, -2, and -4 induce surface exposure of phosphatidylserine on activated (but not naïve) neutrophils without inducing cell death by apoptosis. Notably, likely due to the high contents of phosphatidylserine on the surface, galectin-1-treated neutrophils are phagocytosed by activated macrophages to a level similar to aged neutrophils. Importantly, critical factors for resolution are not only the induction of neutrophil apoptosis, but also the rapid clearance. If this clearance is not efficient enough, apoptotic neutrophils eventually lose their membrane potentials, leading to necrosis and release of intracellular components, including alarmins, and toxic substances (like a product of myeloperoxidase, hypochlorous acid). Indeed, the clearance of neutrophils is rapid, as the neutrophil number usually decreases to 50% of its maximum within an interval of only 8 h (153). Thus, the unique function of galectins in the clearance of activated neutrophils plays a critical role during the resolution period.

The clearance of aged and/or phosphatidylserine-exposed neutrophils by macrophages also results in the secretion of anti-inflammatory cytokines, such as TGF-β and IL-10 (157). Thus, some functions of inflammatory macrophages are turned off, called deactivated macrophages (158). Like T cells, it is now known that in inflamed tissues, the precursors of the elicited macrophages acquire distinct phenotypes and physiological activities. For example, when stimulated with IFN-γ, macrophages show high microbicidal activity, secreting proinflammatory cytokines and reactive oxygen/nitrogen species (ROS/RNS) and increasing antigen presentation (normally classified as "classically activated macrophages"). IL-4 or IL-13 differentiate macrophages into "alternatively activated macrophages," which secrete anti-inflammatory cytokines, such as IL-10 and TGF-β, lower the RNS production by inducing arginase production, and release enzymes and matrix components required for tissue repair (158). Galectin-1 increases arginase activity and decreases LPS-induced RNS production by macrophages (159). In addition, MHC-II expression and antigen-presenting activity by macrophages was decreased when galectin-1 is present *in vivo* and *in vitro* (160). Thus, galectin-1 appears to skew macrophages to states of "alternative activation" or "deactivation" (160).

When the initial immune response fails to clear the insult, subsequent adaptive immune response follows. Since several cytokines, such as IL-10 and TGF-β, and regulatory cells ($CD4^+CD25^+FoxP3^+$ T regulatory cells, IL-10-producing $FoxP3^-$ regulatory T cells, and alternatively activated macrophages) are known to suppress or regulate subsequent immune responses (158, 161–164), these regulatory components of the adaptive immune response may also participate in the resolution phase to provide tissue repair and serve as a homeostatic mechanism to return to basal conditions. In this regard, galectin-1 has been documented to negatively regulate cell growth and apoptosis of activated but not resting T-cells, suggesting the potential role of this protein in restoring T-cell homeostasis (36, 38, 165–171) by apoptosis.

Activation of naïve $CD4^+$ T helper cells results in the development of functionally distinct T-cell effectors, which are responsible for orchestrating adaptive immune responses. While Th1 cells are essential for the clearance of intracellular pathogens and are traditionally associated with the initiation of autoimmunity and cancer, Th2 cells are critical for the control of extracellular microorganisms. Moreover, recent evidence has pointed out the emergence of a novel T helper cell subsets—Th17 cells—which are implicated in the response to certain types of bacteria and fungi and are responsible for the perpetuation of autoimmune inflammation (172). However, the extent of tissue damage by Th1 or Th17 response is sometimes beyond the tissue healing capacity. Thus, appropriate regulation of Th1 and Th17 responses is also an essential domain of immunity. Through a mechanism based on differential glycosylation of T helper cells, we recently demonstrated that galectin-1 can modulate the survival of Th1 and Th17 effector cells (99). At lower doses, which do not induce apoptosis, galectin-1 also suppresses production of T-cell receptor-induced IL-2, TNF-α, and IFN-γ (173–175). *In vivo*, galectin-1 also regulates Th1 and Th17 effector cells and skews the cytokine balance toward Th2-mediated responses, with decreased amounts of IFN-γ and IL-2 (99, 148, 176–179) and increased levels of IL-10 (180, 181).

Recently, Matzinger, who originally proposed the "danger model," argued that pathogen- or insult-related immunity might be regulated in a tissue-specific manner (182). Induction of Th1 responses often leads to delayed-type hypersensitivity response, which may introduce damages. This response may be the most efficient way to eliminate intracellular pathogens in mucous membrane or in skin where the quick replacement of these tissues is possible without any loss of their functionality. However, in more specialized vulnerable tissues, such as the eyes and lungs, this type of destruction can be extremely dangerous. For example, in lungs, thin layers of functional alveolar epithelium could be replaced by fibrosis or tissue integrity could be destroyed, leading to the penetration of microorganism into blood (182). Indeed, a recent report suggests that lung macrophages are influenced by alveolar epithelium in a way to preserve lung function (183). As galectin-1 is abundantly expressed in various tissues, and those reported activity require relatively high concentrations, it could be speculated that suppression of DTH-like Th1 by galectin-1 become active after massive release of this lectin from damaged tissue. In addition, galectin-2, -3, and -9 have also been reported to induce T-cell apoptosis (69, 184–186). Moreover, Kuchroo and colleagues (187) recently reported that galectin-9 can specifically regulate Th1 responses through binding to the Th1-specific type membrane protein, Tim-3.

Recent DNA microarray analysis revealed the high expression of galectin-1 and galectin-10 by naturally occurring regulatory $CD4^{+}CD25^{+}$ T cells (Treg) (70, 71, 188). Treg can suppress the activation of immune response possibly through direct cell–cell contact and/or indirect suppression of IL-2 and interferon-γ (164, 189, 190). Furthermore, galectin-1 has been shown to be a key mediator of the suppressive activity of Tregs (70). Very recent findings revealed that galectin-1 plays a key role in the induction and maintenance of fetal survival *in vivo* through modulation of multiple tolerogenic mechanisms (191). Progesterone-regulated galectin-1 instructs dendritic cells to differentiate into IL-10-producing tolerogenic dendritic cells, which in turn modulate the Th1/Th2 cytokine balance and promote the generation and/or expansion of $CD4^{+}CD25^{+}$ Tregs *in vivo* (191). The ability of galectin-1 to promote the expansion of Tregs has also been found *in vitro* in cocultures of Hodgkin lymphoma cells and activated T cells (192) and *in vivo* in an experimental model of autoimmune uveitis (177).

GALECTINS IN THE REGULATION OF HOST–TUMOR INTERACTIONS

Understanding why the immune system is initially unable to detect transformed cells and becomes subsequently tolerant to tumor growth is critical for developing effective antitumor responses (193). According to the currently accepted model, different stress signals released by tumor cells undergoing damage (probably alarmins) can activate dendritic cells and foster adaptive immunity (193). Dendritic cells capture antigens secreted or shed by tumor cells and migrate to draining lymph nodes to stimulate antigen-specific T cells. Ultimately, activated antigen-specific $CD8^{+}$ T cells differentiate into effector cytotoxic T lymphocytes, which can specifically recognize and destroy tumor cells. Such extensive progress in defining the cells that regulate the immune network has bolstered our belief in the potential power of tumor immunotherapy.

Our renewed enthusiasm in cancer immunotherapy is basically derived from the convergence of four different but partially overlapping streams of research: (a) the appreciation that danger signals and stress-induced ligands connect innate and adaptive immunity via dendritic cells; (b) the identification of immunogenic tumor-associated antigens and cytotoxic T-cell responses in cancer patients; and (c) the discovery of negative regulatory pathways and novel immune escape strategies that influence the magnitude of antitumor responses by counteracting tumor effector mechanisms (194).

However, a fundamental change in our view of tumor immunology occurred in the early 1990s, following the surprising observations that most antigens expressed by tumor cells were not necessarily neoantigens uniquely present in cancer cells, but rather tissue differentiation antigens also expressed in normal cells (195, 196). Therefore, and following the "danger model," one would expect that tissue damage would promote the release of "alarmin-like molecules" to initiate or modulate tumor immunity. Interestingly, Zitvogel and colleagues (197) recently showed exciting results demonstrating that the activation of tumor antigen-specific T-cell immunity involves the secretion of HMGB1, an alarmin, by dying tumor cells and the action of HMGB1 on Toll-like receptor 4 (TLR4) expressed by dendritic cells. In this regard, it is expected that galectins might be secreted by tumors following stress signals to initiate or modulate tumor immunity.

The potential role of galectins in tumor progression was proposed in the mid-1980s, when a differential expression of endogenous β-galactoside-binding proteins on the surface of nontumorigenic, tumorigenic, and metastatic cells (198, 199) was reported. Since then, expression of galectins has been well documented in cancer cells and cancer-associated stromal cells of different tumor types including glioblastoma, melanoma, prostate, thyroid, breast, and ovary carcinomas and also in hematological malignancies (90–92). In addition, galectins are overexpressed in endothelial cells at sites of tumor growth and metastasis (200). Interestingly, in most cases, this expression correlates with the aggressiveness of these tumors and the acquisition of metastatic phenotype (93). In this regard, many laboratories have made considerable efforts to provide the "poor prognosis signature" to identify tumor- and metastasis-related genes. Interestingly, gene and protein expression profiles using microarrays or proteomic analysis have recurrently led to the identification of galectins, particularly galectin-1 and -3, whose expression is upregulated in a plethora of tumors and metastatic lesions, as compared to their nontransformed counterparts (201, 202).

The ability of galectin-1 to negatively regulate T-cell survival, skew the balance of the immune response toward a Th2 cytokine profile, and promote the differentiation and/or expansion of Tregs and alternatively activated macrophages prompted us to investigate the ability of this protein to confer a status of immune privilege to tumor microenvironments. By a combination of *in vitro* and *in vivo* experiments, a link between galectin-1-mediated immunoregulation and tumor immune escape was established (203). Human and mouse melanoma cells were found to secrete substantial levels of galectin-1 that contributes to the immunosuppressive activity of tumor cells. Remarkably, blockade of the inhibitory effects of galectin-1 within the tumor tissue resulted in reduced tumor size and potentiation of tumor-specific T-cell responses *in vivo*. This effect required intact $CD4^+$ and $CD8^+$ T-cell compartments, as the simultaneous depletion of these T-cell subpopulations *in vivo* completely abrogated the antitumor response induced by knockdown transfectants (203). Furthermore, tumor-secreted galectin-1 induced apoptosis of tumor-infiltrating lymphocytes and interruption of this proapoptotic pathway within the tumor microenvironment resulted in the generation of a Th1-specific response in tumor-draining lymph nodes (203).

Interestingly, these findings were subsequently confirmed in human tumor samples; Le and colleagues (204) found a strong inverse correlation between the galectin-1 expression and the recruitment of T cells in tumor sections from head and neck squamous cell carcinoma patients, supporting a role for galectin-1 in negative regulation of antitumor responses. Most recently, Juszczynski et al. (192) found that Reed Sternberg cells in classical Hodgkin lymphoma selectively overexpress galectin-1 through an AP-1-dependent enhancer. In cocultures of activated T cells and Hodgkin cell lines, interference RNA-mediated silencing of galectin-1 increased T-cell viability and restored the Th1/Th2 cytokine balance. Galectin-1 treatment also favored the expansion of $CD4^+CD25^+FoxP3^+$ Tregs (192). In addition, Ghandi et al. (205) showed that exposure to galectin-1 inhibited proliferation and IFN-γ production by Epstein–Barr virus-specific T cells. These results suggest a critical role of galectin-1 in conferring a status of immune privilege to classical Hodgkin lymphoma. In addition, as galectin-2, -3, and -9 can also impair T-cell functions by modulating cytokine production and T-cell survival (69, 184–186), these lectins may also contribute to tumor cell evasion of immune responses. The role of these galectins in tumor immune escape *in vivo* merits further consideration.

In addition to the role of galectins in tumor immunity, these glycan-binding proteins can have important roles in cancer by contributing to neoplastic transformation, tumor cell migration, angiogenesis, and metastasis (extensively revised by Liu and Rabinovich (93)). Therefore, the overall effects of galectins on tumor progression may result not only from the modulation of immune-mediated mechanisms, but also from the control of other biological functions, including homotypic and heterotypic cell aggregation, cell adhesion, migration, proliferation, and angiogenesis. The functional interplay between immune and nonimmune mechanisms by which galectins may affect tumor progression, escape and metastasis remains to be further defined.

CONCLUSIONS: SEEING STRANGERS OR ANNOUNCING DANGER?

As illustrated in this chapter, galectins are induced under stress (microbial infection and tumor growth) and can modulate the development and resolution of innate and adaptive immune responses. In this regard, these glycan-binding proteins fulfill the essential requirements to be considered members of the growing family of alarmins. These include their striking cytoplasmic localization, atypical secretory pathway, and the ability to trigger and/or modulate immune responses. Galectins participate in a plethora of immune responses by acting intracellularly and extracellularly, as cytokines, growth inhibitory factors, death triggers, and survival inducers. Under distinct physiological or pathological conditions, galectin-1 may provide inhibitory or stimulatory signals to control immune cell homeostasis and regulate inflammation following an antigenic challenge. In addition, galectins may act at different steps of tumor progression to modulate tumor growth and metastasis, including tumor cell evasion of immune responses, cell adhesion, migration, and angiogenesis. Although a great deal remains to be learned about the functional roles of galectins during inflammatory episodes, infectious processes, and tumor dissemination, sufficient evidence has been accumulated demonstrating their relevance *in vivo*, and new therapeutic approaches directed toward modulating their activities are being developed.

ACKNOWLEDGMENTS

We thank members (C. St-Pierre, P. Bhaumik, and J. Nieminen) of the Sato laboratory and (M. Toscano, J. Ilarregui, G. Bianco, M. Salatino, and D. Croci) of the Rabinovich laboratory for critical discussions. Work in the Sato's laboratory is supported by the Canadian Institutes for Health Research, Canadian Foundation for Innovation, Mizutani Foundation for Glycoscience, and Fonds de la recherche en santé Québec. Work in the Rabinovich's laboratory is supported by the Cancer Research Institute (New York), the Mizutani Foundation for Glycoscience (Japan), the Argentina National Agency for Promotion of Science and Technology, the University of Buenos Aires, the John Simon Guggemheim Memorial Foundation (New York), and the Argentina National Research Council (CONICET).

REFERENCES

1. Nowak TP, Kobiler D, Roel LE, Barondes SH. Developmentally regulated lectin from embryonic chick pectoral muscle. Purification by affinity chromatography. J Biol Chem 1977;252:6026–6030.
2. Nowak TP, Haywood PL, Barondes SH. Developmentally regulated lectin in embryonic chick muscle and a myogenic cell line. Biochem Biophys Res Commun 1976; 68:650–657.
3. de Waard A, Hickman S, Kornfeld S. Isolation and properties of beta-galactoside binding lectins of calf heart and lung. J Biol Chem 1976;251:7581–7587.
4. Childs RA, Feizi T. Calf heart lectin reacts with blood group Ii antigens and other precursor chains of the major blood group antigens. FEBS Lett 1979;99:175–179.
5. Childs RA, Feizi T. Beta-galactoside-binding muscle lectins of man and monkey show antigenic cross-reactions with those of bovine origin. Biochem J 1979;183:755–758.
6. Powell JT. Purification and properties of lung lectin. Rat lung and human lung beta-galactoside-binding proteins. Biochem J 1980;187:123–129.

7. Raz A, Lotan R. Lectin-like activities associated with human and murine neoplastic cells. Cancer Res 1981;41:3642–3647.
8. Pitts MJ, Yang DC. Mitogenicity and binding properties of beta-galactoside-binding lectin from chick-embryo kidney. Biochem J 1981;195:435–439.
9. Lee RT, Myers RW, Lee YC. Further studies on the binding characteristics of rabbit liver galactose/*N*-acetylgalactosamine-specific lectin. Biochemistry 1982;21:6292–6298.
10. Roff CF, Wang JL. Endogenous lectins from cultures cells: isolation and characterization of carbohydrate binding proteins from 3T3 fibroblasts. J Biol Chem 1983;258: 10657–10663.
11. Roff CF, Rosevear PR, Wang JL, Barker R. Identification of CBP from mouse and human fibroblasts. Biochem J 1983;211:625–629.
12. Oda Y, Kasai K. Purification and characterization of beta-galactoside-binding lectin from chick embryonic skin. Biochim Biophys Acta 1983;761:237–245.
13. Harrison FL, FitzGerald JE, Catt JW. Endogenous beta-galactoside-specific lectins in rabbit tissues. J Cell Sci 1984;72:147–162.
14. Gabius HJ, Engelhardt R, Cramer F, Batge R, Nagel GA. Pattern of endogenous lectins in a human epithelial tumor. Cancer Res 1985;45:253–257.
15. Merkle RK, Zhou Q, Schultz TK, Harper WB, Cummings RD. Characterization of an S-type lectin purified from porcine heart. Arch Biochem Biophys 1989;274:404–416.
16. Foddy L, Stamatoglou SC, Hughes RC. An endogenous carbohydrate-binding protein of baby hamster kidney (BHK21 C13) cells. J Cell Sci 1990;97:139–148.
17. Woo HJ, Shaw LM, Messier JM, Mercurio AM. The major non-integrin laminin binding protein of macrophages is identical to carbohydrate binding protein 35 (Mac-2). J Biol Chem 1990;265:7097–7099.
18. Castronovo V, Luyten F, van den Brule F, Sobel ME. Identification of a 14-kDa laminin binding protein (HLBP14) in human melanoma cells that is identical to the 14-kDa galactoside binding lectin. Arch Biochem Biophys 1992;297:132–138.
19. Poirier F, Timmons PM, Chan CT, Guenet JL, Rigby PW. Expression of the L14 lectin during mouse embryogenesis suggests multiple roles during pre- and post-implantation development. Development 1992;115:143–155.
20. Ho M-K, Springer TA. MAC-2, a novel 32,000 M_r mouse macrophage subpopulation-specific antigen defined by monoclonal antibodies. J Immunol 1982;128:1221–1227.
21. Liu F-T, Albrandt K, Mendel E, Kulczycki A, Orida NK. Identification of an IgE binding protein by molecular cloning. Proc Natl Acad Sci USA 1985;82:4100–4104.
22. Ohyama Y, Hirabayashi J, Oda Y, Ohno S, Kawasaki H, Suzuki K, Kasai K. Nucleotide sequence of chick 14K beta-galactoside-binding lectin mRNA. Biochem Biophys Res Commun 1986;134:51–56.
23. Oda Y, Hirabayashi J, Kasai K. Nucleotide sequence of chick 14K beta-galactoside-binding lectin mRNA. J Biochem (Tokyo) 1986;99:1063–1070.
24. Albrandt K, Orida NK, Liu FT. An IgE-binding protein with a distinctive repetitive sequence and homology with an IgG receptor. Proc Natl Acad Sci USA 1987;84: 6859–6863.
25. Jia S, Mee RP, Morford G, Agrwal N, Voss PG, Moutsatsos IK, Wang JL. Carbohydrate-binding protein 35: molecular cloning and expression of a recombinant polypeptide with lectin activity in *Escherichia coli*. Gene 1987;60:197–204.

26. Raz A, Avivi A, Pazerini G, Carmi P. Cloning and expression of cDNA for two endogenous UV-2237 fibrosarcoma lectin genes. Exp Cell Res 1987;173:109–116.
27. Cherayil BJ, Weiner SJ, Pillai S. The Mac-2 antigen is a galactose-specific lectin that binds IgE. J Exp Med 1989;170:1959–1972.
28. Drickamer K. Two distinct classes of carbohydrate-recognition domains in animal lectins. J Biol Chem 1988;263:9557–9560.
29. Hirabayashi J, Kasai K. The family of metazoan metal-independent beta-galactoside-binding lectins: structure, function and molecular evolution. Glycobiology 1993;3: 297–304.
30. Barondes SH, Castronovo V, Cooper DN, Cummings RD, Drickamer K, Feizi T, Gitt MA, Hirabayashi J, Hughes C, Kasai K, Leffler H, Liu FT, Lotan R, Mercurio AM, Monsigny M, Pillai S, Poirier F, Raz A, Rigby PWJ, Rini JM, Wang JL. Galectins: a family of animal beta-galactoside-binding lectins. Cell 1994;76:597–598.
31. Wang JL, Gray RM, Haudek KC, Patterson RJ. Nucleocytoplasmic lectins. Biochim Biophys Acta 2004;1673:75–93.
32. Hsu DK, Liu FT. Regulation of cellular homeostasis by galectins. Glycoconj J 2004;19:507–515.
33. Cooper DN, Barondes SH. Evidence for export of a muscle lectin from cytosol to extracellular matrix and for a novel secretory mechanism. J Cell Biol 1990;110: 1681–1691.
34. Sato S, Hughes RC. Regulation of secretion and surface expression of Mac-2, a galactoside-binding protein of macrophages. J Biol Chem 1994;269:4424–4430.
35. Sato S, Hughes RC. Control of Mac-2 surface expression on murine macrophage cell lines. Eur J Immunol 1994;24:216–221.
36. Blaser C, Kaufmann M, Muller C, Zimmermann C, Wells V, Mallucci L, Pircher H. Beta-galactoside-binding protein secreted by activated T cells inhibits antigen-induced proliferation of T cells. Eur J Immunol 1998;28:2311–2319.
37. Fuertes MB, Molinero LL, Toscano MA, Ilarregui JM, Rubinstein N, Fainboim L, Zwirner NW, Rabinovich GA. Regulated expression of galectin-1 during T-cell activation involves Lck and Fyn kinases and signaling through MEK1/ERK, p38 MAP kinase and p70S6 kinase. Mol Cell Biochem 2004;267:177–185.
38. Rabinovich GA, Iglesias MM, Modesti NM, Castagna LF, Wolfenstein-Todel C, Riera CM, Sotomayor CE. Activated rat macrophages produce a galectin-1-like protein that induces apoptosis of T cells: biochemical and functional characterization. J Immunol 1998;160:4831–4840.
39. Witko-Sarsat V, Rieu P, Descamps-Latscha B, Lesavre P, Halbwachs-Mecarelli L. Neutrophils: molecules, functions, and pathophysiological aspects. Lab Invest 2000; 80:617.
40. Kohl J. Self, non-self, and danger: a complementary view. Adv Exp Med Biol 2006; 586:71–94.
41. Janeway Jr CA. Approaching the asymptote? Evolution and revolution in immunology. Cold Spring Harb Symp Quant Biol 1989;54:1–13.
42. Janeway Jr CA, Medzhitov R. Innate immune recognition. Annu Rev Immunol 2002;20: 197–216 (Epub 2001 Oct 2004).
43. Medzhitov R, Janeway Jr CA. Decoding the patterns of self and nonself by the innate immune system. Science 2002;296:298–300.

44. Matzinger P. Tolerance, danger, and the extended family. Annu Rev Immunol 1994; 12:991–1045.
45. Matzinger P. Essay 1: the Danger model in its historical context. Scand J Immunol 2001;54:4–9.
46. Matzinger P. The danger model: a renewed sense of self. Science 2002;296:301–305.
47. Scaffidi P, Misteli T, Bianchi ME. Release of chromatin protein HMGB1 by necrotic cells triggers inflammation. Nature 2002;418:191–195.
48. Bianchi ME. Significant (re)location: how to use chromatin and/or abundant proteins as messages of life and death. Trends Cell Biol 2004;14:287–293.
49. Roth J, Vogl T, Sorg C, Sunderkotter C. Phagocyte-specific S100 proteins: a novel group of proinflammatory molecules. Trends Immunol 2003;24:155–158.
50. Foell D, Wittkowski H, Vogl T, Roth J. S100 proteins expressed in phagocytes: a novel group of damage-associated molecular pattern molecules. J Leukoc Biol 2007;81: 28–37.
51. Schmitt E, Gehrmann M, Brunet M, Multhoff G, Garrido C. Intracellular and extracellular functions of heat shock proteins: repercussions in cancer therapy. J Leukoc Biol 2007;81:15–27.
52. Wallin RP, Lundqvist A, More SH, von Bonin A, Kiessling R, Ljunggren HG. Heat-shock proteins as activators of the innate immune system. Trends Immunol 2002;23:130–135.
53. Munoz LE, Franz S, Pausch F, Furnrohr B, Sheriff A, Vogt B, Kern PM, Baum W, Stach C, von Laer D, Brachvogel B, Poschl E, Herrmann M, Gaipl US. The influence on the immunomodulatory effects of dying and dead cells of Annexin V. J Leukoc Biol 2007;81:6–14.
54. Bianchi ME. DAMPs, PAMPs and alarmins: all we need to know about danger. J Leukoc Biol 2007;81:1–5.
55. Bernardes ES, Silva NM, Ruas LP, Mineo JR, Loyola AM, Hsu DK, Liu FT, Chammas R, Roque-Barreira MC. *Toxoplasma gondii* infection reveals a novel regulatory role for galectin-3 in the interface of innate and adaptive immunity. Am J Pathol 2006;168: 1910–1920.
56. del Pozo V, Rojo M, Rubio ML, Cortegano I, Cardaba B, Gallardo S, Ortega M, Civantos E, Lopez E, Martin-Mosquero C, Peces-Barba G, Palomino P, Gonzalez-Mangado N, Lahoz C. Gene therapy with galectin-3 inhibits bronchial obstruction and inflammation in antigen-challenged rats through interleukin-5 gene downregulation. Am J Respir Crit Care Med 2002;166:732–737.
57. Sano H, Hsu DK, Apgar JR, Yu L, Sharma BB, Kuwabara I, Izui S, Liu FT. Critical role of galectin-3 in phagocytosis by macrophages. J Clin Invest 2003;112: 389–397.
58. Chen HY, Sharma BB, Yu L, Zuberi R, Weng IC, Kawakami Y, Kawakami T, Hsu DK, Liu FT. Role of galectin-3 in mast cell functions: galectin-3-deficient mast cells exhibit impaired mediator release and defective JNK expression. J Immunol 2006;177: 4991–4997.
59. Cortegano I, del Pozo V, Cardaba B, de Andres B, Gallardo S, del Amo A, Arrieta I, Jurado A, Palomino P, Liu FT, Lahoz C. Galectin-3 down-regulates IL-5 gene expression on different cell types. J Immunol 1998;161:385–389.
60. Leffler H. Special issue on galectins. Glycoconj J 2002;19:433–630.

61. Ahmad N, Gabius HJ, Andre S, Kaltner H, Sabesan S, Roy R, Liu B, Macaluso F, Brewer CF. Galectin-3 precipitates as a pentamer with synthetic multivalent carbohydrates and forms heterogeneous cross-linked complexes. J Biol Chem 2004; 279:10841–10847 (Epub 2003 Dec 12).
62. Nieminen J, Kuno A, Hirabayashi J, Sato S. Visualization of galectin-3 oligomerization on the surface of neutrophils and endothelial cells using fluorescence resonance energy transfer. J Biol Chem 2007;282:1374–1383.
63. Liu FT. Regulatory roles of galectins in the immune response. Int Arch Allergy Immunol 2005;136:385–400.
64. Rabinovich GA, Liu FT, Hirashima M, Anderson A. An emerging role for galectins in tuning the immune response: lessons from experimental models of inflammatory disease, autoimmunity and cancer. Scand J Immunol 2007;66:143–158.
65. Rabinovich G, Castagna L, Landa C, Riera CM, Sotomayor C. Regulated expression of a 16-kD galectin-like protein in activated rat macrophages. J Leukoc Biol 1996;59: 363–370.
66. Baum LG, Pang M, Perillo NL, Wu T, Delegeane A, Uittenbogaart CH, Fukuda M, Seilhamer JJ. Human thymic epithelial cells express an endogenous lectin, galectin-1, which binds to core 2 *O*-glycans on thymocytes and T lymphoblastoid cells. J Exp Med 1995;181:877–887.
67. Baum LG, Seilhamer JJ, Pang M, Levine WB, Beynon D, Berliner JA. Synthesis of an endogenous lectin, galectin-1, by human endothelial cells is up-regulated by endothelial cell activation. Glycoconj J 1995;12:63–68.
68. Zuniga E, Rabinovich GA, Iglesias MM, Gruppi A. Regulated expression of galectin-1 during B-cell activation and implications for T-cell apoptosis. J Leukoc Biol 2001; 70:73–79.
69. Stillman BN, Hsu DK, Pang M, Brewer CF, Johnson P, Liu FT, Baum LG. Galectin-3 and galectin-1 bind distinct cell surface glycoprotein receptors to induce T cell death. J Immunol 2006;176:778–789.
70. Garin MI, Chu CC, Golshayan D, Cernuda-Morollon E, Wait R, Lechler RI. Galectin-1: a key effector of regulation mediated by $CD25^+CD4^+$ T cells. Blood 2007;109: 2058–2065.
71. Sugimoto N, Oida T, Hirota K, Nakamura K, Nomura T, Uchiyama T, Sakaguchi S. Foxp3-dependent and -independent molecules specific for $CD25^+CD4^+$ natural regulatory T cells revealed by DNA microarray analysis. Int Immunol 2006;18: 1197–1209.
72. Sato S. Galectin as a molecule of danger signal, which could evoke immune response to infection. Trends Glycosci Glycotechnol 2002;14:285–301.
73. Sato S, Nieminen J. Seeing strangers or announcing "danger": Galectin-3 in two models of innate immunity. Glycoconj J 2004;19:583–591.
74. Hirashima M, Kashio Y, Nishi N, Yamauchi A, Imaizumi TA, Kageshita T, Saita N, Nakamura T. Galectin-9 in physiological and pathological conditions. Glycoconj J 2004;19:593–600.
75. Nieminen J, St-Pierre C, Sato S. Galectin-3 interacts with naive and primed neutrophils, inducing innate immune responses. J Leukoc Biol 2005;78:1127–1135.
76. Rabinovich GA, Baum LG, Tinari N, Paganelli R, Natoli C, Liu FT, Iacobelli S. Galectins and their ligands: amplifiers, silencers or tuners of the inflammatory response? Trends Immunol 2002;23:313–320.

77. Ouellet M, Mercier S, Pelletier I, Bounou S, Roy J, Hirabayashi J, Sato S, Tremblay MJ. Galectin-1 acts as a soluble host factor that promotes HIV-1 infectivity through stabilization of virus attachment to host cells. J Immunol 2005;174:4120–4126.

78. Reichert F, Saada A, Rotshenker S. Peripheral nerve injury induces Schwann cells to express two macrophage phenotypes: phagocytosis and the galactose-specific lectin MAC-2. J Neurosci 1994;14:3231–3245.

79. Reichert F, Rotshenker S. Galectin-3/MAC-2 in experimental allergic encephalomyelitis. Exp Neurol 1999;160:508–514.

80. Lalancette-Hebert M, Gowing G, Simard A, Weng YC, Kriz J. Selective ablation of proliferating microglial cells exacerbates ischemic injury in the brain. J Neurosci 2007;27:2596–2605.

81. Sato S, Ouellet N, Pelletier I, Simard M, Rancourt A, Bergeron MG. Role of galectin-3 as an adhesion molecule for neutrophil extravasation during streptococcal pneumonia. J Immunol 2002;168:1813–1822.

82. Zuniga E, Gruppi A, Hirabayashi J, Kasai KI, Rabinovich GA. Regulated expression and effect of galectin-1 on *Trypanosoma cruzi*-infected macrophages: modulation of microbicidal activity and survival. Infect Immun 2001;69:6804–6812.

83. Vray B, Camby I, Vercruysse V, Mijatovic T, Bovin NV, Ricciardi-Castagnoli P, Kaltner H, Salmon I, Gabius HJ, Kiss R. Up-regulation of galectin-3 and its ligands by *Trypanosoma cruzi* infection with modulation of adhesion and migration of murine dendritic cells. Glycobiology 2004;14:647–657.

84. van den Berg TK, Honing H, Franke N, van Remoortere A, Schiphorst WE, Liu FT, Deelder AM, Cummings RD, Hokke CH, van Die I. LacdiNAc-glycans constitute a parasite pattern for galectin-3-mediated immune recognition. J Immunol 2004;173: 1902–1907.

85. Fowler M, Thomas RJ, Atherton J, Roberts IS, High NJ. Galectin-3 binds to *Helicobacter pylori* O-antigen: it is upregulated and rapidly secreted by gastric epithelial cells in response to *H. pylori* adhesion. Cell Microbiol 2006;8:44–54.

86. Kasamatsu A, Uzawa K, Shimada K, Shiiba M, Otsuka Y, Seki N, Abiko Y, Tanzawa H. Elevation of galectin-9 as an inflammatory response in the periodontal ligament cells exposed to *Porphylomonas gingivalis* lipopolysaccharide *in vitro* and *in vivo*. Int J Biochem Cell Biol 2005;37:397–408.

87. Fogel S, Guittaut M, Legrand A, Monsigny M, Hebert E. The tat protein of HIV-1 induces galectin-3 expression. Glycobiology 1999;9:383–387.

88. Hsu DK, Hammes SR, Kuwabara I, Greene WC, Liu FT. Human T lymphotropic virus-I infection of human T lymphocytes induces expression of the beta-galactoside-binding lectin, galectin-3. Am J Pathol 1996;148:1661–1670.

89. Schroder HC, Ushijima H, Theis C, Seve AP, Hubert J, Muller WE. Expression of nuclear lectin carbohydrate-binding protein 35 in human immunodeficiency virus type 1-infected Molt-3 cells. J Acquir Immune Defic Syndr Hum Retrovirol 1995;9: 340–348.

90. Camby I, Le Mercier M, Lefranc F, Kiss R. Galectin-1: a small protein with major functions. Glycobiology 2006;16:137R–157R.

91. Danguy A, Camby I, Kiss R. Galectins and cancer. Biochim Biophys Acta 2002;1572: 285–293.

92. van den Brule F, Califice S, Castronovo V. Expression of galectins in cancer: a critical review. Glycoconj J 2004;19:537–542.
93. Liu FT, Rabinovich GA. Galectins as modulators of tumour progression. Nat Rev Cancer 2005;5:29–41.
94. Sparrow CP, Leffler H, Barondes SH. Multiple soluble beta-galactoside-binding lectins from human lung. J Biol Chem 1987;262:7383–7390.
95. Barondes SH, Cooper DN, Gitt MA, Leffler H. Galectins. Structure and function of a large family of animal lectins. J Biol Chem 1994;269:20807–20810.
96. Hirabayashi J, Hashidate T, Arata Y, Nishi N, Nakamura T, Hirashima M, Urashima T, Oka T, Futai M, Muller WE, Yagi F, Kasai K. Oligosaccharide specificity of galectins: a search by frontal affinity chromatography. Biochim Biophys Acta 2002;1572: 232–254.
97. Amano M, Galvan M, He J, Baum LG. The ST6Gal I sialyltransferase selectively modifies *N*-glycans on CD45 to negatively regulate galectin-1-induced CD45 clustering, phosphatase modulation, and T cell death. J Biol Chem 2003;278:7469–7475.
98. Sato S, Hughes RC. Binding specificity of a baby hamster kidney lectin for H type I and II chains, polylactosamine glycans, and appropriately glycosylated forms of laminin and fibronectin. J Biol Chem 1992;267:6983–6990.
99. Toscano MA, Bianco GA, Ilarregui JM, Croci DO, Correale J, Hernandez JD, Zwirner NW, Poirier F, Riley EM, Baum LG, Rabinovich GA. Differential glycosylation of Th1, Th2 and Th17 effector cells selectively regulates susceptibility to cell death. Nat Immunol 2007;8:825–834.
100. Dumic J, Dabelic S, Flogel M. Galectin-3: an open-ended story. Biochim Biophys Acta 2006;1760:616–635.
101. Brewer CF, Miceli MC, Baum LG. Clusters, bundles, arrays and lattices: novel mechanisms for lectin–saccharide-mediated cellular interactions. Curr Opin Struct Biol 2002;12:616–623.
102. Leppanen A, Stowell S, Blixt O, Cummings RD. Dimeric galectin-1 binds with high affinity to alpha2,3-sialylated and non-sialylated terminal *N*-acetyllactosamine units on surface-bound extended glycans. J Biol Chem 2005;280:5549–5562.
103. Nieminen J, St-Pierre C, Bhaumik P, Sato S, Role of galectin-3 in neutrophil recruitment in pneumococcal pneumonia. J Immunol 2008;180: 2466–2473.
104. Partridge EA, Le Roy C, Di Guglielmo GM, Pawling J, Cheung P, Granovsky M, Nabi IR, Wrana JL, Dennis JW. Regulation of cytokine receptors by Golgi *N*-glycan processing and endocytosis. Science 2004;306:120–124.
105. Lau KS, Partridge EA, Grigorian A, Silvescu CI, Reinhold VN, Demetriou M, Dennis JW. Complex *N*-glycan number and degree of branching cooperate to regulate cell proliferation and differentiation. Cell 2007;129:123–134.
106. Demetriou M, Granovsky M, Quaggin S, Dennis JW. Negative regulation of T-cell activation and autoimmunity by Mgat5 N-glycosylation. Nature 2001;409:733–739.
107. Ohtsubo K, Takamatsu S, Minowa MT, Yoshida A, Takeuchi M, Marth JD. Dietary and genetic control of glucose transporter 2 glycosylation promotes insulin secretion in suppressing diabetes. Cell 2005;123:1307–1321.
108. Hughes RC. Secretion of the galectin family of mammalian carbohydrate-binding proteins. Biochim Biophys Acta 1999;1473:172–185.

109. Sato S, Burdett I, Hughes RC. Secretion of the baby hamster kidney 30-kDa galactose-binding lectin from polarized and nonpolarized cells: a pathway independent of the endoplasmic reticulum–Golgi complex. Exp Cell Res 1993;207:8–18.

110. Nickel W. The mystery of nonclassical protein secretion, A current view on cargo proteins and potential export routes. Eur J Biochem 2003;270:2109–2119.

111. Nickel W. Unconventional secretory routes: direct protein export across the plasma membrane of mammalian cells. Traffic 2005;6:607–614.

112. Oppenheim JJ, Tewary P, de la Rosa G, Yang D. Alarmins initiate host defense. Adv Exp Med Biol 2007;601:185–194.

113. Rabinovich GA, Ramhorst RE, Rubinstein N, Corigliano A, Daroqui MC, Kier-Joffe EB, Fainboim L. Induction of allogenic T-cell hyporesponsiveness by galectin-1-mediated apoptotic and non-apoptotic mechanisms. Cell Death Differ 2002;9:661–670.

114. Matsumoto R, Matsumoto H, Seki M, Hata M, Asano Y, Kanegasaki S, Stevens RL, Hirashima M. Human ecalectin, a variant of human galectin-9, is a novel eosinophil chemoattractant produced by T lymphocytes. J Biol Chem 1998;273:16976–16984.

115. Gil CD, La M, Perretti M, Oliani SM. Interaction of human neutrophils with endothelial cells regulates the expression of endogenous proteins annexin 1, galectin-1 and galectin-3. Cell Biol Int 2006;30:338–344.

116. Bergeron Y, Ouellet N, Deslauriers AM, Simard M, Olivier M, Bergeron MG. Cytokine kinetics and other host factors in response to pneumococcal pulmonary infection in mice. Infect Immun 1998;66:912–922.

117. Fillion I, Ouellet N, Simard M, Bergeron Y, Sato S, Bergeron MG. Role of chemokines and formyl peptides in pneumococcal pneumonia-induced monocyte/macrophage recruitment. J Immunol 2001;166:7353–7361.

118. Mercier S, St-Pierre C, Pelletier I, Ouellet M, Tremblay MJ, Sato S, Galectin-1 promotes HIV-1 infectivity in macrophages through stabilization of viral adsorption. Virology 2008;371:121–129.

119. Levroney EL, Aguilar HC, Fulcher JA, Kohatsu L, Pace KE, Pang M, Gurney KB, Baum LG, Lee B. Novel innate immune functions for galectin-1: galectin-1 inhibits cell fusion by Nipah virus envelope glycoproteins and augments dendritic cell secretion of proinflammatory cytokines. J Immunol 2005;175:413–420.

120. Mey A, Leffler H, Hmama Z, Normier G, Revillard JP. The animal lectin galectin-3 interacts with bacterial lipopolysaccharides via two independent sites. J Immunol 1996;156:1572–1577.

121. John CM, Jarvis GA, Swanson KV, Leffler H, Cooper MD, Huflejt ME, Griffiss JM. Galectin-3 binds lactosaminylated lipooligosaccharides from *Neisseria gonorrhoeae* and is selectively expressed by mucosal epithelial cells that are infected. Cell Microbiol 2002;4:649–662.

122. Pelletier I, Hashidate T, Urashima T, Nishi N, Nakamura T, Futai M, Arata Y, Kasai K, Hirashima M, Hirabayashi J, Sato S. Specific recognition of *Leishmania major* poly-beta-galactosyl epitopes by galectin-9: possible implication of galectin-9 in interaction between *L. major* and host cells. J Biol Chem 2003;278:22223–22230.

123. Pelletier I, Sato S. Specific recognition and cleavage of galectin-3 by *Leishmania major* through species-specific polygalactose epitope. J Biol Chem 2002;277:17663–17670.

124. van Die I, Cummings RD. Glycans modulate immune responses in helminth infections and allergy. Chem Immunol Allergy 2006;90:91–112.

125. Kleshchenko YY, Moody TN, Furtak VA, Ochieng J, Lima MF, Villalta F. Human galectin-3 promotes *Trypanosoma cruzi* adhesion to human coronary artery smooth muscle cells. Infect Immun 2004;72:6717–6721.

126. Hohl TM, Pamer EG. Cracking the fungal armor. Nat Med 2006;12:730–732.

127. Taylor PR, Tsoni SV, Willment JA, Dennehy KM, Rosas M, Findon H, Haynes K, Steele C, Botto M, Gordon S, Brown GD. Dectin-1 is required for beta-glucan recognition and control of fungal infection. Nat Immunol 2007;8:31–38.

128. Saijo S, Fujikado N, Furuta T, Chung SH, Kotaki H, Seki K, Sudo K, Akira S, Adachi Y, Ohno N, Kinjo T, Nakamura K, Kawakami K, Iwakura Y. Dectin-1 is required for host defense against *Pneumocystis carinii* but not against *Candida albicans*. Nat Immunol 2007;8:39–46.

129. Fradin C, Poulain D, Jouault T. Beta-1,2-linked oligomannosides from *Candida albicans* bind to a 32-kilodalton macrophage membrane protein homologous to the mammalian lectin galectin-3. Infect Immun 2000;68:4391–4398.

130. Jouault T, El Abed-El Behi M, Martinez-Esparza M, Breuilh L, Trinel PA, Chamaillard M, Trottein F, Poulain D. Specific recognition of *Candida albicans* by macrophages requires galectin-3 to discriminate *Saccharomyces cerevisiae* and needs association with TLR2 for signalling. J Immunol 2006;177:4679–4687.

131. Kohatsu L, Hsu DK, Jegalian AG, Liu FT, Baum LG. Galectin-3 induces death of *Candida* species expressing specific beta-1,2-linked mannans. J Immunol 2006;177: 4718–4726.

132. Sorg C. The calcium binding proteins MRP8 and MRP14 in acute and chronic inflammation. Behring Inst Mitt :1992;91:126–137.

133. Oppenheim JJ, Biragyn A, Kwak LW, Yang D. Roles of antimicrobial peptides such as defensins in innate and adaptive immunity. Ann Rheum Dis 2003;62 (Suppl 2):ii17–ii21.

134. Yang D, Oppenheim JJ. Antimicrobial proteins act as "alarmins" in joint immune defense. Arthritis Rheum 2004;50:3401–3403.

135. Yamaoka A, Kuwabara I, Frigeri LG, Liu FT. A human lectin, galectin-3 (epsilon bp/ Mac-2), stimulates superoxide production by neutrophils. J Immunol 1995;154: 3479–3487.

136. Karlsson A, Follin P, Leffler H, Dahlgren C. Galectin-3 activates the NADPH-oxidase in exudated but not peripheral blood neutrophils. Blood 1998;91:3430–3438.

137. Schroder AK, Uciechowski P, Fleischer D, Rink L. Crosslinking of CD66B on peripheral blood neutrophils mediates the release of interleukin-8 from intracellular storage. Hum Immunol 2006;67:676–682.

138. Wang J, Barke RA, Charboneau R, Roy S. Morphine impairs host innate immune response and increases susceptibility to *Streptococcus pneumoniae* lung infection. J Immunol 2005;174:426–434.

139. Hsu DK, Yang RY, Pan Z, Yu L, Salomon DR, Fung-Leung WP, Liu FT. Targeted disruption of the galectin-3 gene results in attenuated peritoneal inflammatory responses. Am J Pathol 2000;156:1073–1083.

140. Colnot C, Ripoche MA, Milon G, Montagutelli X, Crocker PR, Poirier F. Maintenance of granulocyte numbers during acute peritonitis is defective in galectin-3-null mutant mice. Immunology 1998;94:290–296.

141. Rabinovich GA, Sotomayor CE, Riera CM, Bianco I, Correa SG. Evidence of a role for galectin-1 in acute inflammation. Eur J Immunol 2000;30:1331–1339.

142. La M, Cao TV, Cerchiaro G, Chilton K, Hirabayashi J, Kasai K, Oliani SM, Chernajovsky Y, Perretti M. A novel biological activity for galectin-1: inhibition of leukocyte–endothelial cell interactions in experimental inflammation. Am J Pathol 2003;163: 1505–1515.

143. Asakura H, Kashio Y, Nakamura K, Seki M, Dai S, Shirato Y, Abedin MJ, Yoshida N, Nishi N, Imaizumi T, Saita N, Toyama Y, Takashima H, Nakamura T, Ohkawa M, Hirashima M. Selective eosinophil adhesion to fibroblast via IFN-gamma-induced galectin-9. J Immunol 2002;169:5912–5918.

144. Imaizumi T, Kumagai M, Sasaki N, Kurotaki H, Mori F, Seki M, Nishi N, Fujimoto K, Tanji K, Shibata T, Tamo W, Matsumiya T, Yoshida H, Cui XF, Takanashi S, Hanada K, Okumura K, Yagihashi S, Wakabayashi K, Nakamura T, Hirashima M, Satoh K. Interferon-gamma stimulates the expression of galectin-9 in cultured human endothelial cells. J Leukoc Biol 2002;72:486–491.

145. Frigeri LG, Zuberi RI, Liu FT. Epsilon BP, a beta-galactoside-binding animal lectin, recognizes IgE receptor (Fc epsilon RI) and activates mast cells. Biochemistry 1993;32:7644–7649.

146. Sano H, Hsu DK, Yu L, Apgar JR, Kuwabara I, Yamanaka T, Hirashima M, Liu FT. Human galectin-3 is a novel chemoattractant for monocytes and macrophages. J Immunol 2000;165:2156–2164.

147. Jeng KC, Frigeri LG, Liu FT. An endogenous lectin, galectin-3 (epsilon BP/Mac-2), potentiates IL-1 production by human monocytes. Immunol Lett 1994;42:113–116.

148. Perone MJ, Larregina AT, Shufesky WJ, Papworth GD, Sullivan ML, Zahorchak AF, Stolz DB, Baum LG, Watkins SC, Thomson AW, Morelli AE. Transgenic galectin-1 induces maturation of dendritic cells that elicit contrasting responses in naive and activated T cells. J Immunol 2006;176:7207–7220.

149. Fulcher JA, Hashimi ST, Levroney EL, Pang M, Gurney KB, Baum LG, Lee B. Galectin-1-matured human monocyte-derived dendritic cells have enhanced migration through extracellular matrix. J Immunol 2006;177:216–226.

150. Breuilh L, Vanhoutte F, Fontaine J, van Stijn CM, Tillie-Leblond I, Capron M, Faveeuw C, Jouault T, van Die I, Gosset P, Trottein F. Galectin 3 modulates immune and inflammatory responses during helminthic infection: impact of galectin 3 deficiency on the functions of dendritic cells. Infect Immun 2007;75:5148–5157.

151. Dai SY, Nakagawa R, Itoh A, Murakami H, Kashio Y, Abe H, Katoh S, Kontani K, Kihara M, Zhang SL, Hata T, Nakamura T, Yamauchi A, Hirashima M. Galectin-9 induces maturation of human monocyte-derived dendritic cells. J Immunol 2005;175: 2974–2981.

152. Henson PM. Dampening inflammation. Nat Immunol 2005;6:1179–1181.

153. Serhan CN, Savill J. Resolution of inflammation: the beginning programs the end. Nat Immunol 2005;6:1191–1197.

154. Dias-Baruffi M, Zhu H, Cho M, Karmakar S, McEver RP, Cummings RD. Dimeric galectin-1 induces surface exposure of phosphatidylserine and phagocytic recognition of leukocytes without inducing apoptosis. J Biol Chem 2003;278: 41282–41293.

155. Karmakar S, Cummings RD, McEver RP. Contributions of Ca^{2+} to galectin-1-induced exposure of phosphatidylserine on activated neutrophils. J Biol Chem 2005;280(31): 28623–28631.

156. Stowell SR, Karmakar S, Stowell CJ, Dias-Baruffi M, McEver RP, Cummings RD. Human galectin-1, -2, and -4 induce surface exposure of phosphatidylserine in activated human neutrophils but not in activated T cells. Blood 2007;109:219–227.

157. Erwig LP, Henson PM. Immunological consequences of apoptotic cell phagocytosis. Am J Pathol 2007;171:2–8.

158. Gordon S, Taylor PR. Monocyte and macrophage heterogeneity. Nat Rev Immunol 2005;5:953–964.

159. Correa SG, Sotomayor CE, Aoki MP, Maldonado CA, Rabinovich GA. Opposite effects of galectin-1 on alternative metabolic pathways of L-arginine in resident, inflammatory, and activated macrophages. Glycobiology 2003;13:119–128.

160. Barrionuevo P, Beigier-Bompadre M, Ilarregui JM, Toscano MA, Bianco GA, Isturiz MA, Rabinovich GA. A novel function for galectin-1 at the crossroad of innate and adaptive immunity: galectin-1 regulates monocyte/macrophage physiology through a nonapoptotic ERK-dependent pathway. J Immunol 2007;178:436–445.

161. Gordon S. Alternative activation of macrophages. Nat Rev Immunol 2003;3:23–35.

162. Mosser DM. The many faces of macrophage activation. J Leukoc Biol 2003;73: 209–212.

163. Belkaid Y, Rouse BT. Natural regulatory T cells in infectious disease. Nat Immunol 2005;6:353–360.

164. von Boehmer H. Mechanisms of suppression by suppressor T cells. Nat Immunol 2005;6:338–344.

165. Goldstone SD, Lavin MF. Isolation of a cDNA clone, encoding a human beta-galactoside binding protein, overexpressed during glucocorticoid-induced cell death. Biochem Biophys Res Commun 1991;178:746–750.

166. Perillo NL, Pace KE, Seilhamer JJ, Baum LG. Apoptosis of T cells mediated by galectin-1. Nature 1995;378:736–739.

167. Rabinovich GA, Modesti NM, Castagna LF, Landa CA, Riera CM, Sotomayor CE. Specific inhibition of lymphocyte proliferation and induction of apoptosis by CLL-I, a beta-galactoside-binding lectin. J Biochem (Tokyo) 1997;122:365–373.

168. He J, Baum LG. Presentation of galectin-1 by extracellular matrix triggers T cell death. J Biol Chem 2004;279:4705–4712.

169. Matarrese P, Tinari A, Mormone E, Bianco GA, Toscano MA, Ascione B, Rabinovich GA, Malorni W. Galectin-1 sensitizes resting human T lymphocytes to Fas (CD95)-mediated cell death via mitochondrial hyperpolarization, budding, and fission. J Biol Chem 2005;280:6969–6985.

170. Ion G, Fajka-Boja R, Toth GK, Caron M, Monostori E. Role of p56lck and ZAP70-mediated tyrosine phosphorylation in galectin-1-induced cell death. Cell Death Differ 2005;12:1145–1147.

171. Ion G, Fajka-Boja R, Kovacs F, Szebeni G, Gombos I, Czibula A, Matko J, Monostori E. Acid sphingomyelinase mediated release of ceramide is essential to trigger the mitochondrial pathway of apoptosis by galectin-1. Cell Signal 2006;18:1887–1896.

172. Steinman L. A brief history of T(H)17, the first major revision in the T(H)1/T(H)2 hypothesis of T cell-mediated tissue damage. Nat Med 2007;13:139–145.

173. Vespa GN, Lewis LA, Kozak KR, Moran M, Nguyen JT, Baum LG, Miceli MC. Galectin-1 specifically modulates TCR signals to enhance TCR apoptosis but inhibit IL-2 production and proliferation. J Immunol 1999;162:799–806.

174. Chung CD, Patel VP, Moran M, Lewis LA, Miceli MC. Galectin-1 induces partial TCR zeta-chain phosphorylation and antagonizes processive TCR signal transduction. J Immunol 2000;165:3722–3729.

175. Rabinovich GA, Ariel A, Hershkoviz R, Hirabayashi J, Kasai KI, Lider O. Specific inhibition of T-cell adhesion to extracellular matrix and proinflammatory cytokine secretion by human recombinant galectin-1. Immunology 1999;97:100–106.

176. Rabinovich GA, Daly G, Dreja H, Tailor H, Riera CM, Hirabayashi J, Chernajovsky Y. Recombinant galectin-1 and its genetic delivery suppress collagen-induced arthritis via T cell apoptosis. J Exp Med 1999;190:385–398.

177. Toscano MA, Commodaro AG, Ilarregui JM, Bianco GA, Liberman A, Serra HM, Hirabayashi J, Rizzo LV, Rabinovich GA. Galectin-1 suppresses autoimmune retinal disease by promoting concomitant Th2- and T regulatory-mediated anti-inflammatory responses. J Immunol 2006;176:6323–6332.

178. Perone MJ, Bertera S, Tawadrous ZS, Shufesky WJ, Piganelli JD, Baum LG, Trucco M, Morelli AE. Dendritic cells expressing transgenic galectin-1 delay onset of autoimmune diabetes in mice. J Immunol 2006;177:5278–5289.

179. Santucci L, Fiorucci S, Rubinstein N, Mencarelli A, Palazzetti B, Federici B, Rabinovich GA, Morelli A. Galectin-1 suppresses experimental colitis in mice. Gastroenterology 2003;124:1381–1394.

180. van der Leij J, van den Berg A, Blokzijl T, Harms G, van Goor H, Zwiers P, van Weeghel R, Poppema S, Visser L. Dimeric galectin-1 induces IL-10 production in T-lymphocytes: an important tool in the regulation of the immune response. J Pathol 2004;204:511–518.

181. van der Leij J, van den Berg A, Harms G, Eschbach H, Vos H, Zwiers P, van Weeghel R, Groen H, Poppema S, Visser L. Strongly enhanced IL-10 production using stable galectin-1 homodimers. Mol Immunol 2007;44:506–513.

182. Matzinger P. Friendly and dangerous signals: is the tissue in control? Nat Immunol 2007;8:11–13.

183. Takabayshi K, Corr M, Hayashi T, Redecke V, Beck L, Guiney D, Sheppard D, Raz E. Induction of a homeostatic circuit in lung tissue by microbial compounds. Immunity 2006;24:475–487.

184. Kashio Y, Nakamura K, Abedin MJ, Seki M, Nishi N, Yoshida N, Nakamura T, Hirashima M. Galectin-9 induces apoptosis through the calcium-calpain-caspase-1 pathway. J Immunol 2003;170:3631–3636.

185. Sturm A, Lensch M, Andre S, Kaltner H, Wiedenmann B, Rosewicz S, Dignass AU, Gabius HJ. Human galectin-2: novel inducer of T cell apoptosis with distinct profile of caspase activation. J Immunol 2004;173:3825–3837.

186. Fukumori T, Takenaka Y, Yoshii T, Kim HR, Hogan V, Inohara H, Kagawa S, Raz A. CD29 and CD7 mediate galectin-3-induced type II T-cell apoptosis. Cancer Res 2003;63:8302–8311.

187. Zhu C, Anderson AC, Schubart A, Xiong H, Imitola J, Khoury SJ, Zheng XX, Strom TB, Kuchroo VK. The Tim-3 ligand galectin-9 negatively regulates T helper type 1 immunity. Nat Immunol 2005;6:1245–1252.

188. Kubach J, Lutter P, Bopp T, Stoll S, Becker C, Huter E, Richter C, Weingarten P, Warger T, Knop J, Mullner S, Wijdenes J, Schild H, Schmitt E, Jonuleit H. Human $CD4^{+}CD25^{+}$ regulatory T cells: proteome analysis identifies galectin-10 as a novel marker essential for their anergy and suppressive function. Blood 2007;110:1550–1558.

189. Jonuleit H, Schmitt E, Stassen M, Tuettenberg A, Knop J, Enk AH. Identification and functional characterization of human CD4(+)CD25(+) T cells with regulatory properties isolated from peripheral blood. J Exp Med 2001;193:1285–1294.

190. Thornton AM, Shevach EM. $CD4^+CD25^+$ immunoregulatory T cells suppress polyclonal T cell activation *in vitro* by inhibiting interleukin 2 production. J Exp Med 1998;188: 287–296.

191. Blois SM, Ilarregui JM, Tometten M, Garcia M, Orsal AS, Cordo-Russo R, Toscano MA, Bianco GA, Kobelt P, Handjiski B, Tirado I, Markett UR, Klapp BF, Poirier F, Szekeres-Bartho J, Arck PC, Rabinovich GA. A pivotal role for galectin-1 in fetomaternal tolerance. Nat Med. 2007;13:1450–1457.

192. Juszczynski P, Ouyang J, Monti S, Rodig SJ, Takeyama K, Abramson J, Chen W, Kutok JL, Rabinovich GA, Shipp MA. The AP1-dependent secretion of galectin-1 by Reed Sternberg cells fosters immune privilege in classical Hodgkin lymphoma. Proc Natl Acad Sci USA 2007;104:13134–13139.

193. Rabinovich GA, Gabrilovich D, Sotomayor EM. Immunosuppressive strategies that are mediated by tumor cells. Annu Rev Immunol 2007;25:267–296.

194. Pure E, Allison JP, Schreiber RD. Breaking down the barriers to cancer immunotherapy. Nat Immunol 2005;6:1207–1210.

195. Rosenberg SA. A new era for cancer immunotherapy based on the genes that encode cancer antigens. Immunity 1999;10:281–287.

196. Boon T, Coulie PG, Van den Eynde BJ, van der Bruggen P. Human T cell responses against melanoma. Annu Rev Immunol 2006;24:175–208.

197. Apetoh L, Ghiringhelli F, Tesniere A, Obeid M, Ortiz C, Criollo A, Mignot G, Maiuri MC, Ullrich E, Saulnier P, Yang H, Amigorena S, Ryffel B, Barrat FJ, Saftig P, Levi F, Lidereau R, Nogues C, Mira JP, Chompret A, Joulin V, Clavel-Chapelon F, Bourhis J, Andre F, Delaloge S, Tursz T, Kroemer G, Zitvogel L. Toll-like receptor 4-dependent contribution of the immune system to anticancer chemotherapy and radiotherapy. Nat Med 2007;13:1050–1059.

198. Gabius HJ, Engelhardt R, Cramer F. Endogenous tumor lectins: overview and perspectives. Anticancer Res 1986;6:573–578.

199. Raz A, Meromsky L, Lotan R. Differential expression of endogenous lectin on the surface of nontumorigenic, tumorigenic and metastatic cells. Cancer Res 1986;46: 3667–3672.

200. He J, Baum LG. Endothelial cell expression of galectin-1 induced by prostate cancer cells inhibits T-cell transendothelial migration. Lab Invest 2006;86:578–590.

201. Jung EJ, Moon HG, Cho BI, Jeong CY, Joo YT, Lee YJ, Hong SC, Choi SK, Ha WS, Kim JW, Lee CW, Lee JS, Park ST. Galectin-1 expression in cancer-associated stromal cells correlates tumor invasiveness and tumor progression in breast cancer. Int J Cancer 2007;120:2331–2338.

202. Kreunin P, Yoo C, Urquidi V, Lubman DM, Goodison S. Proteomic profiling identifies breast tumor metastasis-associated factors in an isogenic model. Proteomics 2007;7: 299–312.

203. Rubinstein N, Alvarez M, Zwirner NW, Toscano MA, Ilarregui JM, Bravo A, Mordoh J, Fainboim L, Podhajcer OL, Rabinovich GA. Targeted inhibition of galectin-1 gene expression in tumor cells results in heightened T cell-mediated rejection: a potential mechanism of tumor-immune privilege. Cancer Cell 2004;5:241–251.

204. Le QT, Shi G, Cao H, Nelson DW, Wang Y, Chen EY, Zhao S, Kong C, Richardson D, O'Byrne KJ, Giaccia AJ, Koong AC. Galectin-1: a link between tumor hypoxia and tumor immune privilege. J Clin Oncol 2005;23:8932–8941.

205. Gandhi MK, Moll G, Smith C, Dua U, Lambley E, Ramuz O, Gill D, Marlton P, Seymour JF, Khanna R. Galectin-1 mediated suppression of Epstein-Barr virus specific T-cell immunity in classic Hodgkin lymphoma. Blood 2007;110:1326–1329.

8

THE ROLE OF GALECTINS IN ORGAN FIBROSIS

NEIL C. HENDERSON
The Queen's Medical Research Institute, 47 Little France Crescent, Edinburgh EH16 4TJ, Scotland, UK

TARIQ SETHI
MRC Centre for Inflammation Research, The Queen's Medical Research Institute, University of Edinburgh Medical School, Royal Infirmary, 51 Little France Crescent, Old Dalkeith Road, Edinburgh EH16 4SA, UK

INTRODUCTION

Chronic inflammation leading to fibrosis with loss of tissue architecture and subsequent organ failure represents a massive health care problem. Tissue fibrosis represents the final common pathway of chronic tissue injury. Inflammation is normally a beneficial protective response to tissue damage promoting healing or repair of affected tissues. However, chronic inflammation with the formation of scar tissue and organ failure is a characteristic feature of the pathogenesis of many human diseases and results in major morbidity and mortality worldwide. Currently, our therapeutic repertoire is limited to immunosuppression and/or organ transplantation (1–3). Therefore, effective alternative therapies are urgently required.

Previous studies of the pathogenesis of tissue fibrosis have identified a number of cells as being important in the evolution of tissue scarring. The fibroblast and myofibroblast have been identified as key cells in the initiation and perpetuation of organ scarring (4). Classically, tissue fibroblasts become activated to a contractile myofibroblast matrix-secreting phenotype, which is central to the fibrotic process.

Galectins, Edited by Anatole A. Klyosov, Zbigniew J. Witczak, and David Platt

However, cells of the innate and adaptive immune system are also known to play an important role in tissue fibrosis, and macrophages are a good example of immune cells that are intrinsically involved in tissue scarring and remodeling (5). Therefore, understanding the complex interplay between immune cells and tissue fibroblasts and also elucidation of the molecular mechanisms that drive the phenotype switch from fibroblasts to myofibroblasts may allow the development of targeted antifibrotic therapies. This chapter will focus on the role of galectins in organ fibrosis with particular emphasis on galectins-1 and -3, as the majority of the data to date have mainly examined the role of these two galectins in tissue fibrosis.

GALECTIN EXPRESSION IN HUMAN ORGAN FIBROSIS

A number of studies have demonstrated upregulation of galectin-3 in different human fibrotic conditions. Hsu et al. (6) demonstrated that although galectin-3 is absent in normal hepatocytes, examination of liver biopsies by immunohistochemistry showed that galectin-3 is upregulated both in hepatitis B induced cirrhosis of the liver and in hepatocellular carcinoma complicating hepatitis B infection. Galectin-3 was abundantly expressed in cirrhotic liver in a peripheral distribution within regenerating nodules. Henderson et al. (7) also found upregulation of galectin-3 in human liver cirrhosis. Furthermore, confirming and extending Hsu's findings (6), we demonstrated that galectin-3 was upregulated in liver cirrhosis secondary to a broad range of different etiologies with quite different mechanisms of injury including viral-induced liver disease (hepatitis B and C), autoimmune, copper or iron overload, primary biliary cirrhosis, and alcohol-induced liver disease. We also found that galectin-3 expression was negligible in the normal liver and increased in the cirrhotic nodules of hepatocytes, particularly at the nodule periphery.

Galectin-1 and galectin-3 expression have been studied in chronic pancreatitis. Wang et al. (8) examined galectin-1 and galectin-3 expression by Northern blot analysis, *in situ* hybridization, immunohistochemistry, and Western blot analysis in normal and chronic pancreatitis tissue. The authors found a significant correlation between galectin-1 and fibrosis and between galectin-3 and fibrosis and the density of ductular complexes within the pancreas.

More recently, the role of galectin-3 in human pulmonary fibrosis has been examined (9). Galectin-3 expression was increased in the bronchoalveolar lavage fluid from patients with idiopathic pulmonary fibrosis and interstitial pneumonia associated with collagen vascular disease. Interestingly, galectin-3 levels appeared to be lower in the bronchoalveolar lavage fluid in patients with idiopathic pulmonary fibrosis and interstitial pneumonia associated with collagen vascular disease who were receiving corticosteroid therapy. Furthermore, alveolar macrophages from idiopathic pulmonary fibrosis patients expressed more galectin-3 compared to those from control.

GALECTIN EXPRESSION IN ANIMAL MODELS OF TISSUE FIBROSIS

To determine the distribution and regulation of galectin-3 during pulmonary injury in an animal model, Casper and Hughes (10) utilized a rat model of irradiation-induced lung inflammation and repair. Immunocytochemistry demonstrated that in normal rat lungs, galectin-3 was localized to alveolar macrophages with weaker staining of bronchial epithelial cells. Shortly after the irradiation-induced lung injury, the total galectin concentration in the lung increased markedly. This upregulation in lung expression of galectin-3 was secondary to an increased population of galectin-3 positive interstitial and alveolar macrophages, and surface expression on the newly formed type I alveolar epithelium and to a lesser extent at the apical surface of type II cells.

Galectin-3 expression has also been examined in the rat kidney following induction of experimental glomerulonephritis by anti-Thy1.1 antibodies (11). Galectin-3 expression, which is sparse in the mature rat kidney and confined to the apical face of some distal tubules, was increased within 1–3 days following antibody administration, with recruitment of glomerular macrophages and pronounced neoexpression in the cytoplasm and at the basal face of distal tubules. In addition, at later time points galectin-3 was detectable in the repopulating mesangial cell mass, preceding the extensive mesangial deposition of laminin and collagen type IV.

Furthermore, we have demonstrated in a murine model of experimental hydronephrosis (unilateral ureteric ligation) a marked upregulation of galectin-3 expression with progressive renal fibrosis (unpublished data). We found that galectin-3 expression, measured by immunohistochemistry and real time PCR, increases progressively following unilateral ureteric ligation (experiments were conducted up to day 14).

We have also examined galectin-3 expression in a murine liver fibrosis model (chronic carbon tetrachloride (CCl_4) induced) (7). After 8 weeks of CCl_4 treatment, galectin-3 expression was increased in the periportal areas and areas of bridging fibrosis in the liver. Dense hepatocyte galectin-3 staining was also present at the periphery of the hepatocyte nodules. Furthermore, when we examined galectin-3 expression in a rat model of reversible CCl_4-induced liver fibrosis, we found that galectin-3 expression was temporally and spatially associated with fibrosis, with expression minimal in normal rat liver, maximal at peak fibrosis, and then virtually absent again at 24 weeks (recovery from fibrosis) (7). This suggests that the development and resolution of fibrosis may be regulated by galectin-3.

FIBROBLASTS AND GALECTINS

Some of the galectins have been shown to be potent mitogens for fibroblasts *in vitro*. Incubation of quiescent cultures of normal human lung fibroblast IMR-90 cells with recombinant galectin-3 stimulated DNA synthesis as well as cell proliferation in a dose-dependent manner (12). This mitogenic activity was dependent on the lectin property of galectin-3 as it was significantly inhibited by lactose. A number of further studies examining primary fibroblasts extracted from different organs have also demonstrated a potent mitogenic effect of different galectins on fibroblast proliferation.

Kristensen et al. (13) performed a proteome analysis on the cellular and secreted proteins of normal (quiescent) and activated rat hepatic stellate cells (liver myofibroblasts). Hepatic stellate cells were activated either *in vitro* (plated on tissue culture plastic) or *in vivo* (by injecting rats with CCl_4 for 8 weeks). Forty-three proteins/polypeptides were identified to have altered their expression levels when the cells were activated *in vivo* and/or *in vitro*. Interestingly, 27 of them showed similar changes *in vivo* and *in vitro* and amongst the proteins upregulated was galectin-1. Upregulation of galectin-1 was directly confirmed in fibrotic liver tissue and Northern blots confirmed upregulation of mRNA for galectin-1 in activated hepatic stellate cells, indicating that the changes were controlled at the mRNA level. The same group then demonstrated that galectin-1 and galectin-3 are capable of stimulating rat hepatic stellate cell proliferation through different intercellular signaling pathways (14).

Furthermore, galectin-1 but not galectin-3 promoted the migration of rat hepatic stellate cells *in vitro*. In addition, we have shown that recombinant galectin-3 drives the proliferation of mouse hepatic stellate cells and that the inhibition of galectin-3 expression by siRNA in murine hepatic stellate cells is able to block myofibroblast activation (7).

Rat kidney mesangial cells in culture do not produce appreciable amounts of galectin-3 but are able to bind and endocytose exogenously added galectin (11). Addition of galectin-3 to primary cultures of rat mesangial cells increased synthesis of collagen type IV and also acted in synergy with a quantitatively similar stimulatory effect of transforming growth factor beta (TGF-β) on matrix synthesis. Furthermore, exogenous galectin-3 prolonged the survival of mesangial cells in serum-free cultures and also protected these cells against the cytotoxic effects of TGF-β.

Using microarray profiling, galectin-3 has been shown to be a robustly over-expressed gene in failing versus functionally compensated hearts from homozygous transgenic TGRmRen2-27 (Ren-2) rats (15). Galectin-3-binding sites were demonstrated in rat cardiac fibroblasts and the extracellular matrix. Recombinant galectin-3-induced cardiac fibroblast proliferation, collagen production, and cyclin D1 expression (15). Furthermore, in a recent study focusing on fibrotic processes in the colon, the examination of primary human colonic lamina propria fibroblasts isolated from human colonic tissue demonstrated that soluble galectin-3 is a strong colonic lamina propria fibroblast activating factor (16).

Pancreatic stellate cells subserve a similar function to hepatic stellate cells in liver fibrosis, that is, they are a major source of extracellular matrix secretion in the pancreas. Fitzner et al. (17) demonstrated that activation of rat pancreatic stellate cells *in vitro* was associated with increased expression of galectin-1, and pancreatic stellate cells exposed to exogenous galectin-1 proliferated at a higher rate and synthesized more collagen than controls (an effect inhibitable by lactose).

Therefore, it is clear that in a number of different organ systems (both human and animal) galectins play a prominent role in the growth promotion and activation of tissue fibroblasts. This suggests that galectins may be an important basic mechanism driving fibrosis in different tissue types and a number of *in vivo* studies (detailed below) have begun to address this concept.

IN VIVO MODELS EXAMINING THE ROLE OF GALECTINS IN ORGAN FIBROSIS

Previous work examining the role of galectin-3 in animal models of tissue fibrosis suggests that galectin-3 can be both protective and profibrotic, depending on both the mechanism of injury and the type of organ fibrosis modeled. For example, in the murine lung, when chronic inflammation was induced by intranasal administration of OVA (ovalbumin) in mice over a period of 12 weeks, treatment with a plasmid encoding galectin-3 (instilled intranasally) led to an improvement in the eosinophil count and the normalization of hyper-responsiveness to methacholine. Furthermore, galectin-3 gene therapy resulted in an improvement in mucus secretion and subepithelial fibrosis in the chronically asthmatic mice, with a reduction in lung collagen (18).

Studies in mouse kidney have shown a protective role for galectin-3 in some models of renal injury. Pugliese et al. (19) rendered galectin-3 knockout and wild-type mice diabetic with streptozotocin. Despite a comparable degree of metabolic derangement, galectin-3 knockout mice developed accelerated glomerulopathy compared to wild-type animals, as evidenced by a more pronounced increase in proteinuria, extracellular matrix gene expression, and mesangial expansion. This was also associated with a more marked renal/glomerular AGE (advanced glycation end product) accumulation, suggesting it was attributable to the lack of galectin-3 AGE receptor function.

Furthermore, a follow-up study demonstrated that galectin-3 ablation is associated with enhanced susceptibility to AGE-induced glomerular disease, with galectin-3 knockout mice demonstrating higher circulating and renal AGE levels and more marked renal functional and structural changes than wild-type mice with significantly higher proteinuria, albuminuria, glomerular and mesangial area, and glomerular sclerosis index (20). Although the above two models [19, 20] of kidney injury do not give rise to major renal fibrosis, they do demonstrate a protective role for galectin-3 in these settings.

However, there is also a strong evidence that galectin-3 plays a profibrotic role in tissue fibrosis. We have previously shown that the disruption of the galectin-3 gene blocks hepatic stellate cell activation (Fig. 8.1, α-smooth muscle actin (α-SMA) was used as a marker of myofibroblast activation) and collagen expression (Fig. 8.2) in the liver, markedly attenuating CCl_4-induced hepatic fibrosis (7). The reduction in liver fibrosis observed in the galectin-3-null mouse occurred despite equivalent liver injury and inflammation, and similar tissue expression of TGF-β. TGF-β failed to

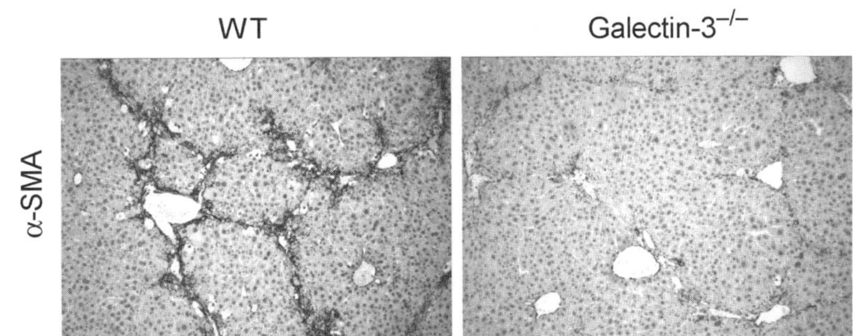

FIGURE 8.1 Liver myofibroblast activation is defective in galectin-3$^{-/-}$ mice. α-SMA (α-smooth muscle actin) staining (marker of myofibroblast activation) of liver tissue after chronic CCl_4 treatment (8 weeks) of WT (wild type) and galectin-3-null mice. This figure is reproduced with permission from Reference (7) © 2006 National Academy of Sciences, USA.

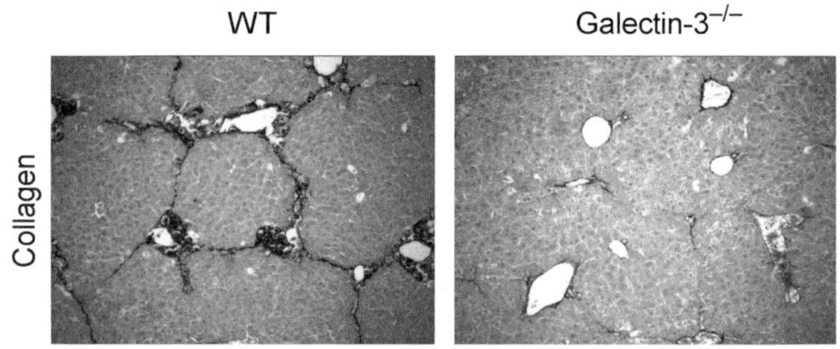

FIGURE 8.2 Liver fibrosis is reduced in galectin-$3^{-/-}$ mice. Collagen staining with picrosirius red of liver tissue after chronic CCl_4 treatment (8 weeks) of WT (wild type) and galectin-3-null mice. This figure is reproduced with permission from Reference (7) ©2006 National Academy of Sciences, USA.

transactivate galectin-3-null hepatic stellate cells, in contrast with wild-type hepatic stellate cells. However, TGF-β stimulated Smad-2 and -3 activation was equivalent. These data suggest that galectin-3 is required for TGF-β mediated myofibroblast activation and matrix production.

Galectin-3 siRNA treatment *in vitro* and *in vivo* also inhibited the hepatic stellate cell activation. Furthermore, preliminary data generated in our laboratory have suggested that galectin-3 may also be important in myofibroblast activation in the kidney. Utilizing a murine model of experimental hydronephrosis (unilateral ureteric obstruction—UUO), we found that myofibroblast activation is markedly reduced in the galectin-3-null mouse compared to wild type following UUO (unpublished data).

As detailed above, Sharma et al. (15) examined galectin-3 expression in a rat model of heart failure. Interestingly, they demonstrated that a 4-week continuous infusion of low dose galectin-3 into the pericardial sac of healthy Sprague–Dawley rats led to left ventricular dysfunction, with a threefold differential increase of collagen I over collagen III. This was in addition to their *in vitro* work, which showed that galectin-3-induced cardiac fibroblast proliferation, collagen production, and cyclin D1 expression.

A recent parasitic infection study investigated the acute and chronic phases of *Schistosomiasis mansoni* infection in wild-type and galectin-3 knockout mice (21). In the absence of galectin-3, chronic-phase granulomas were smaller in diameter, displaying thinner collagen fibers with a loose orientation. The authors suggested a role for galectin-3 in the organization, collagen distribution, and mobilization of inflammatory cells to chronic phase granulomas, as well as interfering with monocyte to macrophage and B cell to plasma cell differentiation.

SUMMARY

The number of studies reporting an important role for galectins in the pathogenesis of tissue fibrosis is steadily increasing. Both profibrotic and protective roles for galectins

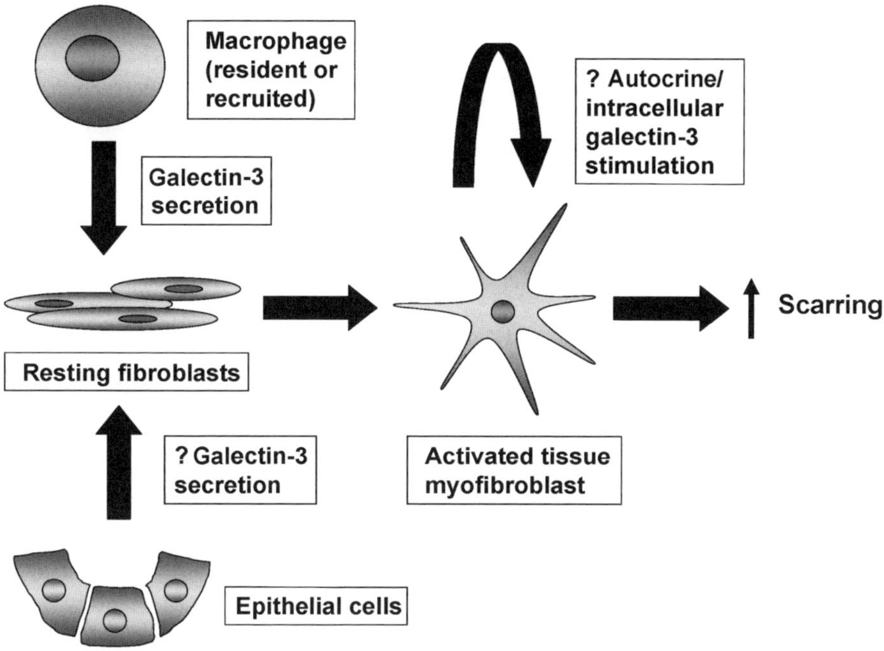

FIGURE 8.3 Potential profibrotic roles of galectin-3 in tissue scarring.

have been reported in models of chronic inflammation, and the effects of galectins do appear to be organ and context dependent. It is interesting to speculate how, for example, galectin-3 may be acting within a tissue when it mediates profibrotic effects (Fig. 8.3). Secondary to their ubiquity, there are various potential cellular sources of galectin within tissues during chronic inflammation, and further studies will be required to fully elucidate how galectins are mechanistically involved in organ fibrosis. Finally, strategies to manipulate galectins during tissue scarring will hopefully lead to the development of novel antifibrotic therapies.

REFERENCES

1. Neuberger J. Liver transplantation. J Hepatol 2000;32:198–207.
2. Simpson KJ, Garden OJ. The indications and implications of liver transplantation. Proc R Coll Phys Edinb 1999;29:144–152.
3. El Nahas AM, Bello AK. Chronic kidney disease: the global challenge. Lancet 2005;365:331–340.
4. Friedman SL. Molecular regulation of hepatic fibrosis, an integrated cellular response to tissue injury. J Biol Chem 2000;275:2247–2250.
5. Duffield JS, Forbes SJ, Constandinou CM, Clay S, Partolina M, Vuthoori S, Wu S, Lang R, Iredale JP. Selective depletion of macrophages reveals distinct, opposing roles during liver injury and repair. J Clin Invest 2005;115:56–65.

6. Hsu DK, Dowling CA, Jeng KC, Chen JT, Yang RY, Liu FT. Galectin-3 expression is induced in cirrhotic liver and hepatocellular carcinoma. Int J Cancer 1999;81: 519–526.
7. Henderson NC, Mackinnon AC, Farnworth SL, Poirier F, Russo FP, Iredale JP, Haslett C, Simpson KJ, Sethi T. Galectin-3 regulates myofibroblast activation and hepatic fibrosis. Proc Natl Acad Sci USA 2006;103:5060–5065.
8. Wang L, Friess H, Zhu Z, Frigeri L, Zimmermann A, Korc M, Berberat PO, Buchler MW. Galectin-1 and galectin-3 in chronic pancreatitis. Lab Invest 2000; 80:1233–1241.
9. Nishi Y, Sano H, Kawashima T, Okada T, Kuroda T, Kikkawa K, Kawashima S, Tanabe M, Goto T, Matsuzawa Y, Matsumura R, Tomioka H, Liu FT, Shirai K. Role of galectin-3 in human pulmonary fibrosis. Allergol Int 2007;56(1):57–65.
10. Kasper M, Hughes RC. Immunocytochemical evidence for a modulation of galectin 3 (Mac-2), a carbohydrate binding protein, in pulmonary fibrosis. J Pathol 1996; 179:309–316.
11. Sasaki S, Bao Q, Hughes RC. Galectin-3 modulates rat mesangial cell proliferation and matrix synthesis during experimental glomerulonephritis induced by anti-Thy1.1 antibodies. J Pathol 1999;187:481–489.
12. Inohara H, Akahani S, Raz A. Galectin-3 stimulates cell proliferation. Exp Cell Res 1998;245:294–302.
13. Kristensen DB, Kawada N, Imamura K, Miyamoto Y, Tateno C, Seki S, Kuroki T, Yoshizato K. Proteome analysis of rat hepatic stellate cells. Hepatology 2000;32(2): 268–277.
14. Maeda N, Kawada N, Seki S, Arakawa T, Ikeda K, Iwao H, Okuyama H, Hirabayashi J, Kasai K, Yoshizato K. Stimulation of proliferation of rat hepatic stellate cells by galectin-1 and galectin-3 through different intracellular signaling pathways. J Biol Chem 2003; 278:18938–18944.
15. Sharma UC, Pokharel S, van Brakel TJ, van Berlo JH, Cleutjens JP, Schroen B, Andre S, Crijns HJ, Gabius HJ, Maessen J, Pinto YM. Galectin-3 marks activated macrophages in failure-prone hypertrophied hearts and contributes to cardiac dysfunction. Circulation 2004;110:3121–3128.
16. Lippert E, Falk W, Bataille F, Kaehne T, Naumann M, Goeke M, Herfarth H, Schoelmerich J, Rogler G. Soluble galectin-3 is a strong, colonic epithelial-cell-derived, lamina propria fibroblast-stimulating factor. Gut 2007;56:43–51.
17. Fitzner B, Walzel H, Sparmann G, Emmrich J, Liebe S, Jaster R. Galectin-1 is an inductor of pancreatic stellate cell activation. Cell Signal 2005;17(10): 1240–1247.
18. Lopez E, del Pozo V, Miguel T, Sastre B, Seoane C, Civantos E, Llanes E, Baeza ML, Palomino P, Cardaba B, Gallardo S, Manzarbeitia F, Zubeldia JM, Lahoz C. Inhibition of chronic airway inflammation and remodeling by galectin-3 gene therapy in a murine model. J Immunol 2006;176(3):1943–1950.
19. Pugliese G, Pricci F, Iacobini C, Leto G, Amadio L, Barsotti P, Frigeri L, Hsu DK, Vlassara H, Liu FT, Di Mario U. Accelerated diabetic glomerulopathy in galectin-3/AGE receptor 3 knockout mice. FASEB J 2001;15(13):2471–2479.
20. Iacobini C, Menini S, Oddi G, Ricci C, Amadio L, Pricci F, Olivieri A, Sorcini M, Di Mario U, Pesce C, Pugliese G. Galectin-3/AGE-receptor 3 knockout mice show accelerated

AGE-induced glomerular injury: evidence for a protective role of galectin-3 as an AGE receptor. FASEB J 2004;18(14):1773–1775.

21. Oliveira FL, Frazao P, Chammas R, Hsu DK, Liu FT, Borojevic R, Takiya CM, El-Cheikh MC. Kinetics of mobilization and differentiation of lymphohematopoietic cells during experimental murine schistosomiasis in galectin-3/mice. J Leukoc Biol 2007;82(2): 300–310.

9

GALECTIN-1, CANCER CELL MIGRATION, ANGIOGENESIS, AND CHEMORESISTANCE

Florence Lefranc,[1,2] Marie Le Mercier,[1] Véronique Mathieu,[1] and Robert Kiss[1]

[1]*Laboratory of Toxicology, Institute of Pharmacy, Université Libre de Bruxelles, Brussels, Belgium*

[2]*Laboratory of Toxicology, Institute of Pharmacy, Department of Neurosurgery, Erasme University Hospital, Université Libre de Bruxelles, Brussels, Belgium*

INTRODUCTION

Although galectin-1 and -3 are the most extensively studied members of this family of proteins in the field of cancer biology (1, 2), and the focus of this chapter is galectin-1, this certainly does not mean that other galectins do not play a major role in the aggressive biological behavior of various types of human cancer (1). Indeed, readers interested by the involvement of specific galectins in cancer biology can refer to References (3–5) for galectin-3, Reference (6) for galectin-4, References 7 and 8 for galectin-7, References 9 and 10 for galectin-8, Reference 11 for galectin-9, and Reference 12 for galectin-12.

Of the above, galectin-3 is the sole member of the family to be found in solution as a monomer with two functional domains. Its carbohydrate-binding properties constitute the basis of cell–cell and cell–matrix interactions as well as of cancer progression and metastasis (13, 14). Galectin-3 plays an important role in cancer biology because it exerts antiapoptotic effects in that it suppresses cytotoxic

Galectins, Edited by Anatole A. Klyosov, Zbigniew J. Witczak, and David Platt

drug-induced apoptosis and anoikis (apoptosis induced by the loss of cell anchorage) (15) that contribute to cancer cell survival (4, 16–18). Conversely, it has also been shown that the galectin-3 secreted by tumor cells induces apoptosis in cancer-infiltrating T cells and so plays a role in the tumor immune escape mechanism (13).

Galectin-1 itself is expressed differentially by various normal and pathological tissues and appears to be functionally multivalent, undertaking a wide range of biological activities (19). Evidence points to a major role played by galectin-1 in skeletal muscle differentiation and regeneration (20, 21), sensory and motoneuron biology (22), nerve regeneration (23), and neurodegenerative diseases (24), none of which are related to cancer. However, galectin-1 also plays a key role in cancer cell biology in that it interacts with major signaling pathways involved in cancer (19), such as p21 (25), p27 (26), ganglioside GM1 receptor related pathways (26), Ras (27), Raf (27), and phosphatidyl inositol 3 kinase (PI3-K) (27).

This chapter will thus focus on the underlying biology of galectin-1 in cancer and elaborate on the development of antigalectin-1 strategies to combat this complex disease.

CANCER CELL MIGRATION

Cell migration in general, and cancer cell migration in particular, is a complex dynamic process that involves multiple biological features (27, 28). Indeed, cell migration involves at least three independent but highly coordinated biological processes: (i) cell adhesion to numerous components of the extracellular matrix (ECM) with resulting modification of ECM molecular composition in the case of cancer cells in general (27, 29–34), (ii) cell motility, which involves the reorganization of the actin cytoskeleton (35, 36) mainly through modification of the components of the integrin – ECM interaction (27, 37, 38), and (iii) invasion that leads to the degradation of matrix proteins by tumor-secreted proteolytic enzymes, mainly serine proteases, cathepsins, and metalloproteinases of the MMP-2, MMP-9, and MMP-14 (MT1-MMP) types (39–41). The involvement of galectin-1 in these three biological processes is elaborated further below.

Adhesion

Cell adhesion involves particular ECM components and specific cell receptors known as adhesion molecules and is mediated through interactions between ECM components, integrins, focal adhesion linked molecules, and the actin cytoskeleton to form the so-called adhesion complex (27). Adhesion molecules themselves can be classified into at least four major groups according to their functional behavior (derived from their amino acid sequence and their receptor ligand(s) interactions) and include the integrins, cadherins, lectins (mainly selectins and galectins), and the immunoglobulin superfamily. While galectins bind to *N*-acetyllactose (LacNAc) moieties (galactose-β1,4-glucose-*N*-acetyl, i.e., Galβ1,4GluNAc), selectins bind to fucosyl-related LacNAc ligands (Lewis antigens) (27). These LacNAc and fucosyl-LacNAc moieties are exhibited by almost all ECM components and by a large number of cell surface proteins.

Thus, whereas integrins employ protein–protein interactions with ECM components, galectins reveal protein–carbohydrate interactions between themselves and ECM glycoproteins, and selectins preferentially demonstrate protein–carbohydrate interactions between themselves and cell surface receptors exhibited by other cell types.

Cancer cell adhesion galectin-1. This line produced 43 references by the search engine PubMed. This is the subject of the present section in which we focus on the part played by galectins and, more specifically, galectin-1 in cancer cell–ECM and cell–cell adhesions. Galectin-1 shows both proadhesive and antiadhesive functions, as it can potentiate or inhibit cell–ECM and cell–cell interactions.

Galectin-1 has been shown to modify (increase or decrease) the adhesion of various normal and cancer cells to the ECM via the cross-linking of glycoproteins (integrins) exposed on the cell surfaces with carbohydrate moieties of ECM components such as laminin and fibronectin (42–46). In addition, galectin-1 can also mediate homotypical cell interactions, so favoring the aggregation of human melanoma cells (47) and heterotypical cell interactions such as those between cancer and endothelial cells, which in turn favor the dispersion of tumor cells (48–50). Table 9.1 shows the various binding glycoprotein partners of galectin-1 in the ECM and in the cellular membrane for the mediation of cell–ECM or cell–cell adhesions in the context of cancer cells.

Galectin-1 in the Extracellular Matrix Galectin-1 binds to a number of ECM components in a dose-dependent and β-galactoside-dependent manner in the following order: laminin > cellular fibronectin > thrombospondin > plasma fibronectin > vitronectin > osteopontin (44, 51). Laminin and cellular fibronectin are glycoproteins that are highly N-glycosylated with bi- and tetraantennary poly-*N*-lactosamines (52, 53). Galectin-1 is also involved in ECM assembly and remodeling as it inhibits the incorporation of vitronectin and chondroitin sulfate B into the ECM of vascular smooth muscle cells (VSMCs) (51). The interaction of galectin-1 with vitronectin seems to depend on the vitronectin conformation, since it preferentially recognizes unfolded vitronectin multimers rather than inactive folded monomers (51).

Positive regulation of cancer cell adhesion to laminin and fibronectin substrates by galectin-1 has been documented for human ovarian carcinoma cells (46, 54, 55) and human melanoma (56). By contrast, galectin-1 antiadhesive functions have also been evidenced for various murine and human tumor cells (42). Additionally, an increase in cell surface galectin-1 induced by experimental manipulations can lead to reduced binding of various tumor cells (42).

In conclusion, galectin-1 can certainly promote cell adhesion to laminin and fibronectin by means of binding bridges between ECM components and cell membrane receptors such as the integrins. However, the galectins secreted by tumor cells can also disrupt adhesion by the interaction between ECM components and integrins (57, 58).

Galectin-1 and Its Cell Surface Binding Partners *Galectin and integrin; galectin-1 and integrin; galectin-1 integrin and adhesion.* These lines produced 92, 21, and 12 references, respectively, by the search engine PubMed.

TABLE 9.1 The Binding Partners of Galectin-1 in the Context of Cancer Progression

Binding Partners	Cell/Tissue Types	Biological Functions	References
Laminin, fibronectin	Various human and murine tumor cells	↓ Adhesion	(42)
	Human ovarian carcinoma cells	↑ Adhesion	(46, 54, 55)
	Human melanoma cells	↑ Adhesion	(56)
Osteopontin	VSMC	↑ Adhesion	(99)
Thrombospondin	VSMC	↑ Adhesion	(99)
Vitronectin	VSMC	↓ ECM assembly	(51)
Glycosaminoglycan, (chondroitin sulfate B, heparan sulfate	VSMC	Modulation of ECM assembly, ↓ adhesion	(45)
β1 integrin	VSMC	Adhesion, FAK activation, migration	(45)
α5β1	Epithelial carcinoma cells	Inhibition Ras-MEK-ERK pathway, increase of p21 and p27, growth inhibition	(24)
LAMP-1 and LAMP-2	Human ovarian and colon carcinomas	↑ Adhesion	(55, 61)
Glycoprotein 90K (MAC-2BP)	Human melanoma cells	↑ Homotypical cell aggregation	(47)
GM1	Human neuroblastoma cells	Negative regulation of cell growth	(99)
CEA	Human colon carcinoma cells	Unknown	(61)
CA125	Human ovarian carcinoma cells	Galectin-1 export?	(62)
CD7, CD43, CD45	T cells	Induction of T-cell apoptosis	(63, 116)

VSMC: vascular smooth muscle cells.

The integrins are the major cell surface binding partners of galectin-1 facilitating cell–ECM or cell–cell adhesion. The activity of integrin adhesion receptors is essential for the normal cellular function and survival (15, 59). N-glycosylations of β-integrins regulate β1 integrin function by modulating their heterodimerization with β chains and ligand-binding activity (60).

Numerous variants of integrin glycoforms have been described in many normal and pathological cell types. Moiseeva et al. have shown that galectin-1 interacts with the β1 subunit of integrin. As a result of its direct binding to β1 integrins (without cross-linking), dimeric galectin-1 increases the amounts of partly activated β1 integrins, but does not induce dimerization with β subunits (45). In the case of vascular smooth muscle cells, this interaction of galectin-1 with $\alpha 1\beta 1$ integrin has been reported as both transiently phosphorylating the focal adhesion kinase (FAK) and modulating the attachment of cells and their spreading and migration on laminin, but not on cellular fibronectin (44, 45, 51). Thus galectin-1 is likely to affect smooth muscle cell adhesion by interacting with β1 integrin on the surface of such cells and inducing outside-in signaling (45) (Fig. 9.1).

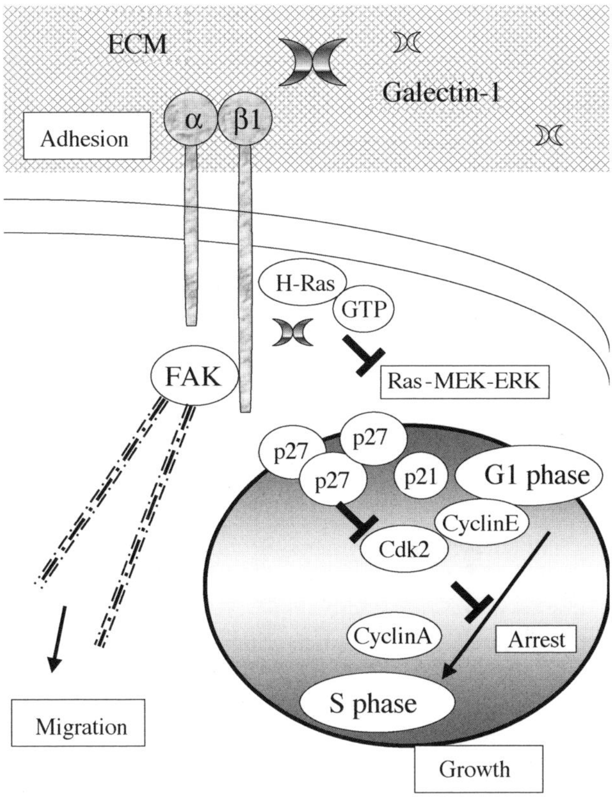

FIGURE 9.1 Galectin-1-integrin partnership and its outside-in signaling. Galectin-1 is present both outside and inside cells, and has both extracellular and intracellular functions. The extracellular functions require the carbohydrate-binding properties of dimeric galectin-1, whereas intracellular activities are associated with carbohydrate-independent interactions between galectin-1 and other proteins. In the case of vascular smooth muscle cells, this interaction of galectin-1 with the $\alpha 1\beta 1$ integrin has been reported both transiently phosphorylating FAK and modulating the attachment of cells and their spreading and migration on laminin, a process involved in angiogenesis. In several carcinoma cell lines, functional interaction of galectin-1 with the fibronectin receptor $\alpha 5\beta 1$ inhibits the Ras-MEK-ERK signaling pathway, resulting in reduced threonine phosphorylation of Sp1, increased Sp1 transactivation and DNA binding, and consecutive induction of p27 gene transcription. Accumulation of p27 inhibits Cdk2 activity and ultimately results in G1 cell cycle arrest and growth inhibition.

Galectin-1 also interacts with α5β1 integrin to restrict epithelial tumor cell growth (24) (Fig. 9.1), a process elaborated upon in the section entitled "Galectin-1 Involvement in Tumor Progression."

In addition to integrins, various membrane glycoproteins have also been identified as binding galectin-1 and thus facilitating cell–ECM or cell–cell adhesion. These include cell membrane receptors such as lysosome-associated membrane proteins 1 and 2 (LAMP-1 and LAMP-2) (61), glycoprotein 90 K (MAC-2BP) (47), cancer antigen CA125 (62), carcinoembryonic antigen (CEA) (61), and GM1 gangliosides CD43, CD45, and CD7 (57, 63) (Table 9.1).

Motility

Cells may peregrinate from one site to another by forming and breaking adherence molecule attachments to the matrix. After attaching themselves, they produce footlike or fingerlike extensions and form new attachments at their front ends. They then release the attachment at the rear. The biological processes regulating cell motility make use of the cell cytoskeleton, a process that includes the dynamic remodeling of the actin cytoskeleton as well as the microtubule network (27, 35, 36, 38).

The extension of both the lamellipodia and the filopodia in response to migratory stimuli is almost universally found in conjunction with local actin polymerization and is under the control of small GTPases, Rac for the lamellipodia, cdc42 for the filopodia, and Rho for the stress fibers. Along with a bias toward membrane extensions at the cell front, attachments preferably tend to form at the leading edge of lamellipodia and filopodia. Actin polymerization and structural organization provide the protrusive force needed to extend the lamellipodia and the filopodia independently of myosin motor activity.

In contrast, the contractile force needed to move a cell forward appears to depend on the active myosin-based motors. Rapid migration requires efficient mechanisms to release adhesions at the rear. While the actin cytoskeleton provides the driving force at the front, the microtubule network assumes a regulatory function in coordinating rear retraction (27, 35, 36, 38).

Galectin motility and cancer; galectin-1 motility and cancer. These lines produced 37 and 13 references, respectively, by the search engine PubMed. Five out of the last 13 papers come from our group. We have investigated the cell migration properties of galectin-1 in human malignant gliomas, human colon cancer, and mouse melanoma cells.

It has been shown that the levels of galectin-1 and galectin-3 expression significantly change during the progression of malignancy in human astrocytic tumors, while that of galectin-8 remained unchanged (64).

Additionally, high grade astrocytic tumors with elevated levels of galectin-1 expression were associated with dismal prognoses as compared to such tumors expressing low levels of galectin-1 (65). Furthermore, galectin-3, galectin-1, and to a lesser extent galectin-8 markedly stimulated glioblastoma (GBM) cell migration *in vitro* (64, 65). In complementary studies, it was shown that the *in vitro* addition of purified galectin-1 to U87 human GBM cells enhanced tumor cell motility in a lactose-inhibiting manner (66). This effect appeared to be related to an increase in polymerized filamentous actin and the expression of small GTPase RhoA (65).

Conversely, knocking down galectin-1 expression in U87 and U373 human GBM cell lines by stable transfection with antisense galectin-1 mRNA impaired motility and increased the survival times of nude mice intracranially grafted with transfected cells (66). Additionally, comparative cDNA microarray analysis of gene expression of stable transfected (with antisense galectin-1 vector), mock-transfected, and wild-type GBM cells revealed that of 631 genes potentially involved in cancer, the expression of 86 genes was increased at least twofold.

Table 9.2 lists some of the overexpressed genes (involved in controlling cell migration), which were found in U87 GBM cells with reduced galectin-1 expression. The expression of $\alpha 9$ integrin was increased in these cells. Major differences in the expression patterns of ADAM-15 and actin stress fiber organization were also observed (66). The ADAM family of membrane anchorage glycoproteins encompasses a catalytically active metalloproteinase domain and a disintegrin domain (67), and may thus be involved both in the proteolytic cleavage of cell surface proteins and in the integrin-mediated cell adhesion (including $\alpha 9\beta 1$/ADAM-15 interactions) via RGD-dependent and -independent binding (68).

The stable knockdown of galectin-1 certainly alters the expression of a number of genes that are either directly or indirectly involved in actin polymerization in GBM cells (66). These results agree with cell invasiveness studies using a proteomic approach for the comparison of highly and poorly invasive mammary carcinoma cells. In these studies, galectin-1 membrane expression was identified as a signature of cell invasiveness (69). In contrast, however, the galectin-1

TABLE 9.2 Some Overexpressed Genes Involved in the Control of Cell Migration in Human GBM U87 Cells with Knocked-Down Galectin-1 Expression (from Reference 66)

Gene Name	Ratio (Expression in Knocked-Down as Compared to Mock-Transfected Cells)
ECM	
Thrombospondin 1	2.64
ECM-1	2.64
Adhesion molecules	
ADAM-15	3.30
Integrin $\alpha 7$	3.57
Integrin $\alpha 9$	2.41
Transmembranar glycoprotein	6.06
Cytoskeleton	
Actin-capping protein	2.36
MAP-2	2.77
Tubulin $\beta 5$	3.15

depletion in B16F10 melanoma cells *in vitro* was not associated with a decrease in cell motility (70).

We have further shown that galectin-1 and galectin-3 are expressed in variable amounts in the epithelial cells and the connective tissue of normal colon from clinical specimens (71). Their expression was found to significantly increase with the degree of dysplasia, suggesting that galectin-1 and galectin-3 are related to malignant progression, whereas galectin-8 has been associated with suppressor activity. A number of galectins, including galectin-1 (71, 72), -3 (71–74), -4 (75), -7 (76), and -8 (77), have been shown to seriously modulate the biological aggressiveness of colon cancers.

Using quantitative computer-assisted phase contrast videomicroscopy on living human colon cancer cells, the direct involvement of galectin-3 (but not of galectin-1) was demonstrated in human colon cancer cell migration. On investigating the migratory effects of galectin-1 on human colorectal cancer cell lines exhibiting galectin-1-binding sites on their surface (HCT-25, LoVo, and CoLo201), the level of migration for the three cell lines was significantly reduced by 0.15 μg/cm^2 of galectin-1, an effect that was partially neutralized by antigalectin-1 antibodies. Galectin-3 also significantly reduced cell migration in all the cell lines (71).

Galectin-1 is also involved in the migratory ability of noncancer cells, that is, dendritic cells (78), liver cells (79), and Schwann cells (80) favoring the process of axonal regeneration. It also promotes the migratory capacity of endothelial cells and therefore the process of angiogenesis, a concept further developed in the context of cancer in the section entitled "Galectin-1 and Angiogenesis." In contrast, it decreases the migratory capacity of (i) leukocytes and thus displays anti-inflammatory properties (81) and (ii) perversely of immune T cells thus favoring immune privilege (82).

Invasion

Invasion involves the degradation of the extracellular matrix components by a number of tumor-secreted proteolytic enzymes (serine proteases, cathepsins, and metalloproteinases such as MMP-2, MMP-9, and MMP-14) (40, 83). This process then creates an intercellular space into which invading cells can migrate by means of an active mechanism that requires receptor turnover (see the Section "Adhesion") and the reorganization of cytoskeleton elements (see the Section "Motility").

Galectin-1, galectin-3, and galectin-8 seem to be involved in tumor astrocyte invasion of the brain parenchyma as their levels of expression are higher in the invasive parts of xenografted glioblastomas than in their less invasive parts (64). The immunohistochemical analysis of galectin-1 expression in human U87 and U373 GBM xenografts from the brains of nude mice revealed a higher level of galectin-1 expression in invasive areas compared to noninvasive areas of the xenografts (65, 84). Figure 9.2 illustrates the involvement of galectin-1 in GBM cell migration. Using a proteomic approach based on the comparison of highly and poorly invasive mammary carcinoma cell lines, Harvey et al. identified the membrane expression of galectin-1 as a signature of cell invasiveness (69).

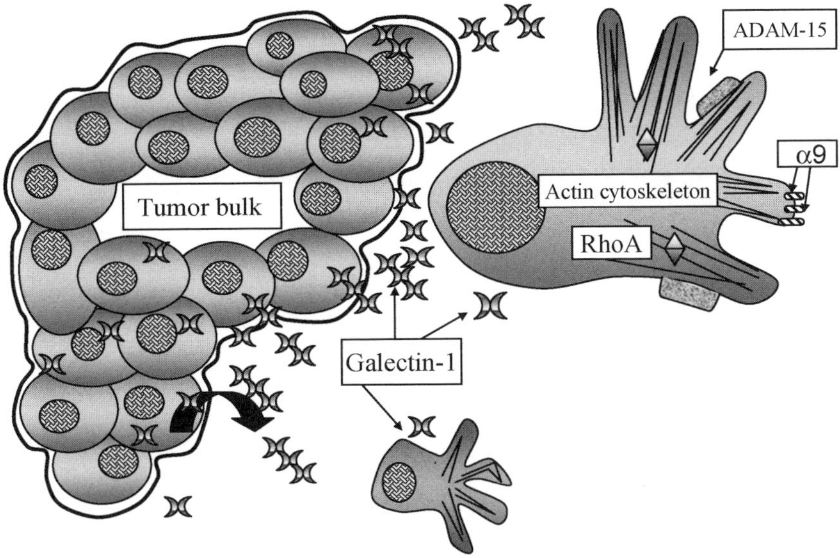

FIGURE 9.2 Involvement of galectin-1 in glioblastoma cell migration. Galectin-1 expression in xenografts of a human GBM reveals a higher level of galectin-1 expression in invasive areas than noninvasive ones (tumor bulk). Galectin-1-induced increase in GBM cell motility is associated with the increased expression of the small GTPase RhoA and modification in the polymerization of the actin cytoskeleton. The levels of expression of the adhesion proteins ADAM15 and $\alpha 9$ integrin are also modified.

THE PLEIOTROPIC BIOLOGICAL ROLES OF GALECTIN-1

Galectin-1 plays an important role in cancer, contributing to neoplastic transformation and tumor progression through tumor cell survival, angiogenesis, and tumor metastasis (1, 13, 58) as indicated in Fig. 9.3.

Furthermore, galectin-1 is also involved in chemoresistance (70, 85). The role of galectin-1 in angiogenesis is demonstrated in Fig. 9.4 and its involvement in chemoresistance is elaborated on in the previous section.

Galectin-1 can also modulate immune and inflammatory responses and may therefore play a key role in helping tumors to escape immune surveillance (1, 13, 81).

Galectin-1 Involvement in Tumor Transformation

It has been recently demonstrated that intracellular galectin-1 may play a role in the initiation of transformed tumor phenotypes. Paz et al. have found that galectin-1 interacts with oncogenic H-Ras and contributes to membrane anchorage of H-Ras (86) and cell transformation (Fig. 9.3).

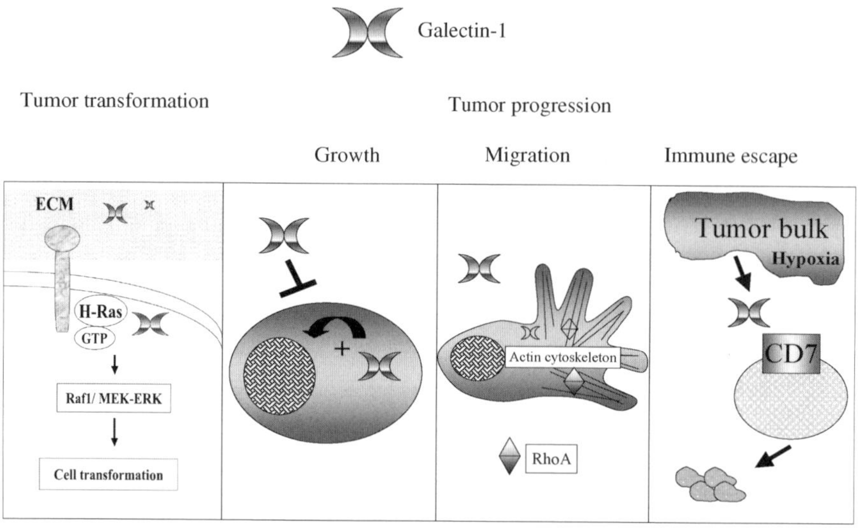

FIGURE 9.3 Contribution of galectin-1 to tumor transformation and tumor progression. Galectin-1 interacts with oncogenic H-Ras and mediates tumor transformation. Galectin-1 is recruited from the cytosol into the cell membrane by H-Ras-GTP in a lactose-independent manner with the resulting stabilization of H-Ras-GTP, the clustering of the H-Ras-GTP and galectin-1 in nonraft microdomains, the subsequent binding to Raf-1 and the activation of the ERK signaling pathway. Galectin-1 also controls tumor growth in a pluripotent fashion. Galectin-1 modulates tumor cell adhesion, motility, and invasiveness, thereby affecting the process of tumor metastases. Furthermore, galectin-1 secreted by tumor cells induces activated T-cell apoptosis, helping the tumor to escape the immune system.

Interaction Between Galectin-1 and Ras Ras genes that are frequently mutated in human tumors promote malignant transformation (86). Ras transformation requires membrane anchorage, which is promoted by Ras farnesylcysteine carboxymethyl ester and a second signal. The overexpression of galectin-1 increases membrane-associated Ras, Ras-GTP, and active extracellular signal-regulated kinase (ERK) and results in cell transformations that are blocked by dominant negative Ras (86). Galectin-1 antisense RNA inhibits transformations by H-Ras and abolishes the membrane anchorage of green fluorescent protein (GFP)-H-Ras but not of GFP-H-Ras wild-type, GFP-K-Ras, and GFP-N-Ras (86).

Thus, H-Ras–galectin-1 interactions establish an essential link between two proteins associated with cell transformation and human malignancies that can be exploited to selectively target oncogenic Ras proteins. In fact, H-Ras-GTP recruits galectin-1 from the cytosol into the cell membrane with the resulting stabilization of H-Ras-GTP, the clustering of H-Ras-GTP and galectin-1 in nonraft microdomains (87), the subsequent binding to Raf-1 (but not to PI3-K), the activation of the

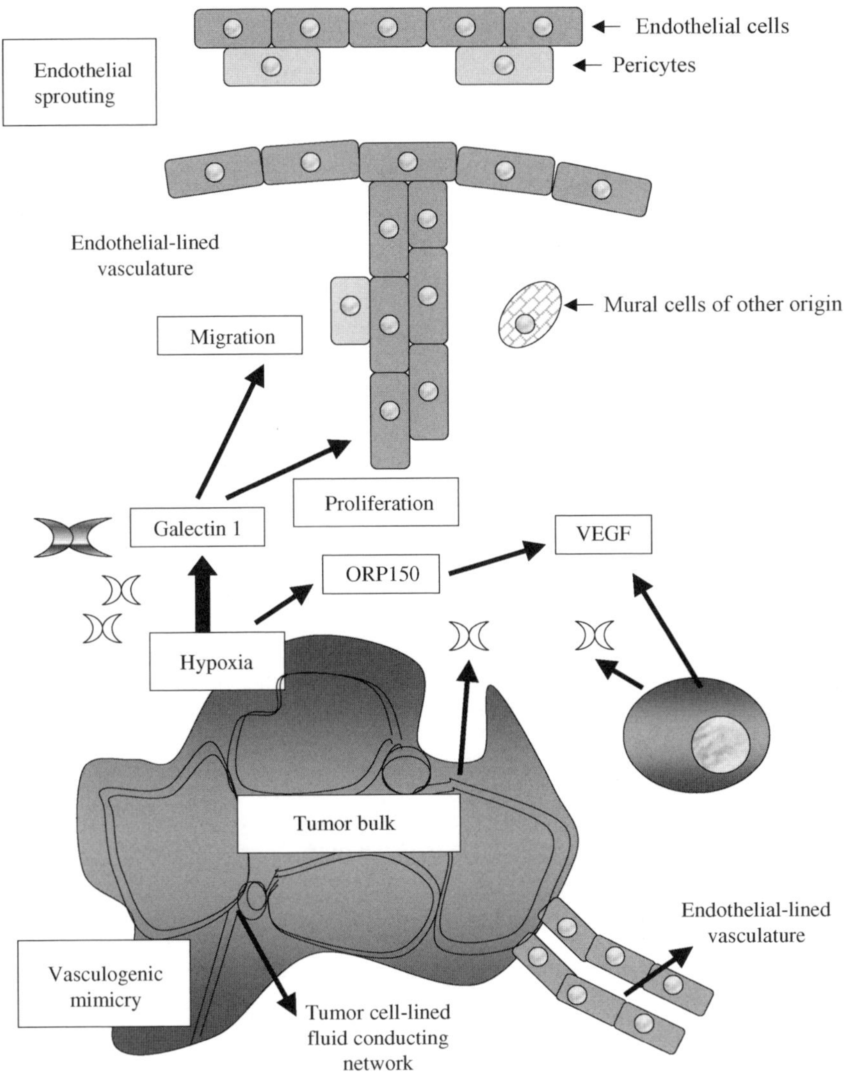

FIGURE 9.4 Scheme of galectin-1 involvement in angiogenesis and vasculogenic mimicry. Tumor vasculature is not necessarily derived from endothelial cell sprouting (upper part of the scheme i.e., angiogenesis) as cancer tissue can acquire vasculature by vasculogenic mimicry (lower part of the figure). Endothelial and vascular smooth muscle cells express galectin-1. Recent evidence indicates that galectin-1 induces angiogenesis in the context of tumor hypoxia. The process is under the control of VEGF and the upstream protein ORP150, also called HYOU1, a hypoxic stress-induced protein.

ERK signaling pathway, and finally increased cell transformation (26) (Fig. 9.3). So, in addition to increasing and prolonging the H-Ras activation, the galectin-1–H-Ras complex renders the activated molecule selective toward Raf-1 but not toward PI3-K (88) (Fig. 9.3).

Rotblat et al. have identified a hydrophobic pocket in galectin-1 analogous to the Cdc42 geranylgeranyl-binding cavity in RhoGDI (89). This pocket possesses homologous isoprenoid-binding residues including the critical L11 (galectin-1), whose RhoGDI L77 homologue changes dramatically on Cdc42 binding. By substituting L11A, Rotblat et al. obtained a dominant interfering galectin-1 that possesses a normal carbohydrate-binding ability, but inhibits the H-Ras GTP loading and extracellular signal-regulated kinase activation, dislodges H-Ras from the cell membrane, and attenuates H-Ras fibroblast transformation (89). Thus, whereas galectin-1 cooperates with Ras independently of carbohydrate binding, galectin-1 (L11A) inhibits it.

Galectin-1 Involvement in Tumor Progression

We have recently reviewed the role of galectin-1 in the different steps of tumor progression to evaluate its potential use as a therapeutic target in cancer (1, 90). Expression of galectin-1 has been well documented in many different tumor types including astrocytoma (64–66, 85, 91), melanoma (56, 70), head and neck (92), prostate (48, 93), thyroid (94, 95), colon (71), bladder, and ovarian carcinomas (46). Table 9.3 illustrates the galectin-1 expression in a variety of human tumors and its correlation with worse prognosis. Interestingly, in most cases such expression correlates with the aggressiveness of these tumors and the acquisition of a metastatic phenotype.

Regulation of Cell Growth Extracellular galectin-1 has no effect on the growth rates of naive T cells (96) or astrocytic (65) or colon (71) tumor cell lines. However, galectin-1 is mitogenic to different normal and pathological rodent and human cells, including murine Thy-1-negative spleen or lymph node cells (63), mammalian vascular cells (97, 98), and rat hepatic stellate cells (79). Exogenously added galectin-1 additionally inhibits the growth of other cell types such as neuroblastoma (99) and stromal bone marrow cells (100).

The knockdown of galectin-1 expression in murine melanomas (70, 101) and human glioma cells (85, 91) does not affect their growth rate *in vitro*. In contrast, Yamaoka et al. showed that inhibition of galectin-1 gene expression in a rat glioma cell line arrests tumor growth (102), suggesting that endogenous galectin-1 has growth-promoting activity in this cell line. Thus, the effects of galectin-1 on cell growth are pluripotent. Interestingly enough, it has been reported that depending on the concentration involved, galectin-1 causes the biphasic modulation of cell growth. While high concentrations ($\sim$1 μM) of recombinant galectin-1 inhibit cell proliferation independently of galectin-1 sugar-binding activity, low levels ($\sim$1 nM) are mitogenic and are susceptible to inhibition by lactose (103, 104). Furthermore, galectin-1 can also regulate cell cycle progression in human mammary tumor cells (105). These seemingly paradoxical positive and negative effects of galectin-1 on cell growth are highly dependent on cell type and cell activation status, and may also be influenced by the relative distribution of monomeric versus dimeric or intracellular versus extracellular forms.

TABLE 9.3 Galectin-1 Expression in Human Tumors

Histological Types	Expression in Tumors as Compared to Normal Tissues	Is Galectin-1 a Diagnostic Marker?	Is Galectin-1 a Prognostic Marker?
Colon carcinomas	↑ In stroma and in epithelial tissues (21, 76, 114)		Yes (76)
Pancreatic ductal adenocarcinomas	↑ (169)	Yes (112) (176)	
Intrahepatic cholangiocarcinomas	↑ in stroma and in epithelial tissues (115)		
Renal cell carcinomas	↑ or ↓ depending on histological grades (170, 171)		
Bladder transitional cell carcinomas	↑ (172)	Yes (172)	
Prostate cancers	↑ In stromal tissues (93)		Yes (93)
Ovarian adenocarcinomas	↑ (46)		
Choriocarcinomas	↑ (173)		
Breast cancers	↑ (174)		Yes
Gliomas	↑ (64–66, 84, 91, 102)		Yes (64, 65)
Nonsmall-cell lung cancers	↑ (175)		
Head and neck squamous cell carcinomas	↑ (82, 92, 113)		Yes (92)
Melanomas	↑ (56, 70)		

Fischer et al. have observed that the antiproliferative potential of galectin-1 in a number of carcinoma cell lines requires functional interaction with the $\alpha 5\beta 1$ integrin (24) (Fig. 9.1). Antiproliferative effects result from the inhibition of the Ras-MEK-ERK pathway and the consecutive transcriptional induction of p27, whose promoter contains two Sp1-binding sites crucial for galectin-1 responsiveness (24) (Fig. 9.1). The inhibition of the Ras-MEK-ERK cascade by galectin-1 increases Sp1 transactivation and DNA binding due to the reduced threonine phosphorylation of Sp1. In addition, galectin-1 induces p21 transcription and selectively increases p27 protein stability, while the galectin-1- mediated accumulation of p27 and p21 inhibits cyclin-dependent kinase 2 activity, a process that ultimately results in G1 cell cycle arrest and growth inhibition (24) (Figs. 9.1 and 9.3).

Regulation of Tumor Cell Metastasis Tumor metastasis is a multistep process that includes changes in cell adhesion, increases in cell motility and invasiveness, with these in turn favoring the detachment of cells from the primary tumor site and their attachment to ECM proteins at distal sites, angiogenesis, and tumor immune escape. We developed further a concept on the role of galectin-1 in angiogenesis.

Regulation of Cell Migration Processes While cell migration is the net result of adhesion, motility, and invasion (27, 28), galectin-1, as indicated above, is known to modify each of these three cell migration-related processes. Galectin-1 is also involved in the migration of nontumor cells, that is, Schwann cells and endothelial cells. Oxidized galectin-1 stimulates the migration of Schwann cells from both the proximal and the distal stumps of transected nerves and promotes axonal regeneration after peripheral nerve injury (80). The effect of galectin-1 on endothelial cell migration is described further in the section entitled "Galectin-1 and Angiogenesis."

Galectin-1 Involvement in Tumor Immune Escape Several mechanisms have been described that potentially contribute to tumor cell evasion from the immune response (106, 107). These mechanisms include the production of immunosuppressive cytokines and other soluble factors, including transforming growth factor-β, interleukin-10, vascular endothelial growth factor (VEGF), and galectin-1. However, a long-standing dilemma in tumor immunology has been the ability of solid tumor cells to escape immune surveillance despite a demonstrable antitumor T-cell response. Recently, studies have highlighted the importance of the host stroma in modulating antitumor cytotoxicity of T lymphocytes (108). From all the studies reported in the literature and summarized in Table 9.3, it is reasonable to assume that galectin-1 expression or overexpression in tumors or the surrounding tissue (the stroma) must be considered a sign of malignant progression and consequently of a poor prognosis for patients. This prognosis is often related to tumor immune escape, to the long-range dissemination of tumoral cells (metastasis), or to their presence in the surrounding normal tissue. Galectins have accordingly attracted the attention of tumor immunologists as novel regulators of the antitumor immune response.

The abundance of proapoptotic galectin-1 in privileged immune sites such as the placenta (109), the brain (110), and the reproductive organs (111), along with the expression of galectin-1 in the stromal tissue around tumors (46, 93, 112–115) or in the endothelial cells from capillaries infiltrating them, rather than in adjacent nontumoral stroma (48), suggest that galectin-1 might trigger the death of infiltrating T cells and protect these sites from the tissue damage induced by T-cell-derived proinflammatory cytokines (Fig. 9.3).

The inhibition of the biological activity of galectin-1 in melanoma tissue results in a reduced tumor mass and stimulates the *in vivo* generation of a tumor-specific T-cell response (101). These observations further support the idea that galectin-1 may contribute to the immune privilege of tumors by modulating the survival or polarization of effector T cells. Additionally, the binding of galectin-1 to CD7 receptors induces major apoptotic features in normal T cells (116). With respect to programmed cell death, galectin-1 induces the inhibition of cell growth and cell cycle arrest and promotes the apoptosis of activated but not resting immune cells (1, 2). The signal transduction events that lead to cell death induced by galectin-1 in activated T cells involve several intracellular mediators including the induction of specific transduction factors (i.e., NFAT and activator protein-1).

Le et al. have also demonstrated in head and neck carcinomas a significant relationship between galectin-1 expression and the presence of hypoxia markers and an inverse correlation with T-cell infiltration (117). The latter fact suggests that hypoxia can affect malignant progression by regulating the secretion by tumor cells of proteins such as galectin-1 that modulate immune privilege. As will be developed later in this chapter, the hypoxia-induced galectin-1 expression is also responsible for angiogenesis stimulation.

The immunomodulatory effects of galectin-1 and the correlation between galectin-1 expression in cancer cells and their aggressiveness suggest the hypothesis that tumor cells may impair T-cell effector functions through the secretion of galectin-1 and that this mechanism may contribute toward tilting the balance in favor of an immunosuppressive environment at a tumor site. Indeed, the immunohistochemical determination of galectin-1 expression is of prognostic value in human squamous laryngeal cancers. Laryngeal squamous cell carcinomas that display high levels of galectin-1 have worse prognoses than those with low level expression. Elevation of galectin-1 levels in laryngeal cancers can contribute to tumor immune escape by killing activated T cells and other protumoral activities such as promoting motility or the activity of oncogenic H-Ras proteins (92).

CANCER CELL MIGRATION AND CHEMORESISTANCE

The development of malignant tumors has been reviewed by different authors who have established the pivotal roles played by cell migration in cancer cell scattering, tissue invasion, and metastasis (118–125). Once a cancer has been diagnosed as having metastasized, the probability that a patient will survive for more than a year after the initial diagnosis is less than 50%. Both this dismal prognosis and the nature of those incurable cancers characterized by dramatic tissue invasion (such as glioblastomas, metastatic melanomas) can at least be partly explained by the fact that while a large majority of the drugs used today to treat cancer are proapoptotic, migratory and metastatic cancer cells are resistant to apoptosis, as indicated by an increasing number of publications (27, 126–131).

Part of the switch between proliferation and migration processes is controlled by the CAS/Crk pathway (132) and part is controlled under the phosphatase and tensin homologue deleted on chromosome 10 gene (PTEN)/Akt/PI3-K/mammalian target of rapamycin (mTOR) pathway, as described below. The CAS/Crk pathway is also involved in the suppression of apoptosis during cell migration (133), a feature that involves the activation of c-Abl tyrosine kinase, at least in the case of carcinoma cells (134). The binding of $\alpha v\beta 3$ and $\alpha v\beta 5$ integrins to vitronectin can also confer resistance to apoptosis in the case of glioma cells migrating at the tumor/brain interface.

McCormick argues that cancer cells have an in-built urge to survive, so that any genetic change that favors survival against adverse conditions will be selected, and the outcome will be that some tumors will survive exposure to even the most potent therapeutic agents (135). The need for survival mechanisms in cancer cells is increased even more dramatically in the case of migrating cancer cells that must resist anoikis during their journeys that culminate in the formation of metastases. Indeed, migratory cells detached from their supports die from anoikis, which has even been suggested to be acting as a physiological barrier to metastasis. Douma et al. report that a resistance to anoikis may permit the survival of cancer cells during their time in the systemic circulation (136), thus facilitating secondary tumor formation in distant organs (28).

Signaling pathways involved in the resistance of migrating cancer cells to apoptosis (and probably to anoikis) include, for example, PI3-K and its downstream targets such as Akt, mTOR, NF-κB, Src, and galectin-3 (27, 28). There are in fact ever more publications available that demonstrate a direct relationship between the active migration of cancer cells and their resistance to apoptosis (27, 91, 126–129, 131, 137, 138).

Glioblastomas are among the most devastating forms of cancer arising from dramatic migration, and no patient has ever been cured to date. In a recent review of the molecular mechanisms behind the resistance of migrating glioblastoma cells to apoptosis, a number of publications have demonstrated that drugs able to reduce the levels of migration of these cells are also likely to confer a certain level of sensitivity to apoptosis in these migration-restricted cells (137, 138). Antimigratory compounds thus not only delay the formation of metastases by cancer cells emerging from primary tumor sites, but also restore a certain level of sensitivity to apoptosis in these slowly migrating cells, which can then be combated using conventional cytotoxic (proapoptotic) drugs. Novel types of antimigratory compounds are thus needed.

Reducing the promigratory roles of specific galectins could potentially reduce the migration of some types of apoptosis-resistant cancer cells and in turn confer/restore a certain level of sensitivity to proapoptotic agents in these migration-restricted cells.

Thus, taken together the data presented hitherto emphasize that

i. galectins are involved in numerous aspects of cancer development and progression, including cancer cell migration,
ii. migrating cancer cells are protected against apoptosis and are resistant to a large majority of the cytotoxic drugs currently used to treat cancer patients as these are designed by and large to be proapoptotic, and
iii. reduction in the migration of cancer cells restores a certain level of sensitivity to apoptosis and so to proapoptotic drugs.

Antigalectin strategies could therefore be helpful in restoring sensitivity to apoptosis in cancer cells by reducing the levels of migration in these populations. Antigalectin strategies could also increase chemosensitivity by means of lysosomal membrane permeabilization (LMP) processes in melanoma models (70) and by impairing the endoplasmic reticulum stress (ERS) response in GBM models (85), as we describe further in the section entitled "Galectin-1 and Chemoresistance."

ANGIOGENESIS AND VASCULOGENIC MIMICRY

Although cancer cells are not generally controlled by normal regulatory mechanisms, tumor growth highly depends on the supply of oxygen, nutrients, and host-derived regulators. It has been over 30 years since Judah Folkman hypothesized that tumor growth depends on angiogenesis (1971) (139).

Angiogenesis is a complex multistep process comprising a series of cellular events that lead to neovascularization from existing blood vessels and is associated with the process of inflammation, wound healing, tumor growth, and metastasis (121, 140) (Fig. 9.4). It is obvious that the formation of new capillary vessels in a tumor is critical for its continuous growth and also provides a gateway for the dissemination of malignant cells.

Determination of microvessel density in a growing cancer has prognostic value for recurrence and survival. For many years, tumor vascularization was explained solely by the ingrowth of new vessels into the tumor from preexisting ones. It is now established that tumor vasculature is not necessarily derived from endothelial cell sprouting, instead cancer tissue can acquire its vasculature by coopting preexisting vessels, intussusceptive microvascular growth, postnatal vasculogenesis, glomeruloid angiogenesis, or vasculogenic mimicry (Fig. 9.4). These different mechanisms may exist concomitantly in the same tumor or may be selectively involved in a specific tumor type or host environment (141).

Vasculogenic mimicry is defined by the unique ability of aggressive melanoma cells to express an endothelial cell phenotype and to form vessel-like networks in three-dimensional cultures, "mimicking" the pattern of embryonic vascular networks and recapitulating the patterned networks seen in patients' aggressive tumors correlating with a poor prognosis (142). Vasculogenic mimicry has been confirmed in breast, prostate, ovarian, chorio- and lung carcinomas (143).

The best-known molecular pathway driving tumor vascularization is the hypoxia-adaptation mechanism (144). However, a broad and diverse spectrum of genetic aberrations is associated with the development of the "angiogenic phenotype." On the basis of this knowledge, novel forms of antivascular modalities have been developed in the past decade. When applying these targeted therapies, the stage of tumor progression, the type of vascularization of the given cancer tissue, and the molecular machinery behind the vascularization process all need to be considered. A further challenge is finding the most appropriate combinations of antivascular therapies and standard radio- and chemotherapies (141).

Galectin-3 is the most extensively studied member of the galectin family of proteins in the field of cancer angiogenesis (145, 146). It has been shown that galectin-3 affects chemotaxis and morphology and stimulates capillary tube formation in human umbilical vein endothelial cells (HUVECs) *in vitro* and angiogenesis *in vivo*. Endothelial cell morphogenesis is a carbohydrate-dependent process, as it is neutralized by specific sugars and antibodies (146). These findings demonstrate that endothelial cell surface carbohydrate recognition event(s) can induce a signaling cascade leading to the differentiation and angiogenesis of endothelial cells. In summary, these results suggest that angiogenesis could be mediated by carbohydrate recognition.

Galectin-3 secreted by tumor cells and macrophages may also act on endothelial cells to induce chemotaxis, facilitating their motility during the initial phase of tube formation. In fact, the galectin-3 C-terminal domain fragment significantly suppresses tumor growth and inhibits metastasis in a mouse model of human breast cancer (147). In addition, peptides specific to the galectin-3 CRD significantly inhibit the adhesion of a human breast carcinoma cell line to endothelial cells *in vitro* (107) and administration of an antigalectin-3 antibody specifically inhibits liver metastasis by adenocarcinoma cell lines (148).

Interestingly, it has been found that citrus pectin, a water-soluble polysaccharide fiber derived from a citrus fruit that specifically inhibits galectin-3, given orally, inhibits carbohydrate-mediated tumor growth, angiogenesis, and metastasis by disrupting the interactions between galectin-3 and its specific carbohydrate ligands (149). Rabinovich et al. reported the synthesis of three novel synthetic lactulose amines as potent and specific inhibitors of the binding of galectin-1 and -3 to the highly glycosylated protein 90 K (150). These inhibitory agents showed selective modulatory effects in critical steps of tumor progression, including galectin-mediated homotypic cell aggregation, tumor cell apoptosis, and endothelial cell morphogenesis, suggesting their potential contribution to the suppression of tumor growth, angiogenesis, and metastasis.

GALECTIN-1 AND ANGIOGENESIS

In this section, the role of galectin-1 in tumor angiogenesis is explored. The evidence is reviewed that galectin-1 is required for tumor angiogenesis and outgrowth of tumors, that galectin-1 is the target for the potent angiogenesis inhibitor anginex, and that decreasing the expression of galectin-1 in tumor cells including glioblastoma and melanoma cells impairs angiogenesis *in vitro* and *in vivo* (Fig. 9.4).

It was first considered possible that galectin-1 could be involved in tumor angiogenesis because both vascular smooth muscle and endothelial cells express the protein (51, 97). Furthermore, although the vessel walls of normal lymphoid tissues do not express galectin-1, the blood vessel walls of lymphomas do so in relation to their vascular density (151). Clausse et al. had also previously shown that galectin-1 was upregulated in capillaries associated with carcinoma cells and that it could mediate interactions between tumors and endothelial cells *in vitro* (48), suggesting a potential role for galectin-1 in modulating angiogenesis. Treatment with galectin-1-specific antisense oligodeoxynucleotides or polyclonal antigalectin-1 antibodies resulted in the inhibition of endothelial cell proliferation and migration, which suggests an essential role for galectin-1 during angiogenesis (152), as originally proposed by pioneering experiments performed by Clausse et al. (48). Thijssen et al. have also described galectin-1 as a receptor for the angiogenesis inhibitor anginex, and that the protein is crucial for tumor angiogenesis (152).

The importance of galectin-1 in angiogenesis has been illustrated in the zebrafish, where its deletion results in impaired vascular guidance and growth of dysfunctional vessels. The role of galectin-1 in tumor angiogenesis is further highlighted in galectin-1-null mice in which tumor growth is markedly impaired because of insufficient tumor angiogenesis. Furthermore, tumor growth in galectin-1-null mice no longer responds to antiangiogenesis treatment with anginex. Thus, galectin-1 regulates tumor angiogenesis and is a target for angiostatic cancer therapy (152).

These authors further suggest that galectin-1 is regulated by hypoxia at both transcriptional and posttranslational levels and that galectin-1 protein accumulation and secretion precede the accumulation of its mRNA. Thus, the critical regulation of galectin-1 by hypoxia appears to occur at the protein level.

We ourselves have also put forward evidence for the role of galectin-1 in the process of angiogenesis using human GBM and murine melanoma models. We used a short interfering RNA (siRNA) approach to decrease the levels of galectin-1 expression *in vitro* in two tumor models, as well as *in vivo* in the brain of adult mice bearing Hs683 GBM model. As described elsewhere (153), a brain-infusion cannula (Alzet) was stereotactically positioned with a view to carrying out infusions from an s.c. implanted minipump into the third ventricule as described by Thakker et al. for siRNA *in vivo* delivery (154). Full genome-wide array analysis followed by Western blotting and immunofluorescence analyses at the proteomic level revealed that galectin-1 depletion in Hs683 GBM cells markedly decreased the expression of DNAJB9/MDG1/ERdj4 (referred to here as the microvascular differentiation gene 1 (MDG1)) and HYOU1/ORP150 (referred to here as ORP150).

ORP150 was first identified and cloned from cultured astrocytes on the basis of their ability to withstand and even produce neurotrophic factors in response to severe hypoxia (155). The expression of ORP150 in cultured human cells is essential for their survival under prolonged hypoxia. Hypoxia confers cellular resistance to conventional chemotherapy and accelerates malignant progression (155–157). This would appear to be consistent with galectin-1 being identified as a hypoxia-regulated protein (117), which plays an important role in endothelial cell biology in general (1) and in endothelial cell morphogenesis in particular (150).

ORP150 has further been shown to modulate angiogenesis via the processing of VEGF (155). Reduced galectin-1 expression in Hs683 glioblastoma cells also leads to sustained decreases in VEGF expression, accompanied *in vivo* by the severe impairment of angiogenesis in Hs683 orthotopic xenografts. The expression of angiogenic factors such as VEGF under conditions of cell stress involves both transcriptional and translational events, as well as an important role for inducible endoplasmic reticulum (ER) chaperones. The *in vivo* reduction in ORP150 expression in C6 rat glioma tumors induces a decrease in angiogenesis in C6 gliomas, while the *in vitro* inhibition of ORP150 expression decreases the release of VEGF into the supernatant of C6 cultured tumor cells. In ORP150 antisense transfected C6 rat glioma cells, VEGF accumulates intracellularly within the ER. Increased levels of ORP150 promote VEGF processing with subsequent transport from the ER to the Golgi followed by secretion from the cell. The cell death mechanism in tumors arising from ORP150 antisense C6 transfectants is mainly based on the inhibition of angiogenesis, which is caused by the retention of VEGF in the ER (155).

As indicated above, galectin-1 reduction also decreases MDG1 expression in Hs683 GBM cells. MDG1 also plays an active role in endothelial cell biology through its modulation of the ERS (158, 159). MDG1 is isolated from differentiating microvascular endothelial cells cultured in collagen type I gels (3D culture) and is upregulated in primary endothelial and mesangial cells when subjected to various stress stimuli (159).

We have also shown that decreasing galectin-1 expression in primary human endothelial cells does not modify their ability to perform capillary networking on Matrigel. In contrast, a decrease in galectin-1 expression in human GBM cells markedly impairs their ability to perform vasculogenic mimicry (Le Mercier et al., 2008, in press). Furthermore, decreasing the expression of galectin-1 by means of siRNA in B16F10 melanoma cells resulted in appreciably lower average weights of lungs excised from animals grafted with transfected B16F10 cells and subsequently treated with temozolomide.

Histopathological analysis revealed that B16F10 lung metastases obtained from mice grafted with transfected cells showed significantly larger necrotic areas than those obtained with scrambled siRNA-transfected cells. Because the levels of angiogenesis in lung tumors revealed a significant decrease in the surface area occupied by blood vessels in those tumors obtained after galectin-1-depleted cells were injected compared to tumors obtained after scrambled siRNA-transfected cells were injected, this marked necrotic process was considered to relate partly to a modification in angiogenesis (70). Whether ORP150 is also involved in melanoma angiogenesis remains to be determined.

GALECTIN-1 AND CHEMORESISTANCE

Galectin-3 has been extensively studied in the field of cancer chemoresistance. Galectin-3 confers chemoresistance to a wide variety of cancer cell types (160). The antiapoptotic molecule galectin-3 was shown to regulate CD95, a member of the tumor necrosis factor (TNF) family of proteins in the apoptotic signaling pathway. Oka et al. suggest that galectin-3 involves Akt as a modulator molecule in protecting bladder carcinoma cells from tumor necrosis factor-related apoptosis inducing ligand (TRAIL)-induced apoptosis (17).

Recent studies have revealed that galectin-3 demonstrates antiapoptotic effects, which contribute to cell survival in several types of cancer cells (4). Intracellular galectin-3 in particular, which contains the NWGR antideath motif of the Bcl-2 family, inhibits cell apoptosis induced by chemotherapeutic agents such as cisplatin and etoposide in some types of cancer cells. It has also been reported that the nuclear export of phosphorylated galectin-3 regulates its antiapoptotic activity in response to chemotherapeutic drugs. Finally, it has been suggested that targeting galectin-3 could improve the efficacy of anticancer drug chemotherapy in several types of cancer (145).

We have recently investigated the roles played by galectin-1 *in vitro* and *in vivo* in the case of B16F10 mouse melanoma and human Hs683 GBM cell resistance to programmed cell death (70, 161). For this purpose, a siRNA approach was used to decrease the level of galectin-1 expression *in vitro* in B16F10 cells (70) and *in vitro* as well as *in vivo* in Hs683 GBMs (161). Decreasing the expression of galectin-1 by means of a siRNA approach in B16F10 melanoma and Hs683 glioblastoma cells sensitizes them to the antitumor effects of temozolomide *in vivo* (70, 85, 91).

Decreasing galectin-1 expression in B16F10 melanoma cells did not induce apoptosis or autophagy in these cells (70). In contrast, this decrease in galectin-1 expression in these melanoma cells appeared to induce marked Hsp70-mediated LMP processes with a significant release of cathepsin B into the cytosol *in vitro* and marked necrosis *in vivo* (70). This, in turn, appeared to synergize the effects of temozolomide. As recently reviewed by Kroemer and Jäättelä, lysosomal alterations are common in cancer cells (162). Indeed, increased expression and altered trafficking of lysosomal enzymes play a key role in tissue invasion, angiogenesis, and sensitization to the lysosomal death pathway, whereas the lysosomal Hsp70 locally prevents LMP.

The same phenomenon was observed on decreasing galectin-1 expression in human Hs683 glioblastoma orthotopic xenografts in the brains of immunodeficient mice (85). In this glioma model, neither apoptotic nor autophagic features were observed *in vitro*. We showed that galectin-1 could be involved in the defense of glioma cells against cytotoxic insults. Indeed, treating human Hs683 GBM cells with subtoxic doses of temozolomide increased galectin-1 expression in Hs683 cell subpopulations (85). In contrast, the *in vivo* delivery of an antigalectin-1 siRNA to Hs683 glioblastoma orthotopic xenograft-bearing immunocompromized mice increased the antitumor effects of this proautophagic drug (85, 91, Le Mercier et al., 2008, in press).

Transiently decreasing galectin-1 expression in Hs683 cells seriously modifies their response to ERS. This appears related to the fact that cancer cells in poorly vascularized areas of solid tumors (including GBMs) are constantly or intermittently exposed to nutrient deprivation, glucose deprivation, hypoxia, redox, and glycosylation reactions, as well as to disturbances in their calcium mobilization (156, 163, 164).

The reduction of galectin-1 in Hs683 cells, even transiently, interferes strongly with ER stress and modifies the expression of genes involved in the unfolded protein response (UPR), and consequently could increase the therapeutic benefits of the proautophagic drug temozolomide. ER stress and UPR-related genes, whose expression decreased most markedly at both genomic and proteomic levels in transiently galectin-1-reduced Hs683 cells, included HERP, DUSP5, ORP150, and MDG1 (see section "Galectin-1 and Angiogenesis," the two last genes).

CONCLUSIONS AND PERSPECTIVES

Galectin-1 is differentially expressed by various normal and pathological tissues and appears to be functionally polyvalent, with a wide range of biological activity. The galectin-1 expression or overexpression in tumors or the tissue surrounding them must be considered a sign of their malignant progression resulting in poor prognoses for large numbers of cancer patients. This is often related to the long-range dissemination of tumoral cells (metastasis), to their ability to induce angiogenic processes, to their dissemination into the surrounding normal tissue, and to tumor immune escape.

Galectin-1 is directly involved in the biological processes of cancer cell migration, at least in nonsmall-cell lung cancers, non-Hodgkin lymphomas, pancreatic adenocarcinomas, head and neck squamous cell carcinomas, melanomas, and gliomas (1). Galectin-1 is also directly implicated in the process of tumor immune escape as it is secreted by tumor cells and kills activated T cells. Very recently, it was shown by Thijssen et al. (152) and ourselves (85) (Le Mercier et al., 2008, in press) that galectin-1 is involved in the process of tumor angiogenesis and that decreasing its expression in tumor cells decreases *in vitro* and *in vivo* angiogenesis in melanoma and glioblastoma models (70) (Le Mercier et al., 2008, in press).

Finally, galectin-1 is involved in chemoresistance and its decreased expression is associated with LMP in a melanoma model and with an impairment of the endoplasmic stress in a glioblastoma model. The decrease in galectin-1 expression by tumor cells increases their sensitivity to chemotherapy in both melanoma and glioblastoma models.

Galectin-1 could constitute a target for the novel treatment of a range of devastating cancers. Indeed, reducing galectin-1 expression in migrating tumor cells, certainly at least in gliomas (85, 91) and melanomas (70), could impair malignancy development through delaying cancer cell migration within the host tissue (as for gliomas in the brain) or at a distance (melanoma metastases), by impairing angiogenesis, by increasing the tumor immune response, or by sensitizing migrating apoptosis-resistant cancer cells to chemotherapy.

Antigalectin-1 compounds are thus required to combat migrating cancer cells and several groups (149, 165, 166) including our own (90, 167) are engaged in this quest. Antigalectin-1 compounds, that is, siRNA antisense oligonucleotide, aptamers, or blocking antibodies able to inhibit or at least markedly decrease the biological activity of galectin-1, could be locally administered to patients suffering from nonsmall-cell lung cancers, non-Hodgkin lymphomas, pancreatic adenocarcinomas, head and neck squamous cell carcinomas, melanomas, or gliomas, which together account for ~20% of all cancers encountered in female patients and up to 37% of all cancers in males. The feasibility and the therapeutic benefit of this approach *in vivo* in preclinical orthotopic models of human glioblastoma have been demonstrated.

Direct *in vivo* delivery of antigalectin-1 siRNA in mice bearing orthotopic xenografts of human glioblastoma is also feasible (Le Mercier et al., 2008, in press). The *in vivo* delivery of antigalectin-1 siRNA in humans could be performed directly in the case of GBMs by means of Ommaya reservoirs, thus minimizing/avoiding systemic release/exposure and the potential for wide spread toxicity. A further source of promise resides in the near future use of Anginex, the antibody targeting galectin-1-related proangiogenic processes, which is currently in clinical trials in cancer patients.

Furthermore, proteins and peptides selected experimentally for high affinity interactions with predetermined target structures are emerging as important molecules, which could serve to extend conventional small drug therapy. Considerable progress has been made in recent years to convert peptides into therapeutically useful molecules, and particularly for targeting cell surface molecules (168). The peptide-related approach has resulted in short synthetic peptides that bind to a galectin carbohydrate recognition domain with a high level of affinity and specificity, and which therefore offer an attractive approach to inhibiting the functioning of galectins and β-galactoside-mediated cell adhesion.

Although there are no data currently available concerning on clinical trials with galectin-targeting antibodies, the development of such biologicals designed to interfere with galectin oligomerization or the interaction of galectins with their ligands has already been described *in vivo* preclinically and demonstrates the potential feasibility of the approach. As early as 1986, Meromsky et al. highlighted the fact that a monoclonal antibody directed against endogenous galactoside-specific lectins was able to inhibit homotypic melanoma and fibrosarcoma cell aggregation and adhesion, and to reduce the colonization of the lungs by melanoma and fibrosarcoma cells in mice. Additionally, antigalectin-3 monoclonal antibodies specifically targeting the adhesive interactions of β-galactoside-mediated tumor endothelial cells inhibit the *in vivo* formation of metastatic breast and prostate carcinoma deposits in mouse lungs and bones by >90% (49).

A large number of compounds with steroid backbones from different marine sponges display antimigratory and cytotoxic effects (167). One of these steroid backbones was selected to initiate a novel antimigratory compound project with the aim of identifying molecules effective both *in vitro* and *in vivo* against various types of cancers (167). To enhance the antimigratory activity of this class, various mono- and disaccharides were added to the steroid backbone. Resulting molecules were found to be associated with a total absence of cytotoxicity and a certain level of antimigratory activity (167). We have recently made the point on antigalectin compounds derived from a wide variety of chemicals (90).

The possibility exists that such antigalectin-1 compounds could be evaluated in clinical trials (in association with cytotoxic agents) in the near future. Furthermore, the availability of synthetic inhibitors might also contribute as basic research tools to dissect the cellular and molecular mechanisms implicated in the biological functions of galectins.

ABBREVIATIONS

CEA	carcinoembryonic antigen
ECM	extracellular matrix
ER	endoplasmic reticulum
ERK	extracellular signal-regulated kinase
ERS	endoplasmic reticulum stress
FAK	focal adhesion kinase

GBM	glioblastoma
LAMP	lysosome-activated membrane protein
LMP	lysosomal membrane permeabilization
MDG1	microvascular differentiation gene 1
mTOR	mammalian target of rapamycin
PI3-K	phosphatidyl inositol 3 kinase
PTEN	phosphatase and tensin homologue deleted on chromosome 10 gene
siRNA	short interfering ribosomal nucleic acid
VEGF	vascular endothelial growth factor
UPR	unfolded protein response

REFERENCES

1. Camby I, Le Mercier M, Lefranc F, Kiss R. Galectin-1: a small protein with major functions. Glycobiology 2006;16:137R–157R.
2. Liu FT, Rabinovich GA. Galectins as modulators of tumour progression. Nat Rev Cancer 2005;5:29–41.
3. Califice S, Castronovo V, Van Den Brule F. Galectin-3 and cancer (Review). Int J Oncol 2004;25:983–992.
4. Nakahara S, Oka N, Raz A. On the role of galectin-3 in cancer apoptosis. Apoptosis 2005;10:267–275.
5. Takenaka Y, Fukumori T, Raz A. Galectin-3 and metastasis. Glycoconj J 2004; 19:543–549.
6. Huflejt ME, Leffler H. Galectin-4 in normal tissues and cancer. Glycoconj J 2004; 20:247–255.
7. Saussez S, Cucu DR, Decaestecker C, Chevalier D, Kaltner H, Andre S, Wacreniez A, Toubeau G, Camby I, Gabius HJ, Kiss R. Galectin-7 (p53-induced gene 1): a new prognostic predictor of recurrence and survival in stage IV hypopharyngeal cancer. Ann Surg Oncol 2006;13:999–1009.
8. Saussez S, Kiss R. Galectin-7. Cell Mol Life Sci 2006;63:686–697.
9. Bidon-Wagner N, Le Pennec JP. Human galectin-8 isoforms and cancer. Glycoconj J 2004;19:557–563.
10. Zick Y, Eisenstein M, Goren RA, Hadari YR, Levy Y, Ronen D. Role of galectin-8 as a modulator of cell adhesion and cell growth. Glycoconj J 2004;19:517–526.
11. Hirashima M, Kashio Y, Nishi N, Yamauchi A, Imaizumi TA, Kageshita T, Saita N, Nakamura T. Galectin-9 in physiological and pathological conditions. Glycoconj J 2004;19:593–600.
12. Yang RY, Hsu DK, Yu L, Ni J, Liu FT. Cell cycle regulation by galectin-12, a new member of the galectin superfamily. J Biol Chem 2001;276:20252–20260.
13. Lahm H, Andre S, Hoeflich A, Kaltner H, Siebert HC, Sordat B, von der Lieth CW, Wolf E, Gabius HJ. Tumor galectinology: insights into the complex network of a family of endogenous lectins. Glycoconj J 2004;20:227–238.
14. van den Brule F, Califice S, Castronovo V. Expression of galectins in cancer: a critical review. Glycoconj J 2004;19:537–542.

15. Frisch SM, Ruoslahti E. Integrins and anoikis. Curr Opin Cell Biol 1997;9:701–706.

16. Fukumori T, Oka N, Takenaka Y, Nangia-Makker P, Elsamman E, Kasai T, Shono M, Kanayama HO, Ellerhorst J, Lotan R, Raz A. Galectin-3 regulates mitochondrial stability and antiapoptotic function in response to anticancer drug in prostate cancer. Cancer Res 2006;66:3114–3119.

17. Oka N, Nakahara S, Takenaka Y, Fukumori T, Hogan V, Kanayama HO, Yanagawa T, Raz A. Galectin-3 inhibits tumor necrosis factor-related apoptosis-inducing ligand-induced apoptosis by activating Akt in human bladder carcinoma cells. Cancer Res 2005;65:7546–7553.

18. Takenaka Y, Fukumori T, Yoshii T, Oka N, Inohara H, Kim HR, Bresalier RS, Raz A. Nuclear export of phosphorylated galectin-3 regulates its antiapoptotic activity in response to chemotherapeutic drugs. Mol Cell Biol 2004;24:4395–4406.

19. Watt DJ, Jones GE, Goldring K. The involvement of galectin-1 in skeletal muscle determination, differentiation, and regeneration. Glycoconj J 2004;19:615–619.

20. Kami K, Senba E. Galectin-1 is a novel factor that regulates myotube growth in regenerating skeletal muscles. Curr Drug Targets 2005;6:395–405.

21. Gaudet AD, Steeves JD, Tetzlaff W, Ramer MS. Expression and functions of galectin-1 in sensory and motoneurons. Curr Drug Targets 2005;6:419–425.

22. Horie H, Kadoya T, Sango K, Hasegawa M. Oxidized galectin-1 is an essential factor for peripheral nerve regeneration. Curr Drug Targets 2005;6:385–394.

23. Kato T, Ren CH, Wada M, Kawanami T. Galectin-1 as a potential therapeutic agent for amyotrophic lateral sclerosis. Curr Drug Targets 2005;6:407–418.

24. Fischer C, Sanchez-Ruderisch H, Welzel M, Wiedenmann B, Sakai T, Andre S, Gabius HJ, Khachigian L, Detjen KM, Rosewicz S. Galectin-1 interacts with the $\alpha 5\beta 1$ fibronectin receptor to restrict carcinoma cell growth via induction of p21 and p27. J Biol Chem 2005;280:37266–37277.

25. Andre S, Kaltner H, Lensch M, Russwurm R, Siebert HC, Fallsehr C, Tajkhorshid E, Heck AJ, von Knebel Doeberitz M, Gabius HJ, Kopitz J. Determination of structural and functional overlap/divergence of five prototype galectins by analysis of the growth-regulatory interaction with ganglioside GM1 *in silico* and *in vitro* on human neuroblastoma cells. Int J Cancer 2005;114:46–57.

26. Elad-Sfadia G, Haklai R, Ballan E, Gabius HJ, Kloog Y. Galectin-1 augments Ras activation and diverts Ras signals to Raf-1 at the expense of phosphoinositide 3-kinase. J Biol Chem 2002;277:37169–37175.

27. Lefranc F, Brotchi J, Kiss R. Possible future issues in the treatment of glioblastomas: special emphasis on cell migration and the resistance of migrating glioblastoma cells to apoptosis. J Clin Oncol 2005;23:2411–2422.

28. Decaestecker C, Debeir O, Van Ham P, Kiss R. Can anti-migratory drugs be screened *in vitro*? A review of 2D and 3D assays for the quantitative analysis of cell migration. Med Res Rev 2007;27:149–176.

29. Bellail AC, Hunter SB, Brat DJ, Tan C, Van Meir EG. Microregional extracellular matrix heterogeneity in brain modulates glioma cell invasion. Int J Biochem Cell Biol 2004;36:1046–1069.

30. Belot N, Rorive S, Doyen I, Lefranc F, Bruyneel E, Dedecker R, Micik S, Brotchi J, Decaestecker C, Salmon I, Kiss R, Camby I. Molecular characterization of cell

substratum attachments in human glial tumors relates to prognostic features. Glia 2001;36:375–390.

31. Bolteus AJ, Berens ME, Pilkington GJ. Migration and invasion in brain neoplasms. Curr Neurol Neurosci Rep 2001;1:225–232.
32. Giese A, Rief MD, Loo MA, Berens ME. Determinants of human astrocytoma migration. Cancer Res 1994;54:3897–3904.
33. Gladson CL. The extracellular matrix of gliomas: modulation of cell function. J Neuropathol Exp Neurol 1999;58:1029–1040.
34. Tonn JC, Goldbrunner R. Mechanisms of glioma cell invasion. Acta Neurochir Suppl 2003;88:163–167.
35. Lauffenburger DA, Horwitz AF. Cell migration: a physically integrated molecular process. Cell 1996;84:359–369.
36. Ridley AJ. Rho GTPases and cell migration. J Cell Sci 2001;114:2713–2722.
37. Palecek SP, Loftus JC, Ginsberg MH, Lauffenburger DA, Horwitz AF. Integrin–ligand binding properties govern cell migration speed through cell–substratum adhesiveness. Nature 1997;385:537–540.
38. Wehrle-Haller B, Imhof BA. Actin, microtubules and focal adhesion dynamics during cell migration. Int J Biochem Cell Biol 2003;35:39–50.
39. Belien AT, Paganetti PA, Schwab ME. Membrane-type 1 matrix metalloprotease (MT1-MMP) enables invasive migration of glioma cells in central nervous system white matter. J Cell Biol 1999;144:373–384.
40. Rao JS. Molecular mechanisms of glioma invasiveness: the role of proteases. Nat Rev Cancer 2003;3:489–501.
41. Rickman DS, Bobek MP, Misek DE, Kuick R, Blaivas M, Kurnit DM, Taylor J, Hanash SM. Distinctive molecular profiles of high-grade and low-grade gliomas based on oligonucleotide microarray analysis. Cancer Res 2001;61:6885–6891.
42. Andre S, Kojima S, Yamazaki N, Fink C, Kaltner H, Kayser K, Gabius HJ. Galectins-1 and -3 and their ligands in tumor biology. Non-uniform properties in cell-surface presentation and modulation of adhesion to matrix glycoproteins for various tumor cell lines, in biodistribution of free and liposome-bound galectins and in their expression by breast and colorectal carcinomas with/without metastatic propensity. J Cancer Res Clin Oncol 1999;125:461–474.
43. Ellerhorst J, Nguyen T, Cooper DN, Lotan D, Lotan R. Differential expression of endogenous galectin-1 and galectin-3 in human prostate cancer cell lines and effects of overexpressing galectin-1 on cell phenotype. Int J Oncol 1999;14: 217–224.
44. Moiseeva EP, Spring EL, Baron JH, de Bono DP. Galectin 1 modulates attachment, spreading and migration of cultured vascular smooth muscle cells via interactions with cellular receptors and components of extracellular matrix. J Vasc Res 1999;36:47–58.
45. Moiseeva EP, Williams B, Goodall AH, Samani NJ. Galectin-1 interacts with beta-1 subunit of integrin. Biochem Biophys Res Commun 2003;310:1010–1016.
46. van den Brule F, Califice S, Garnier F, Fernandez PL, Berchuck A, Castronovo V. Galectin-1 accumulation in the ovary carcinoma peritumoral stroma is induced by ovary carcinoma cells and affects both cancer cell proliferation and adhesion to laminin-1 and fibronectin. Lab Invest 2003;83:377–386.

47. Tinari N, Kuwabara I, Huflejt ME, Shen PF, Iacobelli S, Liu FT. Glycoprotein 90K/MAC-2BP interacts with galectin-1 and mediates galectin-1-induced cell aggregation. Int J Cancer 2001;91:167–172.

48. Clausse N, van den Brule F, Waltregny D, Garnier F, Castronovo V. Galectin-1 expression in prostate tumor-associated capillary endothelial cells is increased by prostate carcinoma cells and modulates heterotypic cell–cell adhesion. Angiogenesis 1999; 3:317–325.

49. Glinskii OV, Huxley VH, Glinsky GV, Pienta KJ, Raz A, Glinsky VV. Mechanical entrapment is insufficient and intercellular adhesion is essential for metastatic cell arrest in distant organs. Neoplasia 2005;7:522–527.

50. Lotan R, Belloni PN, Tressler RJ, Lotan D, Xu XC, Nicolson GL. Expression of galectins on microvessel endothelial cells and their involvement in tumour cell adhesion. Glycoconj J 1994;11:462–468.

51. Moiseeva EP, Williams B, Samani NJ. Galectin 1 inhibits incorporation of vitronectin and chondroitin sulfate B into the extracellular matrix of human vascular smooth muscle cells. Biochim Biophys Acta 2003;1619:125–132.

52. Carsons S, Lavietes BB, Slomiany A, Diamond HS, Berkowitz E. Carbohydrate heterogeneity of fibronectins. Synovial fluid fibronectin resembles the form secreted by cultured synoviocytes but differs from the plasma form. J Clin Invest 1987; 80:1342–1349.

53. Fujiwara S, Shinkai H, Deutzmann R, Paulsson M, Timpl R. Structure and distribution of N-linked oligosaccharide chains on various domains of mouse tumour laminin. Biochem J 1988;252:453–461.

54. Allen HJ, Sucato D, Woynarowska B, Gottstine S, Sharma A, Bernacki RJ. Role of galaptin in ovarian carcinoma adhesion to extracellular matrix *in vitro*. J Cell Biochem 1990;43:43–57.

55. Skrincosky DM, Allen HJ, Bernacki RJ. Galaptin-mediated adhesion of human ovarian carcinoma A121 cells and detection of cellular galaptin-binding glycoproteins. Cancer Res 1993;53:2667–2675.

56. van den Brule FA, Buicu C, Baldet M, Sobel ME, Cooper DN, Marschal P, Castronovo V. Galectin-1 modulates human melanoma cell adhesion to laminin. Biochem Biophys Res Commun 1995;209:760–767.

57. Elola MT, Chiesa ME, Alberti AF, Mordoh J, Fink NE. Galectin-1 receptors in different cell types. J Biomed Sci 2005;12:13–29.

58. Elola MT, Wolfenstein-Todel C, Troncoso MF, Vasta GR, Rabinovich GA. Galectins: matricellular glycan-binding proteins linking cell adhesion, migration, and survival. Cell Mol Life Sci 2007;64:1679–1700.

59. Stupack DG, Cheresh DA. Get a ligand, get a life: integrins, signaling and cell survival. J Cell Sci 2002;115:3729–3738.

60. Gu J, Taniguchi N. Regulation of integrin functions by *N*-glycans. Glycoconj J 2004;21:9–15.

61. Ohannesian DW, Lotan D, Lotan R. Concomitant increases in galectin-1 and its glycoconjugate ligands (carcinoembryonic antigen, lamp-1, and lamp-2) in cultured human colon carcinoma cells by sodium butyrate. Cancer Res 1994;54:5992–6000.

62. Seelenmeyer C, Wegehingel S, Lechner J, Nickel W. The cancer antigen CA125 represents a novel counter receptor for galectin-1. J Cell Sci 2003;116:1305–1318.

63. Symons A, Cooper DN, Barclay AN. Characterization of the interaction between galectin-1 and lymphocyte glycoproteins CD45 and Thy-1. Glycobiology 2000; 10:559–563.

64. Camby I, Belot N, Rorive S, Lefranc F, Maurage CA, Lahm H, Kaltner H, Hadari Y, Ruchoux MM, Brotchi J, Zick Y, Salmon I, Gabius HJ, Kiss R. Galectins are differentially expressed in supratentorial pilocytic astrocytomas, astrocytomas, anaplastic astrocytomas and glioblastomas, and significantly modulate tumor astrocyte migration. Brain Pathol 2001;11:12–26.

65. Camby I, Belot N, Lefranc F, Sadeghi N, de Launoit Y, Kaltner H, Musette S, Darro F, Danguy A, Salmon I, Gabius HJ, Kiss R. Galectin-1 modulates human glioblastoma cell migration into the brain through modifications to the actin cytoskeleton and levels of expression of small GTPases. J Neuropathol Exp Neurol 2002;61:585–596.

66. Camby I, Decaestecker C, Lefranc F, Kaltner H, Gabius HJ, Kiss R. Galectin-1 knocking down in human U87 glioblastoma cells alters their gene expression pattern. Biochem Biophys Res Commun 2005;335:27–35.

67. Primakoff P, Myles DG. The ADAM gene family: surface proteins with adhesion and protease activity. Trends Genet 2000;16:83–87.

68. Eto K, Puzon-McLaughlin W, Sheppard D, Sehara-Fujisawa A, Zhang XP, Takada Y. RGD-independent binding of integrin $\alpha9\beta1$ to the ADAM-12 and -15 disintegrin domains mediates cell–cell interaction. J Biol Chem 2000;275:34922–34930.

69. Harvey S, Zhang Y, Landry F, Miller C, Smith JW. Insights into a plasma membrane signature. Physiol Genomics 2001;5:129–136.

70. Mathieu V, Le Mercier M, De Neve N, Sauvage S, Gras T, Roland I, Lefranc F, Kiss R. Galectin-1 knockdown increases sensitivity to temozolomide in a B16F10 mouse metastatic melanoma model. J Invest Dermatol 2007;127:2399–2410.

71. Hittelet A, Legendre H, Nagy N, Bronckart Y, Pector JC, Salmon I, Yeaton P, Gabius HJ, Kiss R, Camby I. Upregulation of galectins-1 and -3 in human colon cancer and their role in regulating cell migration. Int J Cancer 2003;103:370–379.

72. Dudas SP, Yunker CK, Sternberg LR, Byrd JC, Bresalier RS. Expression of human intestinal mucin is modulated by the beta-galactoside binding protein galectin-3 in colon cancer. Gastroenterology 2002;123:817–826.

73. Ideo H, Seko A, Yamashita K. Galectin-4 binds to sulfated glycosphingolipids and carcinoembryonic antigen in patches on the cell surface of human colon adenocarcinoma cells. J Biol Chem 2005;280:4730–4737.

74. Song S, Byrd JC, Mazurek N, Liu K, Koo JS, Bresalier RS. Galectin-3 modulates MUC2 mucin expression in human colon cancer cells at the level of transcription via AP-1 activation. Gastroenterology 2005;129:1581–1591.

75. Ueda S, Kuwabara I, Liu FT. Suppression of tumor growth by galectin-7 gene transfer. Cancer Res 2004;64:5672–5676.

76. Nagy N, Bronckart Y, Camby I, Legendre H, Lahm H, Kaltner H, Hadari Y, Van Ham P, Yeaton P, Pector JC, Zick Y, Salmon I, Danguy A, Kiss R, Gabius HJ. Galectin-8 expression decreases in cancer compared with normal and dysplastic human colon tissue and acts significantly on human colon cancer cell migration as a suppressor. Gut 2002;50:392–401.

77. Nangia-Makker P, Conklin J, Hogan V, Raz A. Carbohydrate-binding proteins in cancer, and their ligands as therapeutic agents. Trends Mol Med 2002;8:187–192.

78. Fulcher JA, Hashimi ST, Levroney EL, Pang M, Gurney KB, Baum LG, Lee B. Galectin-1-matured human monocyte-derived dendritic cells have enhanced migration through extracellular matrix. J Immunol 2006;177:216–226.
79. Maeda N, Kawada N, Seki S, Arakawa T, Ikeda K, Iwao H, Okuyama H, Hirabayashi J, Kasai K, Yoshizato K. Stimulation of proliferation of rat hepatic stellate cells by galectin-1 and galectin-3 through different intracellular signaling pathways. J Biol Chem 2003;278:18938–18944.
80. Fukaya K, Hasegawa M, Mashitani T, Kadoya T, Horie H, Hayashi Y, Fujisawa H, Tachibana O, Kida S, Yamashita J. Oxidized galectin-1 stimulates the migration of Schwann cells from both proximal and distal stumps of transected nerves and promotes axonal regeneration after peripheral nerve injury. J Neuropathol Exp Neurol 2003;62:162–172.
81. Rabinovich GA, Sotomayor CE, Riera CM, Bianco I, Correa SG. Evidence of a role for galectin-1 in acute inflammation. Eur J Immunol 2000;30:1331–1339.
82. He J, Baum LG. Presentation of galectin-1 by extracellular matrix triggers T cell death. J Biol Chem 2004;279:4705–4712.
83. Friedl P. Prespecification and plasticity: shifting mechanisms of cell migration. Curr Opin Cell Biol 2004;16:14–23.
84. Rorive S, Belot N, Decaestecker C, Lefranc F, Gordower L, Micik S, Maurage CA, Kaltner H, Ruchoux MM, Danguy A, Gabius HJ, Salmon I, Kiss R, Camby I. Galectin-1 is highly expressed in human gliomas with relevance for modulation of invasion of tumor astrocytes into the brain parenchyma. Glia 2001;33:241–255.
85. Le Mercier M, Kondo Y, Mathieu V, Camby I, Decaestecker C, Kiss R, Lefranc F. Decreasing galectin-1 expression in human experimental glioblastomas impairs the endoplasmic reticulum stress (ERS) response and increases temozolomide antitumor effects. In: 98th Annual Meeting of the American Association for Cancer Research (AACR). Los Angeles;2007.
86. Paz A, Haklai R, Elad-Sfadia G, Ballan E, Kloog Y. Galectin-1 binds oncogenic H-Ras to mediate Ras membrane anchorage and cell transformation. Oncogene 2001; 20: 7486–7493.
87. Prior IA, Muncke C, Parton RG, Hancock JF. Direct visualization of Ras proteins in spatially distinct cell surface microdomains. J Cell Biol 2003;160:165–170.
88. Ashery U, Yizhar O, Rotblat B, Elad-Sfadia G, Barkan B, Haklai R, Kloog Y. Spatiotemporal organization of Ras signaling: rasosomes and the galectin switch. Cell Mol Neurobiol 2006;26:471–495.
89. Rotblat B, Niv H, Andre S, Kaltner H, Gabius HJ, Kloog Y. Galectin-1(L11A) predicted from a computed galectin-1 farnesyl-binding pocket selectively inhibits Ras-GTP. Cancer Res 2004;64:3112–3118.
90. Ingrassia L, Camby I, Lefranc F, Mathieu V, Nshimyumukiza P, Darro F, Kiss R. Antigalectin compounds as potential anticancer drugs. Curr Med Chem 2006; 13:3513–3527.
91. Camby I, Lefranc F, Kiss R. *In vivo* siRNA-induced decreases of galectin-1 expression in experimental models of human orthotopic glioblastoma xenografts contributes therapeutic benefits. In: 97th Annual Meeting of the American Association for Cancer Research (AACR). Washington, DC;2006.

92. Saussez S, Decaestecker C, Lorfevre F, Cucu DR, Mortuaire G, Chevalier D, Wacreniez A, Kaltner H, Andre S, Toubeau G, Camby I, Gabius HJ, Kiss R. High level of galectin-1 expression is a negative prognostic predictor of recurrence in laryngeal squamous cell carcinomas. Int J Oncol 2007;30:1109–1117.

93. van den Brule FA, Waltregny D, Castronovo V. Increased expression of galectin-1 in carcinoma-associated stroma predicts poor outcome in prostate carcinoma patients. J Pathol 2001;193:80–87.

94. Chiariotti L, Berlingieri MT, Battaglia C, Benvenuto G, Martelli ML, Salvatore P, Chiappetta G, Bruni CB, Fusco A. Expression of galectin-1 in normal human thyroid gland and in differentiated and poorly differentiated thyroid tumors. Int J Cancer 1995;64:171–175.

95. Xu XC, el-Naggar AK, Lotan R. Differential expression of galectin-1 and galectin-3 in thyroid tumors. Potential diagnostic implications. Am J Pathol 1995;147:815–822.

96. Endharti AT, Zhou YW, Nakashima I, Suzuki H. Galectin-1 supports survival of naive T cells without promoting cell proliferation. Eur J Immunol 2005;35:86–97.

97. Moiseeva EP, Javed Q, Spring EL, de Bono DP. Galectin 1 is involved in vascular smooth muscle cell proliferation. Cardiovasc Res 2000;45:493–502.

98. Sanford GL, Harris-Hooker S. Stimulation of vascular cell proliferation by beta-galactoside specific lectins. FASEB J 1990;4:2912–2918.

99. Kopitz J, von Reitzenstein C, Andre S, Kaltner H, Uhl J, Ehemann V, Cantz M, Gabius HJ. Negative regulation of neuroblastoma cell growth by carbohydrate-dependent surface binding of galectin-1 and functional divergence from galectin-3. J Biol Chem 2001;276:35917–35923.

100. Andersen H, Jensen ON, Moiseeva EP, Eriksen EF. A proteome study of secreted prostatic factors affecting osteoblastic activity: galectin-1 is involved in differentiation of human bone marrow stromal cells. J Bone Miner Res 2003;18:195–203.

101. Rubinstein N, Alvarez M, Zwirner NW, Toscano MA, Ilarregui JM, Bravo A, Mordoh J, Fainboim L, Podhajcer OL, Rabinovich GA. Targeted inhibition of galectin-1 gene expression in tumor cells results in heightened T cell-mediated rejection: a potential mechanism of tumor-immune privilege. Cancer Cell 2004;5:241–251.

102. Yamaoka K, Mishima K, Nagashima Y, Asai A, Sanai Y, Kirino T. Expression of galectin-1 mRNA correlates with the malignant potential of human gliomas and expression of antisense galectin-1 inhibits the growth of 9 glioma cells. J Neurosci Res 2000;59:722–730.

103. Adams L, Scott GK, Weinberg CS. Biphasic modulation of cell growth by recombinant human galectin-1. Biochim Biophys Acta 1996;1312:137–144.

104. Vas V, Fajka-Boja R, Ion G, Dudics V, Monostori E, Uher F. Biphasic effect of recombinant galectin-1 on the growth and death of early hematopoietic cells. Stem Cells 2005;23:279–287.

105. Wells V, Davies D, Mallucci L. Cell cycle arrest and induction of apoptosis by beta galactoside binding protein (beta GBP) in human mammary cancer cells. A potential new approach to cancer control. Eur J Cancer 1999;35:978–983.

106. Dunn GP, Old LJ, Schreiber RD. The immunobiology of cancer immunosurveillance and immunoediting. Immunity 2004;21:137–148.

107. Zou W. Immunosuppressive networks in the tumour environment and their therapeutic relevance. Nat Rev Cancer 2005;5:263–274.

108. Spiotto MT, Schreiber H. Rapid destruction of the tumor microenvironment by CTLs recognizing cancer-specific antigens cross-presented by stromal cells. Cancer Immun 2005;5:8.

109. Hirabayashi J, Ayaki H, Soma G, Kasai K. Cloning and nucleotide sequence of a full-length cDNA for human 14 kDa beta-galactoside-binding lectin. Biochim Biophys Acta 1989;1008:85–91.

110. Joubert R, Kuchler S, Zanetta JP, Bladier D, Avellana-Adalid V, Caron M, Doinel C, Vincendon G. Immunohistochemical localization of a beta-galactoside-binding lectin in rat central nervous system. I. Light- and electron-microscopical studies on developing cerebral cortex and corpus callosum. Dev Neurosci 1989;11:397–413.

111. Wollina U, Schreiber G, Gornig M, Feldrappe S, Burchert M, Gabius HJ. Sertoli cell expression of galectin-1 and -3 and accessible binding sites in normal human testis and Sertoli cell only-syndrome. Histol Histopathol 1999;14:779–784.

112. Berberat PO, Friess H, Wang L, Zhu Z, Bley T, Frigeri L, Zimmermann A, Buchler MW. Comparative analysis of galectins in primary tumors and tumor metastasis in human pancreatic cancer. J Histochem Cytochem 2001;49:539–549.

113. Gillenwater A, Xu XC, el-Naggar AK, Clayman GL, Lotan R. Expression of galectins in head and neck squamous cell carcinoma. Head Neck 1996;18:422–432.

114. Sanjuan X, Fernandez PL, Castells A, Castronovo V, van den Brule F, Liu FT, Cardesa A, Campo E. Differential expression of galectin 3 and galectin 1 in colorectal cancer progression. Gastroenterology 1997;113:1906–1915.

115. Shimonishi T, Miyazaki K, Kono N, Sabit H, Tuneyama K, Harada K, Hirabayashi J, Kasai K, Nakanuma Y. Expression of endogenous galectin-1 and galectin-3 in intrahepatic cholangiocarcinoma. Hum Pathol 2001;32:302–310.

116. Roberts AA, Amano M, Felten C, Galvan M, Sulur G, Pinter-Brown L, Dobbeling U, Burg G, Said J, Baum LG. Galectin-1-mediated apoptosis in mycosis fungoides: the roles of CD7 and cell surface glycosylation. Mod Pathol 2003;16:543–551.

117. Le QT, Shi G, Cao H, Nelson DW, Wang Y, Chen EY, Zhao S, Kong C, Richardson D, O'Byrne KJ, Giaccia AJ, Koong AC. Galectin-1: a link between tumor hypoxia and tumor immune privilege. J Clin Oncol 2005;23:8932–8941.

118. Birchmeier C, Birchmeier W, Gherardi E, Vande Woude GF. Met, metastasis, motility and more. Nat Rev Mol Cell Biol 2003;4:915–925.

119. Brinckerhoff CE, Matrisian LM. Matrix metalloproteinases: a tail of a frog that became a prince. Nat Rev Mol Cell Biol 2002;3:207–214.

120. Egeblad M, Werb Z. New functions for the matrix metalloproteinases in cancer progression. Nat Rev Cancer 2002;2:161–174.

121. Folkman J. Fundamental concepts of the angiogenic process. Curr Mol Med 2003;3:643–651.

122. Hanahan D, Weinberg RA. The hallmarks of cancer. Cell 2000;100:57–70.

123. Onn A, Fidler IJ. Metastatic potential of human neoplasms. In Vivo 2002;16:423–429.

124. Ruoslahti E. Specialization of tumour vasculature. Nat Rev Cancer 2002;2:83–90.

125. Vogelstein B, Kinzler KW. Cancer genes and the pathways they control. Nat Med 2004;10:789–799.

126. Bockhorn M, Roberge S, Sousa C, Jain RK, Munn LL. Differential gene expression in metastasizing cells shed from kidney tumors. Cancer Res 2004;64:2469–2473.

127. Buttiglieri S, Deregibus MC, Bravo S, Cassoni P, Chiarle R, Bussolati B, Camussi G. Role of Pax2 in apoptosis resistance and proinvasive phenotype of Kaposi's sarcoma cells. J Biol Chem 2004;279:4136–4143.

128. Goswami S, Wang W, Wyckoff JB, Condeelis JS. Breast cancer cells isolated by chemotaxis from primary tumors show increased survival and resistance to chemotherapy. Cancer Res 2004;64:7664–7667.

129. Haga A, Funasaka T, Niinaka Y, Raz A, Nagase H. Autocrine motility factor signaling induces tumor apoptotic resistance by regulations Apaf-1 and Caspase-9 apoptosome expression. Int J Cancer 2003;107:707–714.

130. Kinoshita K, Taupin DR, Itoh H, Podolsky DK. Distinct pathways of cell migration and antiapoptotic response to epithelial injury: structure–function analysis of human intestinal trefoil factor. Mol Cell Biol 2000;20:4680–4690.

131. Sekharam M, Zhao H, Sun M, Fang Q, Zhang Q, Yuan Z, Dan HC, Boulware D, Cheng JQ, Coppola D. Insulin-like growth factor 1 receptor enhances invasion and induces resistance to apoptosis of colon cancer cells through the Akt/Bcl-x(L) pathway. Cancer Res 2003;63:7708–7716.

132. Klemke RL, Leng J, Molander R, Brooks PC, Vuori K, Cheresh DA. CAS/Crk coupling serves as a "molecular switch" for induction of cell migration. J Cell Biol 1998; 140:961–972.

133. Cho SY, Klemke RL. Extracellular-regulated kinase activation and CAS/Crk coupling regulate cell migration and suppress apoptosis during invasion of the extracellular matrix. J Cell Biol 2000;149:223–236.

134. Kain KH, Gooch S, Klemke RL. Cytoplasmic c-Abl provides a molecular "Rheostat" controlling carcinoma cell survival and invasion. Oncogene 2003;22:6071–6080.

135. McCormick F. Cancer: survival pathways meet their end. Nature 2004;428:267–269.

136. Douma S, Van Laar T, Zevenhoven J, Meuwissen R, Van Garderen E, Peeper DS. Suppression of anoikis and induction of metastasis by the neurotrophic receptor TrkB. Nature 2004;430:1034–1039.

137. Lefranc F, James S, Camby I, Gaussin JF, Darro F, Brotchi J, Gabius J, Kiss R. Combined cimetidine and temozolomide, compared with temozolomide alone: significant increases in survival in nude mice bearing U373 human glioblastoma multiforme orthotopic xenografts. J Neurosurg 2005;102:706–714.

138. Lefranc F, Yeaton P, Brotchi J, Kiss R. Cimetidine, an unexpected antitumor agent, and its potential for the treatment of glioblastoma (review)Int J Oncol 2006;28:1021–1030.

139. Folkman J. Tumor angiogenesis: therapeutic implications. N Engl J Med 1971; 285:1182–1186.

140. Bergers G, Benjamin LE. Tumorigenesis and the angiogenic switch. Nat Rev Cancer 2003;3:401–410.

141. Dome B, Hendrix MJ, Paku S, Tovari J, Timar J. Alternative vascularization mechanisms in cancer: pathology and therapeutic implications. Am J Pathol 2007;170:1–15.

142. Maniotis AJ, Folberg R, Hess A, Seftor EA, Gardner LM, Pe'er J, Trent JM, Meltzer PS, Hendrix MJ. Vascular channel formation by human melanoma cells *in vivo* and *in vitro*: vasculogenic mimicry. Am J Pathol 1999;155:739–752.

143. Hendrix MJ, Seftor EA, Hess AR, Seftor RE. Vasculogenic mimicry and tumour-cell plasticity: lessons from melanoma. Nat Rev Cancer 2003;3:411–421.

144. Pugh CW, Ratcliffe PJ. Regulation of angiogenesis by hypoxia: role of the HIF system. Nat Med 2003;9:677–684.
145. Fukumori T, Kanayama HO, Raz A. The role of galectin-3 in cancer drug resistance. Drug Resist Updat 2007;10:101–108.
146. Nangia-Makker P, Honjo Y, Sarvis R, Akahani S, Hogan V, Pienta KJ, Raz A. Galectin-3 induces endothelial cell morphogenesis and angiogenesis. Am J Pathol 2000; 156:899–909.
147. John CM, Leffler H, Kahl-Knutsson B, Svensson I, Jarvis GA. Truncated galectin-3 inhibits tumor growth and metastasis in orthotopic nude mouse model of human breast cancer. Clin Cancer Res 2003;9:2374–2383.
148. Inufusa H, Nakamura M, Adachi T, Aga M, Kurimoto M, Nakatani Y, Wakano T, Miyake M, Okuno K, Shiozaki H, Yasutomi M. Role of galectin-3 in adenocarcinoma liver metastasis. Int J Oncol 2001;19:913–919.
149. Nangia-Makker P, Hogan V, Honjo Y, Baccarini S, Tait L, Bresalier R, Raz A. Inhibition of human cancer cell growth and metastasis in nude mice by oral intake of modified citrus pectin. J Natl Cancer Inst 2002;94:1854–1862.
150. Rabinovich GA, Cumashi A, Bianco GA, Ciavardelli D, Iurisci I, D'Egidio M, Piccolo E, Tinari N, Nifantiev N, Iacobelli S. Synthetic lactulose amines: novel class of anticancer agents that induce tumor-cell apoptosis and inhibit galectin-mediated homotypic cell aggregation and endothelial cell morphogenesis. Glycobiology 2006;16:210–220.
151. D'Haene N, Maris C, Sandras F, Dehou MF, Remmelink M, Decaestecker C, Salmon I. The differential expression of Galectin-1 and Galectin-3 in normal lymphoid tissue and non-Hodgkin's and Hodgkin's lymphomas. Int J Immunopathol Pharmacol 2005; 18:431–443.
152. Thijssen VL, Postel R, Brandwijk RJ, Dings RP, Nesmelova I, Satijn S, Verhofstad N, Nakabeppu Y, Baum LG, Bakkers J, Mayo KH, Poirier F, Griffioen AW. Galectin-1 is essential in tumor angiogenesis and is a target for antiangiogenesis therapy. Proc Natl Acad Sci USA 2006;103:15975–15980.
153. Lefranc F, Sadeghi N, Metens T, Brotchi J, Salmon I, Kiss R. Characterization of gastrin-induced cytostatic effect on cell proliferation in experimental malignant gliomas. Neurosurgery 2003;52:881–890; discussion 890–881.
154. Thakker DR, Natt F, Husken D, Maier R, Muller M, van der Putten H, Hoyer D, Cryan JF. Neurochemical and behavioral consequences of widespread gene knockdown in the adult mouse brain by using nonviral RNA interference. Proc Natl Acad Sci USA 2004;101:17270–17275.
155. Ozawa K, Tsukamoto Y, Hori O, Kitao Y, Yanagi H, Stern DM, Ogawa S. Regulation of tumor angiogenesis by oxygen-regulated protein 150, an inducible endoplasmic reticulum chaperone. Cancer Res 2001;61:4206–4213.
156. Kim R, Emi M, Tanabe K, Murakami S. Role of the unfolded protein response in cell death. Apoptosis 2006;11:5–13.
157. Shannon AM, Bouchier-Hayes DJ, Condron CM, Toomey D. Tumour hypoxia, chemotherapeutic resistance, and hypoxia-related therapies. Cancer Treat Rev 2003;29:297–307.
158. Berger BJ, Muller TS, Buschmann IR, Peters K, Kirsch M, Christ B, Prols F. High levels of the molecular chaperone Mdg1/ERdj4 reflect the activation state of endothelial cells. Exp Cell Res 2003;290:82–92.

159. Prols F, Mayer MP, Renner O, Czarnecki PG, Ast M, Gassler C, Wilting J, Kurz H, Christ B. Upregulation of the cochaperone Mdg1 in endothelial cells is induced by stress and during *in vitro* angiogenesis. Exp Cell Res 2001;269:42–53.

160. Buttery R, Monaghan H, Salter DM, Sethi T. Galectin-3: differential expression between small-cell and non-small-cell lung cancer. Histopathology 2004;44:339–344.

161. Le Mercier M, Lefranc F, Mijatovic T, Debeir O, Haibe-Kains B, Bontempi G, Decaestecker C, Kiss R, Mathieu V. Evidence of galectin-1 involvement in glioma chemoresistance. Toxicol Appl Pharmacol 2008; epub ahead of print.

162. Kroemer G, Jäättelä M. Lysosomes and autophagy in cell death control. Nat Rev Cancer 2005;5:886–897.

163. Hori O, Ichinoda F, Yamaguchi A, Tamatani T, Taniguchi M, Koyama Y, Katayama T, Tohyama M, Stern DM, Ozawa K, Kitao Y, Ogawa S. Role of Herp in the endoplasmic reticulum stress response. Genes Cells 2004;9:457–469.

164. Kurisu J, Honma A, Miyajima H, Kondo S, Okumura M, Imaizumi K. MDG1/ERdj4, an ER-resident DnaJ family member, suppresses cell death induced by ER stress. Genes Cells 2003;8:189–202.

165. Andre S, Pieters RJ, Vrasidas I, Kaltner H, Kuwabara I, Liu FT, Liskamp RM, Gabius HJ. Wedgelike glycodendrimers as inhibitors of binding of mammalian galectins to glycoproteins, lactose maxiclusters, and cell surface glycoconjugates. ChemBioChem 2001;2:822–830.

166. Sorme P, Kahl-Knutsson B, Wellmar U, Magnusson BG, Leffler H, Nilsson UJ. Design and synthesis of galectin inhibitors. Methods Enzymol 2003;363:157–169.

167. Ingrassia L, Nshimyumukiza P, Dewelle J, Lefranc F, Wlodarczak L, Thomas S, Dielie G, Chiron C, Zedde C, Tisnes P, van Soest R, Braekman JC, Darro F, Kiss R. A lactosylated steroid contributes *in vivo* therapeutic benefits in experimental models of mouse lymphoma and human glioblastoma. J Med Chem 2006;49:1800–1807.

168. Hamm CW. Antiintegrin therapy. Annu Rev Med 2003;54:425–435.

169. Shen J, Person MD, Zhu J, Abbruzzese JL, Li D. Protein expression profiles in pancreatic adenocarcinoma compared with normal pancreatic tissue and tissue affected by pancreatitis as detected by two-dimensional gel electrophoresis and mass spectrometry. Cancer Res 2004;64:9018–9026.

170. François C, van Velthovan R, De Lathouwer O, Moreno C, Peltier A, Kaltner H, Salmon I, Gabius HJ, Danguy A, Decaestecker C, Kiss R. Galectin-1 and galectin-3 binding pattern expression in renal cell carcinomas. Am J Clin Pathol 1999;112:194–203.

171. Saussez S, Nonclercq D, Laurent G, Wattiez R, André S, Kaltner H, Gabius HJ, Kiss R, Toubeau G. Toward functional glycomics by localization of tissue lectins: immunohistochemical galectin fingerprinting during diethylstilbestrol-induced kidney tumorigenesis in male Syrian hamster. Histochem Cell Biol 2005;123:29–41.

172. Cindolo L, Benvenuto G, Salvatore P, Pero R, Salvatore G, Mirone V, Prezioso D, Altieri V, Bruni CB, Chiariotti L. Galectin-1 and galectin-3 expression in human bladder transitional-cell carcinomas. Int J Cancer 1999;84:39–43.

173. Bozić M, Petronijević M, Milenković S, Atanacković J, Lazić J, Vićovac Lj. Galectin-1 and galectin-3 in the trophoblast of the gestational trophoblastic desease. Placenta 2004;25:797–802.

174. Jung EJ, Moon HG, Cho BI, Jeong CY, Joo YT, Lee YJ, Hong SC, Choi SK, Ha WS, Kim JW, Lee CW, Lee JS, Park ST. Galectin-1 expression in cancer-associated stromal cells

correlates tumor invasiveness and tumor progression in breast cancer. Int J Cancer 2007;120:2331–2338.

175. Szoke T, Kayser K, Baumhakel JD, Trojan I, Furak J, Tiszlavicz L, Horvath A, Szluha K, Gabius HJ, Andre S. Prognostic significance of endogenous adhesion/growth-regulatory lectins in lung cancer. Oncology 2005;69:167–174.

176. Fitzner B, Walzel H, Sparmann G, Emmrich J, Liebe S, Jaster R. Galectin-1 is an inductor of pancreatic stellate cell activation. Cell Signal 2005;17:1240–1247.

10

GALECTIN-3 IN THE PROGRESSION AND METASTASIS OF COLORECTAL NEOPLASIA

JAMES C. BYRD AND ROBERT S. BRESALIER

Department of Gastroenterology, Hepatology, and Nutrition, The University of Texas MD Anderson Cancer Center, Houston, TX 77030-4009, USA

INTRODUCTION

Colorectal cancer is the second most common cause of cancer-related death in the United States. Colorectal cancer related mortality is due in large part to metastasis, a complex multistage process by which tumor cells escape the primary tumor and establish secondary foci at distant sites. Galectin-3, an endogenous lectin, and glycoproteins that bind to galectin-3 have each been associated with the metastatic potential of human colon cancer cells. Galectin-3 may be localized in the nucleus, cytoplasm, and cell surface or be secreted and may therefore act in several ways to promote metastasis. Direct evidence in support of particular functions for galectin-3 is now emerging through the identification of interacting ligands and distinct structural domains that contribute to its pleiotropic functions. This chapter will focus on work from our laboratory (1–14) on the role of galectin-3 in neoplastic progression in the colon (1–3), the interaction between galectin-3 and colon cancer mucin (4–7), the functional consequences of galectin-3 phosphorylation (5, 8–11), and the diagnostic and therapeutic potential of galectin-3 and galectin-3-binding glycoproteins (12–14).

Galectins, Edited by Anatole A. Klyosov, Zbigniew J. Witczak, and David Platt

GALECTIN-3 IN COLON CANCER

The galectins are widely distributed and evolutionarily conserved carbohydrate-binding proteins characterized by their binding affinity for beta-galactosides and by conserved sequence elements in the carbohydrate-binding site (15). Among at least 14 mammalian galectins that have been identified, galectin-3 is of the most interest with regard to colon cancer (1, 2, 5, 6, 16, 17). Galectin-3 is a chimeric gene product with a monomer subunit of about 30,000 Da that undergoes noncovalent homodimerization (18).

Of all the galectins reported so far, galectin-3 is unique in that it shares the carbohydrate-binding domain with other galectins and possesses a 20-residue-long N-terminal and a domain consisting of Pro-Gly-Tyr-rich repeating units. Galectin-3 has pleiotropic biological functions and has been implicated in cell growth, differentiation, adhesion, RNA processing, and malignant transformation (19–21). It is found in the cytoplasm, on the cell surface, in the nucleus, and is secreted by tumor and inflammatory cells. While many of its biological functions appear to be based on its role as a carbohydrate-binding protein, several intracellular ligands may interact with galectin-3 via protein–protein, rather than lectin–glycoconjugate interactions.

Expression of Galectin-3 Correlates with Neoplastic Progression in the Colon

In our initial study (1), galectin-3 expression was studied in 153 human colonic tissue specimens including 29 adenomas containing early cancer, 66 colon carcinomas of known Dukes' stage with available long-term patient survival data, and 23 additional primary carcinomas with 35 associated metastases. An immunohistochemical scoring system was used that considers tumor heterogeneity and yields an integrated numeric score subject to statistical analysis. Adenomatous tissue demonstrated weak but homogeneous cytoplasmic staining for galectin-3, while associated carcinomas demonstrated moderate to strong staining (Fig. 10.1). Galectin-3 expression was significantly higher in high grade dysplasia and early invasive cancers than in the adenomatous tissues from which they evolved ($p = 0.008$).

Furthermore, metastases expressed a higher level of galectin-3 than the primary cancers from which they evolved ($p < 0.005$). Galectin-3 expression also varied according to Dukes' stage (Table 10.1). Scores were compared among five separate stages (Dukes' A, B1, B2, C, and D). A statistically significant difference was detected for the comparison of the five groups ($p = 0.016$) as well as for comparison of Dukes' A and B1 combined versus the remainder of the group ($p = 0.006$). Enhanced expression also correlated with decreased long-term patient survival ($p = 0.021$). Also, cultured cells of high metastatic capability had a higher level of galectin-3 compared to their counterparts with low metastatic potential. Tissue expression of galectin-3 in colonic mucosa therefore appears to be related to neoplastic transformation and metastatic progression.

Most studies from other laboratories have also found a positive correlation between total galectin-3 and colon cancer progression (22–24). Others have not found an increase in total galectin-3, but have found that cytoplasmic galectin-3 is increased and nuclear galectin-3 is decreased in colon cancers as compared to normal

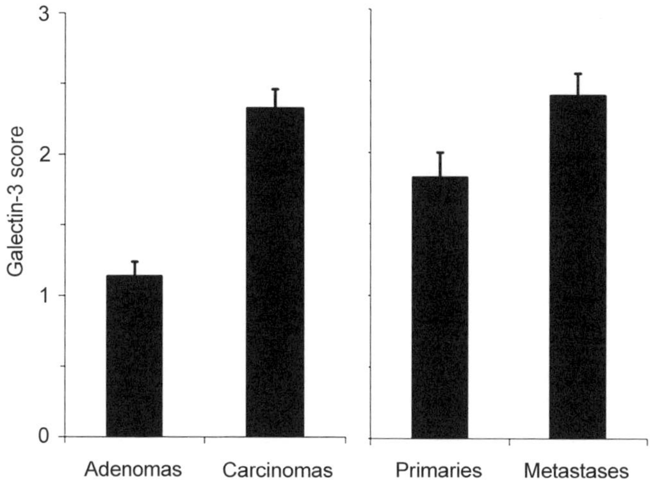

FIGURE 10.1 (Left panel) Galectin-3 staining (mean score and SEM) of adenomatous polyps and foci of adenocarcinoma within the same polyps ($P = 0.001$). (Right panel) Galectin-3 expression in metastases and the primary tumors from which they were derived ($P < 0.005$).

colon (16, 17). Furthermore, galectin-3 concentrations have been found to be increased in sera from colorectal cancer patients (12) and to be higher in those with metastatic disease than in patients with localized tumors (25).

Metastasis of Human Colon Cancer is Altered by Modifying Expression of Galectin-3

The correlation of galectin-3 expression with neoplastic progression in the colon suggests but does not demonstrate a role of galectin-3 in colon cancer metastasis. To more directly establish the role of galectin-3 in colon cancer metastasis, we used human colon cancer cell lines that could be experimentally manipulated to alter their levels of galectin-3 expression. Earlier studies from our laboratory established experimental models for colon cancer metastasis (25–31) and allowed detailed examination of variant cell lines that differed in their metastatic capacity. LS-LiM6, a derivative of LS174T with high liver-metastasizing ability during cecal growth, was established by serially selecting cells that metastasized from cecum to liver. Western analyses (1) showed that galectin-3 protein was expressed to an eightfold greater

TABLE 10.1 Galectin-3 Expression According to Dukes' Stage

Dukes' Stage	Number of Specimens	Staining Score (Mean ± SD)
A	12	1.92 ± 0.86
B1	8	1.63 ± 1.0
B2	14	2.14 ± 0.59
C	18	2.54 ± 0.85
D	14	2.71 ± 0.56

extent in the highly metastatic LS-LiM6 line compared to its less metastatic parent LS174T, consistent with a role of galectin-3 in colon cancer progression.

Galectin-3 levels were manipulated in human colon cancer cells using eukaryotic expression constructs designed to express the complete galectin-3 complementary DNA in either the sense or antisense orientation (2). LS-LiM6 cells, as well as HM7, another highly metastatic variant, were stably transfected with antisense galectin-3. Conversely, parental LS174T colon cancer cells, with low metastatic potential and low levels of native galectin-3, were transfected with a plasmid containing the complete 881-base pair (bp) coding sequence of galectin-3 designed to confer expression of galectin-3 mRNA.

Introduction of galectin-3 antisense into metastatic colon cancer cells (LS-LiM6, HM7) resulted in a significant reduction in galectin-3-specific messenger RNA and total and cell surface galectin-3 protein (Fig. 10.2). Conversely, a stable integration of

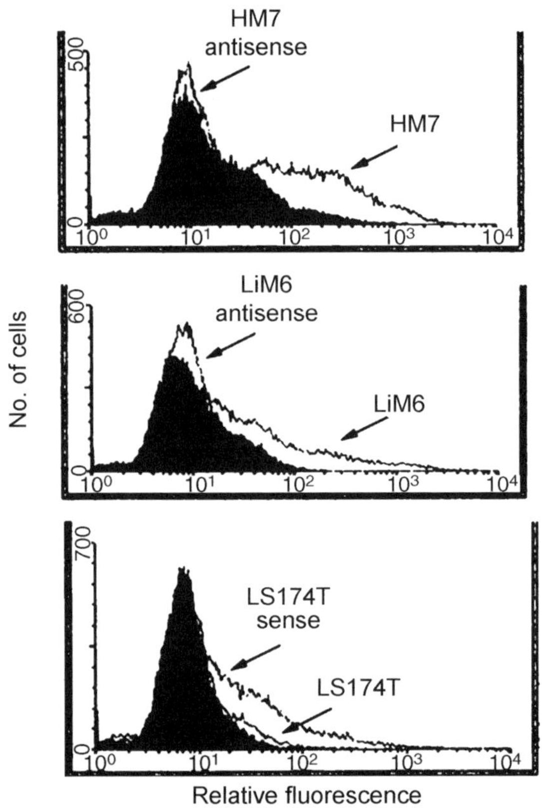

FIGURE 10.2 Alterations in cell surface galectin-3 after stable transfection. Single-cell suspensions of colon cancer cell lines were exposed to antigalectin 3 MAb TIB-166 and submitted to flow analysis. Introduction of galectin-3 antisense into metastatic cell lines HM7 and LS-LiM6 resulted in reductions in cell surface galectin-3, whereas the introduction of galectin-3 sense into low-metastatic LS174T resulted in increased cell surface expression. Solid areas depict controls (second antibody only). Reproduced from Reference 2 with permission.

TABLE 10.2 Effect of Galectin-3 on Liver Colonization by Colon Cancer Cells

Cell Line	Liver Weight, g (±SD)	Tumor Nodules	% Liver Replaced (±SD)
HM7 control	3.33 ± 0.98	>500	70.24 ± 14.51
HM7 galectin-3 antisense	1.06 ± 0.16	0–38	0.67 ± 1.25
LS-LiM6 control	3.36 ± 1.17	>500	51.08 ± 11.53
LS-LiM6 galectin-3 antisense	1.05 ± 0.04	0–12	2.02 ± 4.05
LS174T control	0.45 ± 0.10	0	0
LS174T galectin-3 sense	3.38 ± 1.02	350–500	57.45 ± 6.07

galectin-3 in the sense orientation resulted in an increase in cellular and cell surface galectin-3 in cells of low metastatic potential (LS174T). Liver colonization was assessed in athymic mice after splenic-portal inoculation or after spontaneous metastasis during cecal growth. Downregulation of galectin-3 expression by antisense transfection resulted in a significant decrease in liver colonizing ability, whereas upregulation of galectin-3 increased metastatic potential (Table 10.2). This provided the first direct evidence that galectin-3 plays a role in colon cancer metastasis.

INTERACTION OF GALECTIN-3 WITH COLON CANCER MUCIN

Alterations in the production of galectin-3 (1, 2) and of MUC2 intestinal mucin (28, 32) have each been correlated with the malignant behavior of human colon cancer cells, and colon cancer cells that differ quantitatively in MUC2 expression may also vary in the expression of galectin-3. Experiments from our laboratory have shown that MUC2 mucin is a major ligand for galectin-3 (4, 5) and that galectin-3 can modulate the expression of MUC2 mucin in human colon cancer cells (6, 7). This may have important implications for understanding the role of galectin-3 in colon cancer metastasis.

Mucins are the major secreted glycoproteins of the gastrointestinal tract and play a role in normal physiology and in pathological processes and in the colon (33–40). While the human intestinal mucin encoded by the MUC2 gene is a major product of goblet cells of the colon and small intestine (41), several studies have shown that there is MUC2 expression in colorectal adenocarcinomas and in mucinous carcinomas (39, 40, 42–45). In immunohistochemical studies comparing the expression of MUC2 apoprotein in colon cancers with normal colon, there was decreased expression of MUC2 apomucin in adenocarcinomas, but mucinous carcinomas (which confer a worse prognosis in most studies) were strongly positive for MUC2 (37, 38, 45–47).

Targeted inactivation of the murine Muc2 gene leads to tumor development in the intestine of aged mice (48), accompanied by increased proliferation, decreased apoptosis, and increased migration of intestinal epithelial cells. It is not known whether these alterations are a primary response to Muc2 absence or a secondary response to the inadequate protection or lubrication of the intestinal epithelium (34). In spite of this uncertainty about the importance of mucin in the early stages of carcinogenesis, there is considerable evidence that mucin-associated glycoprotein structures, including those carried on the MUC2 apoprotein, play a role in the later and lethal stages of tumor progression and metastasis.

The conclusion that mucin production by human colonic carcinoma cells correlates with their metastatic potential in animal models of colon cancer metastasis was based on several lines of evidence: cells selected for high metastatic capacity had more mucin; cells selected for high mucin content were more metastatic; inhibition of mucin glycosylation decreased liver colonization; a mucin-specific glycosylation mutant was less metastatic; and inhibition of MUC2 expression (by antisense) decreased metastasis (28, 32, 49).

Aside from alterations in the amount of mucin protein, adenocarcinomas express aberrant forms of mucins. Earlier studies from our laboratory documented a role of mucin-associated carbohydrate structures in the metastasis of colon cancer cells to the liver (28, 49–52). In particular, the expression of sialylated mucin-associated carbohydrates is higher in liver metastases than in associated primary colon cancers. The presence of a large number of oligosaccharide chains (estimated 29 N-linked and 1751 O-linked in MUC2 mucin (53)) with a variety of distinct oligosaccharide structures provides an enormous range of potential ligands for interactions with carbohydrate-binding proteins such as galectin-3.

Binding of Galectin-3 to Colon Cancer Mucin

Colon Cancer Mucin is a Major Ligand for Galectin-3 Galectin-3 binds poly-*N*-acetyllactosamine structures on glycoproteins, but its natural ligands remain to be fully defined. When serum from a patient with metastatic colorectal cancer was chromatographed on Superose 6, >70% of circulating galectin-3 ligand was at the void volume, coincident with mucins (Fig. 10.3). Galectin-3 bound to purified native and desialylated colon cancer mucin in a concentration-dependent manner, which was completely inhibited by 0.1 M lactose, the competitive inhibitory sugar for this protein.

Mucin purified from highly metastatic LS-LiM6 human colon cancer cells bound galectin-3 to a twofold greater extent than mucin from low metastatic parental cell line LS174T. Desialylation increased binding to mucin >4-fold. Mucin purified from LS-B colon cancer cells is fully glycosylated and bound >40-fold more galectin-3 than mucin purified from clonal cell line LS-C, which produces mucin lacking peripheral carbohydrate structures. Endogenous galectin-3 was detected by Western analysis in all these cell lines, and its expression was related to mucin production and metastatic capacity. This study (4) identified colon cancer mucin as a ligand for galectin-3 and showed that binding of galectin-3 to mucins depends on peripheral carbohydrate structures. Binding of this endogenous lectin to mucins may influence cellular interactions that play a role in colon cancer metastasis.

Phosphorylation of Galectin-3 Downregulates Carbohydrate-Dependent Binding to Mucin A better understanding of the function of galectin-3 in normal and neoplastic epithelium requires determination of the distinct role of different portions

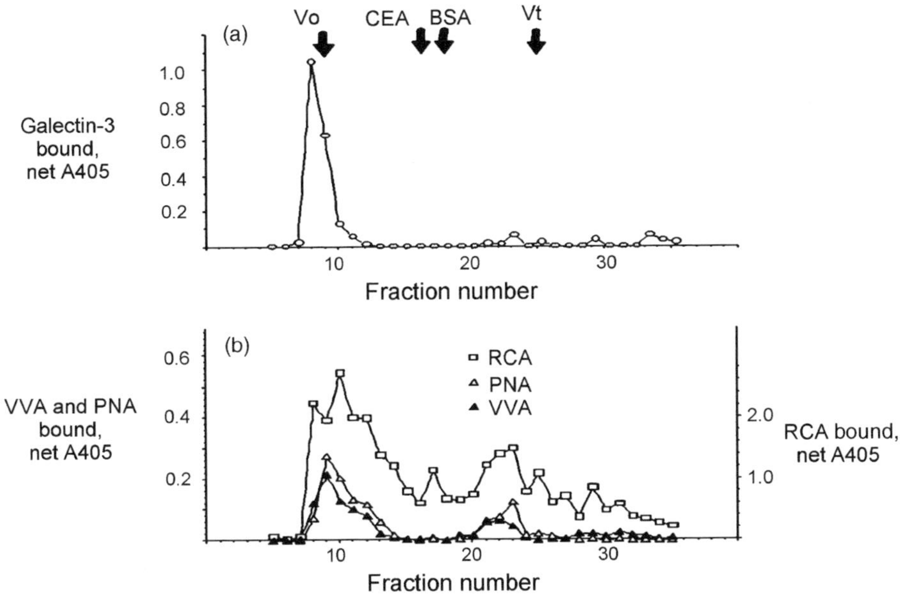

FIGURE 10.3 Demonstration of high molecular weight galectin-3 ligand in colon cancer serum. (a) Serum from a patient with metastatic colon cancer was chromatographed on Superose 6 in 50 mM NH_4HCO_3. Arrows, left to right, show the void volume (Blue Dextran 2000), elution volumes of carcinoembryonic antigen (180 kDa) and serum albumin (67 kDa), and the total column volume. Aliquots of 0.15 mL were adsorbed to polystyrene microtiter plates and assayed for binding of galectin-3. (b) Fractions were assayed for binding of *Ricinus communis* agglutinin, peanut agglutinin, and *Vicia villosa* agglutinin B4. Reproduced from Reference 4 with permission.

of the molecule in ligand–lectin interactions. Much of our understanding of distinct functions of different structural domains of galectin-3 has been derived from the construction of galectin-3 mutations (Table 10.3).

Galectin-3 may be phosphorylated at amino terminal Ser6 (major) and Ser12 (minor) and the major acidic residues on both sides of Ser6 make this a likely substrate for casein kinase I and/or casein kinase II (54–56). Phosphorylated galectin-3 has been demonstrated to be present in the cytosolic and nuclear fractions of 3T3 fibroblasts and in cultured polarized canine epithelial (Madin–Darby canine kidney) cells (54, 55). Metabolic labeling in our laboratory confirmed the presence of phosphorylated galectin-3 *in vivo* in the cytosol of human colon cancer cells (5). It has been suggested that phosphorylation may modulate the intracellular function and translocation of galectin-3, but the functional role of phosphorylation in determining interactions of this endogenous lectin with its ligands remains to be determined (54).

We studied the effect of phosphorylation on binding of galectin-3 to two of its ligands, laminin and purified colon cancer mucin (Fig. 10.4). Human recombinant galectin-3 was phosphorylated *in vitro* by casein kinase I. Galectin-3 and its phosphorylated form were separated by native isoelectric focusing for use in

TABLE 10.3 Functional Consequences of Galectin-3 Mutations

Mutation	Effects on Function	References
Del 2–12	↓ Secretion, ↓ nuclear localization	8
Ser6 → Glu	No phosphorylation, unchanged beta-Gal binding, unchanged secretion, ↓ antiapoptotic functions	57, 58
Ser6 → Ala		
Arg144 → Ser	↓ Binding of NeuAc-alpha-3-Gal	59
Trp181 → Leu	↓ Binding of beta-Gal	60
Gly182 → Ala	↓ Antiapoptotic functions	61
Arg224 → Ala	↓ Importin binding	62
Ile240 → Ala	↓ Nuclear import	63
Leu242 → Ala		
Thr243 → Ala		

Sequence of human galectin-3

MADNF**S**LHDA LSGSGNPNPQ GWPGAWGNQP AGAGGYPGAS YPGAYPGQAP
PGAYPGQAPP GAYHGAPGAY PGAPAPGVYP GPPSGPGAYP SSGQPSAPGA
YPATGPYGAP AGPLIVPYNL PLPGGVVPRM LITILGTVKP NAN**R**IALDFQ
RGNDVAFHFN PRFNENNRRV IVCNTKLDNN **WG**REERQSVF PFESGKPFKI
QVLVEPDHFK VAVNDAHLLQ YNH**R**VKKLNE ISKLGISGD**I** D**LT**SASYTMI

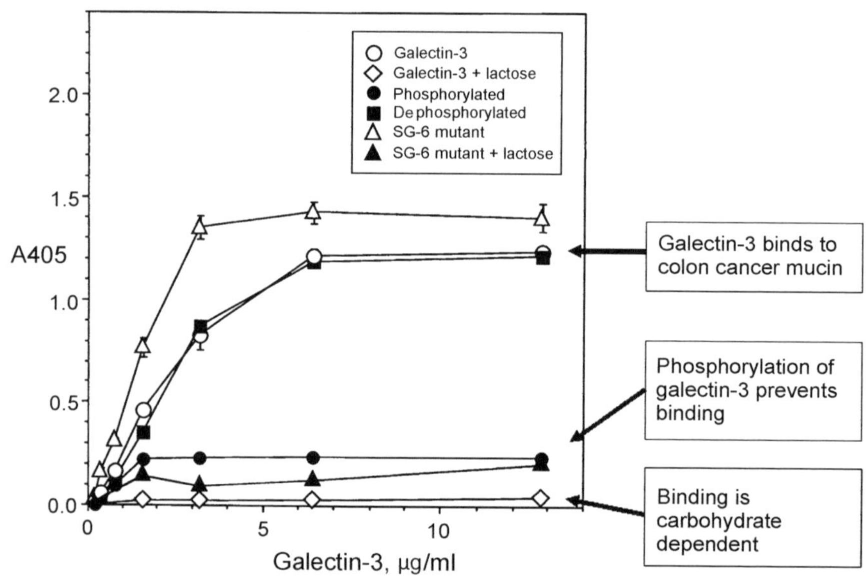

FIGURE 10.4 Saturation binding of galectin-3 and phosphorylated galectin-3 to immobilized asialomucin. Serial dilutions of galectin-3, phosphorylated galectin-3, galectin-3-P dephosphorylated with protein phosphatase type I, and galectin-3 mutated (Ser6Glu) at serine 6 (SG-6 mutant) in the presence or absence of 0.3 M lactose were added to microtiter wells coated with mucin in a solid-phase binding assay. Bars: SE of quadruple analyses.

solid-phase binding assays. Galectin-3 bound to laminin and asialomucin in a dose-dependent manner with half-maximal binding at 1.5 μg/mL (0.06 μM). Phosphorylation reduced saturation binding to each ligand by >85%. Ligand binding was fully restored by dephosphorylation with protein phosphatase type 1. Mutation of galectin-3 at N-terminal Ser6 (Ser to Glu) to introduce a negative charge similar to phosphoserine did not reduce ligand binding, but resulted in increased ligand binding capacity somewhat compared to native galectin-3. Thus, the binding of galectin-3 to colon cancer mucins is not only carbohydrate dependent, but is also influenced by the N-terminal domain, because phosphorylation of serine residue 6 reversibly inhibits binding. Phosphorylation of galectin-3, acting as an "on–off switch" for protein–carbohydrate interactions, may have important implications for understanding the biological functions of this protein.

Galectin-3 Induces MUC2 Expression

Effect of Alterations in Galectin-3 Levels on MUC2 Expression Since the production of MUC2 intestinal mucin and the β-galactoside-binding protein galectin-3 have each been independently correlated with the malignant behavior of human colon cancer cells, we sought to study the relationship between galectin-3 production and MUC2 mucin synthesis. Galectin-3 and MUC2 levels in human colon cancer cells were determined by Western blot analysis of total cell homogenates (Fig. 10.5, left panel). It was observed that colon cancer cell lines with high levels of galectin-3 also demonstrated high levels of MUC2 mucin while those with low galectin-3 levels had low mucin levels. For example, LM12, a poorly metastatic low mucin variant of LS174T, had low levels of galectin-3, while LS-LiM6, a high mucin metastatic variant of LS174T, had higher levels of galectin-3.

The effect of galectin-3 levels on MUC2 production was assessed by stable transfection of sense and antisense constructs of galectin-3 into human colon cancer cells LS174T, LS-LiM6, and HM7. LS174T (with low basal galectin-3 expression) was transfected stably with the galectin-3 sense plasmid, and a galectin-3 sense clone

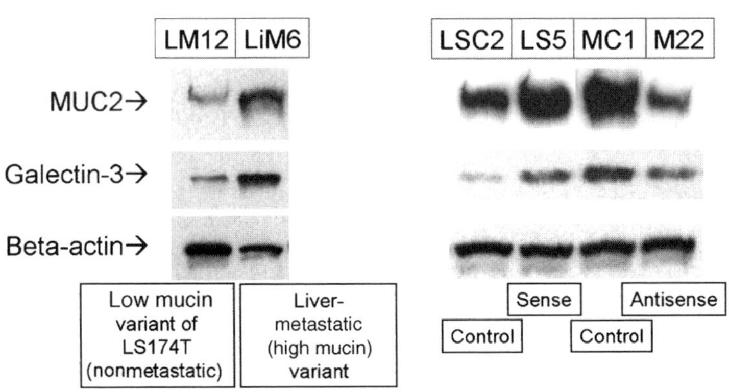

FIGURE 10.5 Western blot analyses of MUC2 in cell lines that vary in expression of galectin-3.

LS5 and a control clone LSC2 were selected. Similarly, LiM6 (with high basal galectin-3 expression) was transfected with the galectin-3 antisense plasmid, and a galectin-3 antisense clone M22 and a control vector clone MC1 were selected (Fig. 10.5, right panel).

Introduction of a galectin-3 antisense construct into LS-LiM6 and HM7 cells resulted in a 35–90% reduction of galectin-3 protein by Western blot analysis with a corresponding 68–93% reduction in immunoreactive MUC2 protein levels. Conversely, introduction of a galectin-3 sense construct into LS174T cells resulted in a 2-fold increase in galectin-3 protein and a 14-fold increase in MUC2 protein levels. There was no change in actin or MUC5AC mucin levels after transfection. Analogous transfection of antisense MUC2 did not alter galectin-3 RNA or protein levels.

The parallel expression of galectin-3 and MUC2 protein could have arisen by chance or could be caused by an unknown factor affecting the expression of both proteins. To better establish a cause-and-effect relationship between galectin-3 and MUC2 expression, the effect of inducible galectin-3 antisense on MUC2 mucin production was assessed *in vitro* and *in vivo* (6). A series of cell lines transfected with an antisense construct of galectin-3 under the control of the tetracycline-inducible promoter was generated to allow controlled manipulation of galectin-3 levels. Induction of galectin-3 antisense by tetracycline resulted in reduction in total cellular galectin-3 and cell surface galectin-3 (30–47% reduction) in three independent cell lines, AG1, AG4, and AG11. This was accompanied by concomitant parallel decreases in total cellular MUC2 protein (AG1, 2.80 ± 1.39-fold; AG11, 1.96 ± 0.25-fold; AG4, reduced to barely detectable levels) as determined by Western blot analysis and in total mucin as determined by metabolic labeling and gel filtration chromatography (42% decrease). The efficiency of secretion was little affected. The inducible reduction in galectin-3 and MUC2 was reversible after withdrawal of tetracycline from the media. Induction of antisense to galectin-3 *in vivo* was associated with decreases in both galectin-3 and MUC2 protein in cecal xenografts (Fig. 10.6). The beta-galactoside-binding protein galectin-3 therefore modulates the expression of its major ligand MUC2 mucin in human colon cancer cells. This may have important implications for understanding the role of galectin-3 in colon cancer metastasis.

Galectin-3 Modulates MUC2 Mucin Expression at the Level of Transcription Via AP-1 Activation Several transcription factors have been reported to regulate MUC2 expression in other systems. For example, MUC2 can be upregulated by lipopolysaccharide, *Haemophilus influenzae*, and PMA via NF-kappa-B (65–67). Transcription factors that have been shown to regulate MUC2 expression in other contexts include SP-1, CDX-2, and GATA-5 (68–70). Recent results from this laboratory have demonstrated that bile acids can induce MUC2 in colon cancer cells through a process involving PKC signaling and AP-1 site in the MUC2 promoter (64). Galectin-3 modulates the expression of MUC2 protein in human colon cancer cells, but the specific regulatory mechanism is unknown.

To elucidate the mechanism by which galectin-3 increases the expression of MUC2 and other genes, we sought to determine whether galectin-3 acts to increase

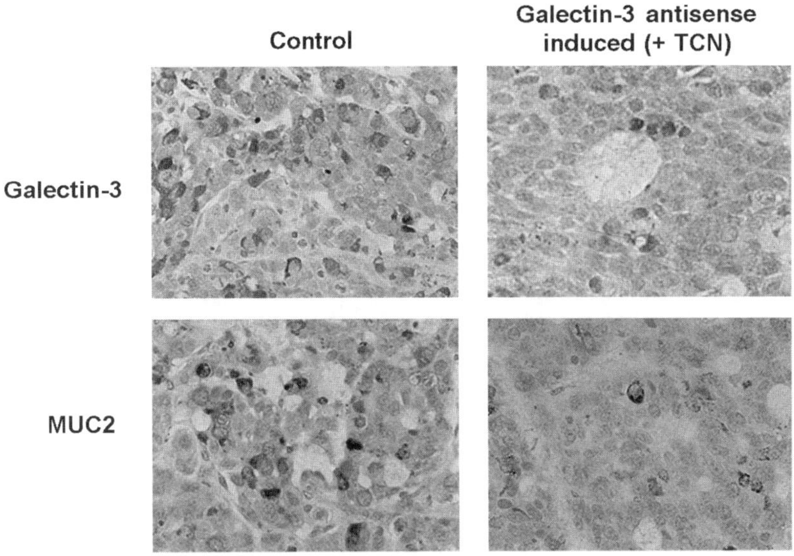

FIGURE 10.6 Effect of induction of galectin-3 antisense on galectin-3 and MUC2 expression *in vivo*. Athymic nude mice were injected in the cecal wall with colon cancer cell line AG4 containing antisense to galectin-3 under control of a tetracycline-inducible promoter. Animals were treated with 1 μg/mL tetracycline in the drinking water for 1 week before injection and continuously for 6 weeks thereafter to induce galectin-3 antisense (right column) or with water minus tetracycline (left column). Cecal xenografts were removed and stained for galectin-3 and MUC2. (See color insert.)

MUC2 expression at the level of transcription or at a post-transcriptional stage, for example, through RNA splicing (7). Promoter reporter constructs comprising sequences 2.2 kb upstream of the transcriptional start site were used to monitor MUC2 transcriptional activity with nuclear extracts from cells engineered to express different levels of galectin-3 (Fig. 10.7). We found that colon cancer cells stably transfected with galectin-3 sense (LS5) had higher levels of MUC2 transcriptional activity than LS174T cells transfected with control vector (LSC2). Conversely, LS-LiM6 cells transfected with galectin-3 antisense (M22) had lower levels of MUC2 transcriptional activity than the control (MC1).

On the basis of deletion analyses, we infer that the region between 1871 and 2185 bp upstream of the MUC2 translation start site is required for the galectin-3-dependent induction of MUC2 transcription (Fig. 10.7). This region contains a single AP-1 site (GCCAGTGACTCCATAGTCGCC). We further examined the correlation of AP-1 activity with MUC2 transcriptional activity and found that both MUC2 transcription and AP-1 activity were markedly higher in the highly metastatic LS-LiM6 cell line than in the related poorly metastatic LM12 cell line. Furthermore, MUC2 transcription was inhibited by cotransfection with a dominant negative AP-1 vector. The levels of both MUC2 transcription and AP-1 activity were found to be up-regulated in LS174T cells transfected with galectin-3 sense (clone LS5) and downregulated in LS-LiM6 cells transfected with galectin-3 antisense (clone M22).

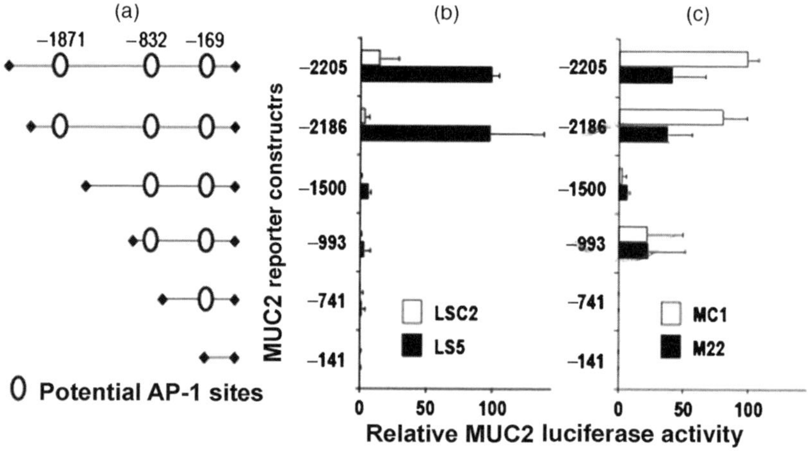

FIGURE 10.7 Deletion analysis of MUC2 promoter activity. (a) Schematic representation of luciferase reporter constructs containing various lengths of the MUC2 promoter. (b) Deletion constructs of the MUC2 promoter were transfected transiently into galectin-3 sense clone LS5 cells (solid bars) and control LSC2 cells (open bars). (c) Deletion constructs of MUC2 promoter were transfected transiently into the galectin-3 antisense clone M22 (open bars) and MC1 vector control (solid bars). Cells were harvested and analyzed for luciferase activity 48 hours after transfection. Reproduced from Reference 7 with permission.

Taken together, these results indicate that galectin-3 upregulation of MUC2 occurs at the level of transcription, rather than at a post-transcriptional stage, and suggest that the activation of AP-1 could be involved in this process.

Further characterization of the effects of galectin-3 on MUC2 transcription indicates that galectin-3 transactivates AP-1 by direct interaction with c-Jun, and Fra-1, two components of AP-1. Electromobility shift assays, coimmunoprecipitation, and chromatin immunoprecipitation analyses demonstrate an association between galectin-3, c-Jun, and Fra-1 in forming a complex at the AP-1 site on the MUC2 promoter (Fig. 10.8). The finding that galectin-3 increases MUC2 transcriptional activity through activation of AP-1 provides a mechanism by which galectin-3 can increase the production of its own ligand, MUC2, in colon cancer cells.

PHOSPHORYLATION OF GALECTIN-3

In addition to the role of phosphorylation of galectin-3 in modulating its binding to carbohydrate ligands, Ser6 phosphorylation has been implicated in other functions of galectin-3 that do not involve glycoprotein binding, including apoptosis and gene induction. Furthermore, the cellular compartmentalization of galectin-3, which is relevant to its distinct functions in glycoprotein binding, apoptosis, and gene induction, may also be influenced by Ser6 phosphorylation. The demonstration of the existence *in vivo* of phosphorylated galectin-3 derived initially from labeling with ^{32}P *in vitro*.

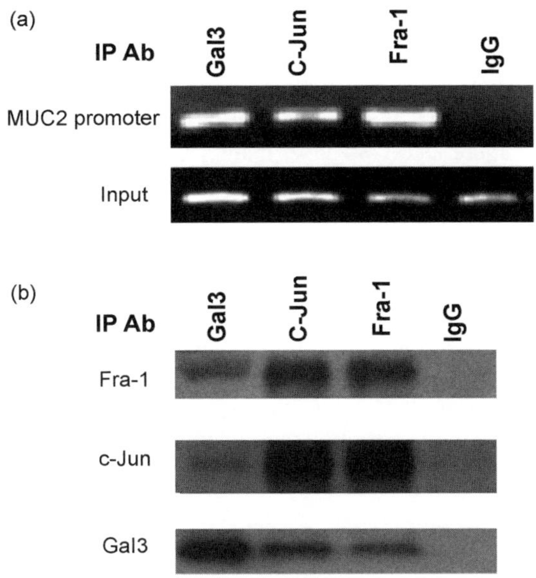

FIGURE 10.8 Nuclear galectin-3 is associated with the MUC2 promoter at the AP-1 site. (a) The association of galectin-3, c-Jun, and Fra-1 with the MUC2 promoter was confirmed in LS5 cells. Chromatin immunoprecipitation was performed using antigalectin-3, anti-c-Jun, or anti-Fra-1 antibodies, and IgG as the negative control. The MUC2 promoter was amplified by PCR. (b) Galectin-3 association with AP-1 components c-Jun and Fra-1 was confirmed by chromatin immunoprecipitation–immunoblot. LS5 cells were fixed, lysed, sonicated, and incubated with either antigalectin-3, anti-c-Jun, or anti-Fra-1 antibodies, and IgG as the negative control. The samples were pulled down by salmon sperm DNA/protein A agarose and immunoblotted with antibodies to Fra-1, c-Jun, or galectin-3. Reproduced from Reference 7 with permission.

Phosphorylated galectin-3 prepared by *in vitro* phosphorylation is distinguishable from nonphosphorylated by SDS-PAGE and by isoelectric focusing. More recently, a phospho-galectin-3-specific antibody has been developed, which should facilitate the detailed examination of alterations in the relative amounts of phospho-galectin-3 as compared to total galectin-3 in cells and tissues. Much of the information about the functional requirements for phosphorylation derives from the use of phospho-mutants, where Ser 6 is substituted with Ala or Glu. Initial studies in this laboratory have been done in galecin-3-null BT-549 breast cancer cells. The extent to which these results apply to colorectal cancer has not yet been determined.

Role of Galectin-3 Phosphorylation in Gene Induction

As mentioned above, galectin-3 induces the expression of MUC2 mucin in colon cancer cells. However, relatively little is known about the role of different structural domains of galectin-3 in gene induction. A recent study in this laboratory examined the effect of wild-type galectin-3 and of a mutant galectin-3 that cannot be phosphorylated on the pattern of gene expression in breast cancer cells (10). We found that phosphorylation of galectin-3 is a prerequisite for the modulation of unique sets of genes. A microarray analysis of 10,000 human genes, comparing BT-549 transfectants expressing wild-type and those expressing phospho-mutant galectin-3, identified genes that were differentially (>2.5-fold) expressed (Fig. 10.9). It is worth noting that the set of 280 genes that were regulated by wild-type galectin-3 was distinct from the set of 313 genes that were regulated by the phospho-mutant galectin-3 constructs. Few genes (34 total) were influenced similarly by both constructs.

Genes regulated by wild-type phosphorylated but not phospho-mutant galectin-3 included those involved in oxidative stress, a novel noncaspase lysosomal apoptotic pathway, cell cycle regulation, transcriptional activation, cytoskeleton remodeling, cell adhesion, and tumor invasion. It is of interest that phospho-galectin-3 suppressed the expression of the transmembrane mucin MUC1 and members of the integrin family and upregulated the secreted gel forming mucin MUC5AC. (MUC1 is common in breast cancer tissues but the expression of MUC5AC and MUC2 expression is rare.)

The reliability of the microarray data was validated by real-time-RT-PCR and Western blot analysis and clinical relevance was evaluated by real-time RT-PCR screening of a panel of matched pairs of breast tumors. Differentially regulated genes in breast cancers that are also predicted to be associated with phospho-galectin-3 in transformed BT-549 cells include C-type lectin 2 (CLEC2B), insulin-like growth factor binding protein 5 (IGFBP5), and cathepsins L2 (CTSL2) and cyclin D1 (CCND1). These data demonstrate the functional diversity of galectin-3 and suggest

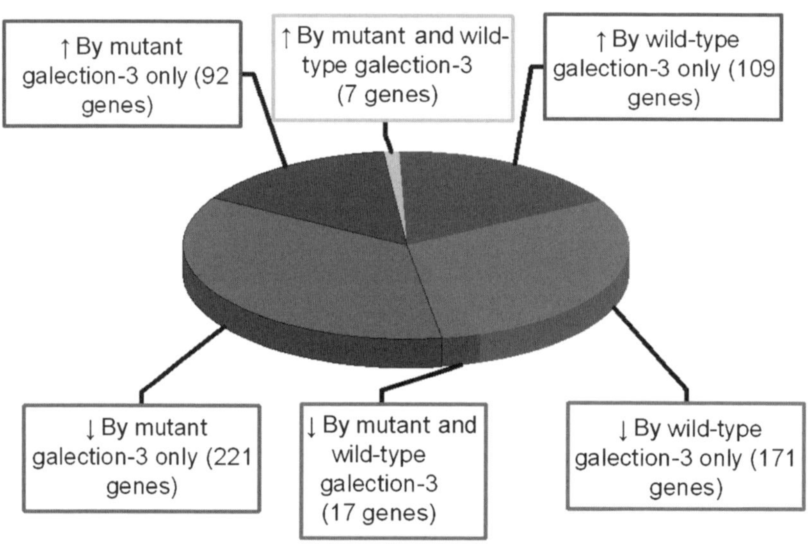

FIGURE 10.9 Number of genes differentially modulated by wild-type galectin-3 and phospho-mutant galectin-3 in BT-549 cells. (See color insert.)

that phosphorylation of protein is necessary for the regulation of a unique set of genes that play a role in malignant transformation. It has not been established whether these results are also applicable to colon cancer cells. It is also not known which of these genes are modulated by galectin-3 through direct transcriptional effects. However, the finding that the set of genes modulated by phospho-mutant galectn-3 is so distinct from the set of genes modulated by wild-type galectin-3 implies that the mechanisms by which these two forms of galectin-3 influence gene expression are fundamentally different.

Cellular Compartmentalization of Phosphorylated Galectin-3

A variety of results from this and other laboratories have found that galectin-3 is present in both the nucleus and the cytosol, and that malignant transformation is accompanied by a shift away from nuclear localization and toward cytosolic localization. The presence of nuclear galectin-3 in many cell types, including colon cancer cells, is well established (70), and its role in the nucleus is an area of active investigation (70–73). The presence of nuclear galectin-3 in colon cancer cells has been demonstrated through immunohistochemistry, subcellular fractionation, and confocal microscopy (7). The dual localization (both cytoplasmic and nuclear) of galectin-3 in the transfectant colon cancer cell clones was established by Western blot analysis of subcellular fractions and immunofluorescence staining (Fig. 10.10).

Nuclear localization of galectin-3 was visualized readily in these cells (Fig. 10.10b) and the expression of both cytoplasmic and nuclear galectin-3 was higher in the sense LS5 transfectant than in LSC2 control and conversely lower in the antisense M22 transfectant (Fig. 10.10a). While an immunohistochemical staining (16) shows a proportional shift in galectin-3 expression to the cytoplasm in primary colon cancers compared to normal mucosa, galectin-3 is never lost from the nucleus.

The structural features of galectin-3 that affect its balance between nuclear import and nuclear export have been studied in breast cancer cells (8, 9). Deletion of the NH_2 terminus resulted in the loss of nuclear localization. When green fluorescent protein was fused to the galectin-3 leader sequence and transiently transfected into BT-549 cells, the uniform cellular distribution of native green fluorescent protein was changed mainly to a nuclear pattern.

To further investigate whether the functional changes observed in a galectin-3 with the 11 NH_2-terminal amino acids deleted were due to the loss of phosphorylation at Ser6, two point mutations were created at this serine: Ser6 $\rightarrow$ Ala and Ser6 $\rightarrow$ Glu. No obvious difference was observed in cellular localization between wild-type and Ser6-mutated transfectants. These results suggest a structural role for the NH_2 terminus leader motif in the nuclear import of galectin-3, independent of phosphorylation (8).

In a subsequent study, BT-549 cell clones expressing wild-type and mutant galectin-3 were exposed to chemotherapeutic anticancer drugs (9). Phosphorylated wild-type galectin-3 was exported from the nucleus to the cytoplasm, where it

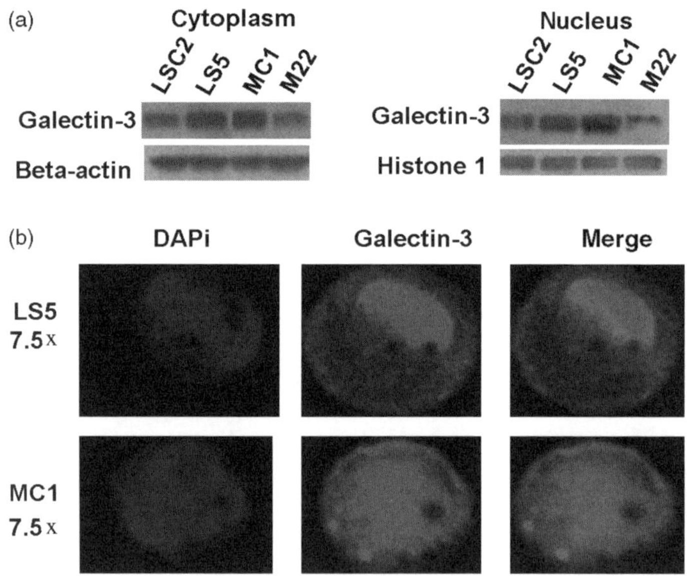

FIGURE 10.10 Subcellular location of galectin-3 in galectin-3 sense and antisense clones of colon cancer cells. (a) Western blots of cytoplasmic and nuclear fractions with antigalectin-3, beta-actin for cytoplasm loading control, and histone 1 for nucleus loading control. (b) Indirect immunofluorescence performed in LS5 and MC1 cells using antigalectin-3 antibody (TIB166, green), followed by DAPI nuclear counterstaining (blue). The merge of galectin-3 (green) with DAPI (blue) also is shown. Reproduced from Reference 7 with permission. (See color insert.)

protected the BT-549 cells from drug-induced apoptosis. In contrast, nonphosphorylated mutant galectin-3 was not exported from the nucleus. Furthermore, leptomycin B, a nuclear export inhibitor, increased the cisplatin-induced apoptosis of galectin-3 expressing BT-549 cells. These results suggest that Ser6 phosphorylation of galectin-3 acts as a molecular switch for its cellular translocation from the nucleus to the cytoplasm and, as a result, regulates the antiapoptotic activity of galectin-3. While it is not known whether the conclusions from these two studies apply to colon cancer cells, the balance between phosphorylation-independent nuclear import and phosphorylation-dependent nuclear export must be considered in understanding the functions of galectin-3 in gene induction and in apoptosis.

Role of Galectin-3 Phosphorylation in Apoptosis

Galectins play a significant role in apoptosis and other cellular processes. In particular, galectin-3 may exert anti- or proapoptotic activity depending on the cell type and the nature of the stimulus (Fig. 10.11). Overexpression of galectin-3 in breast carcinoma cells renders them resistant to chemotherapeutic drugs (75), but inactivates Akt and sensitizes them to tumor necrosis factor-related apoptosis-inducing ligand (TRAIL)-induced apoptosis (11, 76).

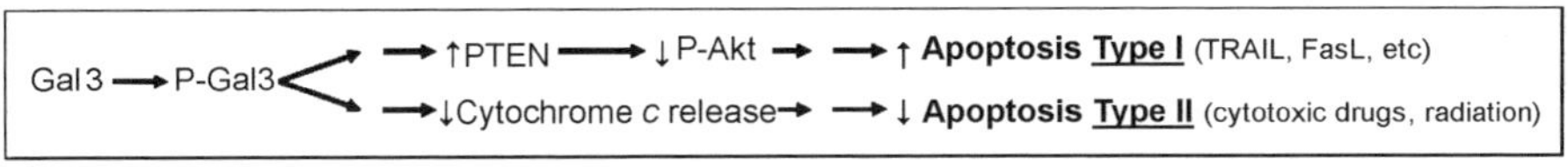

FIGURE 10.11 Antiapoptotic and proapoptotic functions of phosphorylated galectin-3. In BT-549 breast cancer cells, the expression of galectin-3 decreases the intrinsic, mitochondrial-dependent, apoptotic response to cytotoxic drugs, but increases the extrinsic, death receptor-mediated, apoptotic pathway.

The mechanisms by which galectin-3 regulates apoptosis are not fully understood. A portion of its antiapoptotic activity may be attributed to the antideath motif that is conserved in the Bcl-2 family (74). Like Bcl-2, phosphorylation controls the antiapoptotic function of galectin-3 (76). To elucidate the role of phosphorylation of galectin-3 in protection from apoptosis, serine to alanine (S6A) and serine to glutamic acid (S6E) galectin-3 mutants were generated and transfected into the galectin-3-null BT-549 breast cancer cell line (9). In response to an apoptotic insult, wild-type galectin-3 was translocated from the nucleus to the cytoplasm and protected BT-549 cells from drug-induced cell death, while nonphosphorylated mutant galectin-3 neither translocated nor protected BT-549 cells from drug-induced apoptosis. These results indicate that the antiapoptotic activity of galectin-3 results, in part, from its post-translational phosphorylation, acting as a molecular switch for its cellular translocation and its antiapoptotic activity (9).

In contrast, phosphorylation of galectin-3 is required for sensitizing human epithelial carcinoma cells to TRAIL-induced apoptosis via a nonclassical caspase activation cascade (11). We found that parental BT-549 breast cancer cells (galectin-3 null) were resistant to TRAIL, but exposure of clones expressing wild-type phosphorylated galectin-3 to 50 ng/mL TRAIL resulted in >80% apoptotic cell death, accompanied by PARP cleavage, caspase activation, and AKT dephosphorylation. The Ser6Glu and Ser6Ala galectin-3 mutants do not show this proapoptotic activity. We conclude that wild-type galectin-3 exerts a proapoptotic activity in response to ligand-bound death receptors such as TRAIL by upregulating PTEN, resulting in dephosphorylation/inactivation of the Akt signaling pathway.

These results revealed a concurrent dual function of galectin-3: It confers chemoresistance to breast carcinoma cells by blocking the intrinsic apoptosis pathway and renders sensitivity to death receptor-mediated apoptosis by triggering a nonconventional extrinsic apoptosis pathway. The dual function, proapoptotic and antiapoptotic, of galectin-3 may not apply for all cell types (72), but has implications for targeted therapy of susceptible tumors (11).

DIAGNOSTIC AND THERAPEUTIC POTENTIAL OF GALECTIN-3 AND GALECTIN-3 LIGANDS IN COLON CANCER

The potential clinical applications of galectin-3 and galectin-3 ligands include the prospect of using galectin-3 binding as a diagnostic marker for colon cancer and the possibility that inhibitors of galectin-3 could influence the biological behavior of colon cancers. One strategy for increasing the sensitivity and specificity of colon cancer markers is to monitor the expression of cancer-associated proteins and carbohydrates in one assay. Particularly for the colon, cell surface lectins are an attractive target for innovative therapeutic approaches because they have relatively well-defined binding sites for carbohydrates that are accessible to circulating drugs or to luminal carbohydrates.

Galectin-3 Ligands as Colon Cancer Markers

The diagnostic potential of galectin-3-binding is based on the specificity of this lectin for a range of oligosaccharide structures that can be present on diverse glycoproteins. Besides binding to colon cancer mucin (4, 5), galectin-3 has also been shown to bind lysosomal membrane-associated glycoproteins 1 and 2, carcinoembryonic antigen, Mac-2-binding protein, and laminin (77–80). Other biologically relevant ligands for galectin-3 include Gemin4 (a splicing complex protein), hensin (associated with epithelial terminal differentiation), IgE (and high affinity IgE receptor), Fc-gamma-RII (CD32) low affinity IgG receptor, and cubulin (a multiligand receptor in the uteroplacental complex) (81–85).

All galectins bind lactose and *N*-acetyllactosamine (Gal-beta-4GlcNAc) and the amino acid residues that interact with the terminal galactose residue are highly conserved between different galectins (15). However, the specificity of galectin-3 is not simply for the beta-galactoside moiety. Affinity for various ligands may be substantially altered by the substitution of specific oligosaccharides (86–88). For example, blood group A antigen is a 32-fold better ligand than *N*-acetyllactosamine. The strong binding to oligosaccharides containing the type 1 backbone (Gal-beta-3GlcNAc) or type 2 backbone (Gal-beta-4GlcNAc) is further increased by the substitution of the terminal galactose with NeuAc-alpha-3-, 3-sulfate, and Fuc-alpha-2- moieties (87). Because cancer cells differ from their normal counterparts not only in the repertoire of proteins produced but also in the structure of oligosaccharides added to these proteins, glycoproteins have the potential to serve as tumor markers.

Since galectin-3 levels in human colon cancer tissues and derivative cell lines correlate with tumor stage, patient survival, and metastatic capacity, detection of galectin-3 and galectin-3 ligands in serum may be clinically useful (12). Sera from patients with colon cancer or adenomas and from normal individuals were analyzed for galectin-3 by sandwich ELISA (Fig. 10.12). There was increased serum galectin-3 in patients with colon cancer (23.9 ± 4.5 ng/mL) as compared to those with adenomas (11.6 ± 1.6, $p = 0.013$) or to normals (5.4 ± 0.5, $p = 0.0002$). While these differences are statistically significant, the overlap between groups limits the clinical utility of assays of total circulating galectin-3. Assays of galectin-3 ligands, rather than galectin-3 itself, are potentially more robust, in that they measure both cancer-associated proteins and carbohydrates.

A recent study in this laboratory (12) has identified a circulating ligand for galectin-3 as a haptoglobin-related glycoprotein elevated in individuals with colon cancer. In a follow-up to our earlier study (4) that demonstrated the binding of galectin-3 to high molecular weight glycoprotein in colon cancer serum, we sought to

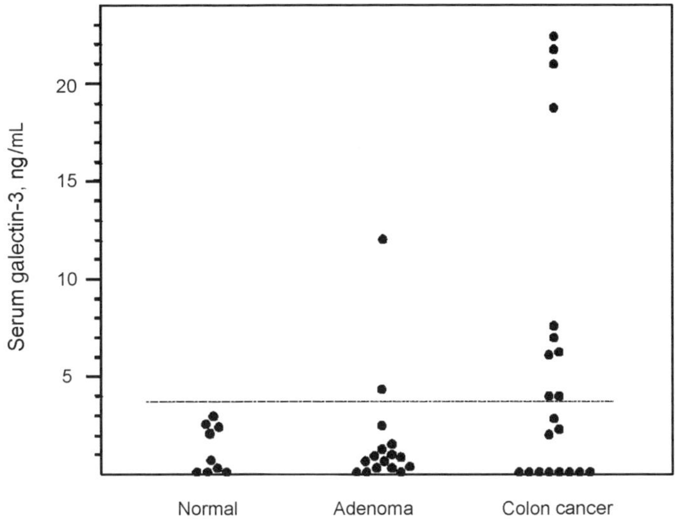

FIGURE 10.12 Levels of total galectin-3 in serum. Sera from normal individuals and patients with adenomatous polyps or with colon cancer were assayed for circulating galectin-3 using a sandwich ELISA.

identify other galectin-3 ligands in serum. When colon cancer sera were desialylated by mild H^+, reduced, separated by SDS-PAGE, and analyzed for binding of biotinylated galectin-3, the major galectin-3 ligand was a 40-kDa band distinct from mucin, CEA, and Mac-2BP. This was purified and several peptide sequences were characteristic of haptoglobin.

Two lines of evidence confirmed that the major galectin-3 ligand in colon cancer sera is a cancer-associated glycoform of haptoglobin. First, on Western blot analysis with antihaptoglobins, purified 40-kDa ligand showed immunological similarily with haptoglobin β-subunit. Second, depletion of haptoglobin by immunoprecipitation also eliminated this 40-kDa ligand.

In the analyses of 76 serum samples, the 40-kDa ligand was significantly higher in patients with colon cancer (68.4 ± 11.2 units, mean $\pm$ SEM) than in those with adenomas (5.1 ± 1.9, $p = 0.0001$), or in normal subjects (2.3 ± 0.3, $p < 0.0001$). Colon cancer sera had ~2-fold more total haptoglobin than normals, versus >30-fold more 40-kDa ligand, suggesting a cancer-associated alteration in glycosylation (Fig. 10.13).

In a subsequent validation study, sera were prospectively obtained from 50 patients representing a spectrum of categories (colorectal cancer, adenomatous polyps, hyperplastic polyps, inflammatory bowel disease, and normal individuals). Sera were assessed in a blinded fashion for galectin-3 ligand (Fig. 10.14). After unbinding, differences between groups were highly significant (normal versus adenoma $p < 0.0077$, normal versus. cancer $p < 0.004$, adenoma versus. cancer $p < 0.0073$). Based on these results, our working hypothesis is that the basis for the colon cancer-associated presence of the 40-kDa galectin-3 ligand in serum is a

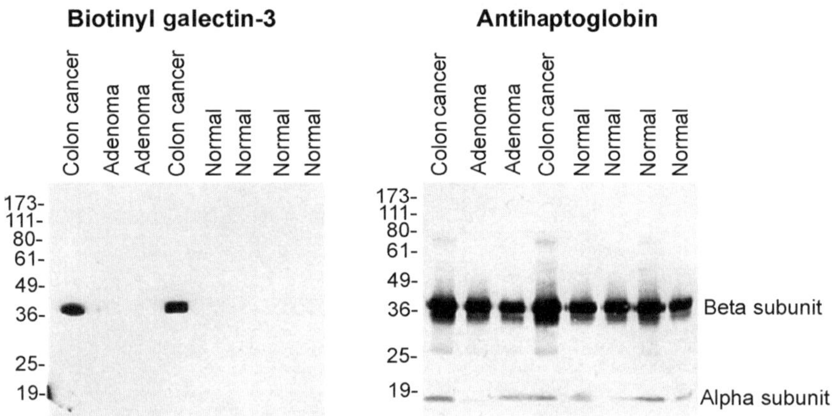

FIGURE 10.13 Lack of correlation between 40-kDa ligand and total haptoglobin. Levels of 40-kDa ligand and total haptoglobin in human serum. Desialylated sera were analyzed with biotinyl-galectin-3 and antihaptoglobin.

combination of tumor-specific ectopic expression of the haptoglobin protein together with a colon-specific glycoprotein biosynthetic pattern.

Mac-2BP, also known as 90 K, is a large oligomeric glycoprotein composed of subunits of approximately 90 kDa, identified as a ligand for the carbohydrate-binding protein, galectin-3 (earlier called Mac-2) (80). The 63-kDa apoprotein (encoded by

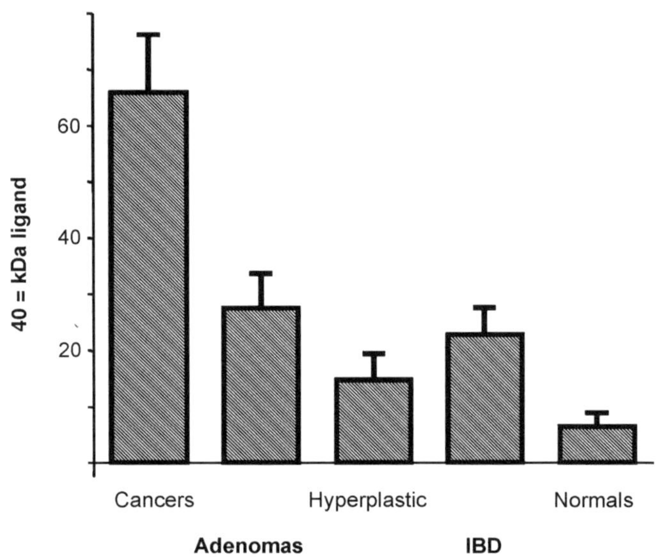

FIGURE 10.14 Validation study by blinded analyses of sera. Serum assays of 40-kDa ligand for 50 sera, presented as mean ± SEM. Coded serum samples (10 per group) obtained through participation in the EDRN (Early Detection Research Network) were assayed for 40-kDa ligand.

the LGALS3BP gene) carries seven N-linked oligosaccharides that are responsible for binding to galectin-3. Previous studies have indicated that serum levels of this protein may have prognostic value in patients with several types of cancer, but immunohistochemical studies of Mac-2BP expression in colon cancer have not been reported.

When 33 sporadic colorectal cancers with 11 associated metastases as well as specimens of normal ($n = 10$) and transitional ($n = 28$) mucosa and colorectal adenomas ($n = 10$) were examined immunohistochemically for the expression of Mac-2BP (Fig. 10.15) and galectin-3, colorectal cancers expressed Mac-2BP in 32 of 33 (97%) cases (mean staining score 1.72 ± 0.8 on 0–3 scale). Expression of this protein in cancers was significantly higher than that in normal mucosa (0.75 ± 0.3, $p < 0.001$), colorectal adenomas (0.3 ± 0.6, $p < 0.001$), and transitional mucosa (0.53 ± 0.7, $p < 0.001$). Metastases expressed higher amounts of Mac-2BP than primary tumors (2.32 ± 0.6; $p = 0.018$). Mac-2BP was expressed in the cytoplasm of neoplastic tissues and in secreted material, with levels of the ligand paralleling that of galectin-3. In keeping with previous reports, normal tissue expressed predominately nuclear galectin-3. Most colorectal carcinomas (85%) lost nuclear localization of galectin-3 and expressed the protein only in the cytoplasm (colocalized with Mac-2BP).

While it is not known whether the carbohydrates present on Mac-2BP of colon cancer cells are different from those of the normal serum Mac-2-BP, these results suggest that the galectin-3-binding protein Mac-2BP is highly expressed in colorectal cancers and may serve as a marker of neoplastic progression (13).

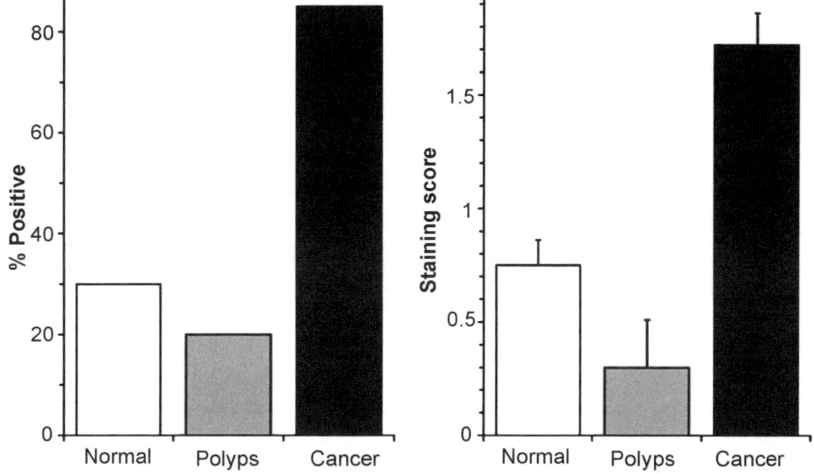

FIGURE 10.15 Expression of Mac-2BP in colon cancer as compared to normal rectal mucosa and adenomatous polyps. Immunohistochemical analysis used polyclonal anti-Mac-2BP antibody, with standard avidin–biotin staining. (Left panel) Percent of cases positive. (Right panel) Colorectal tissues were evaluated for staining intensity and area and scored on a 0 to 3+ scale. Mean and SEM of staining score are shown.

Galectin-3 as a Therapeutic Target

Because galectin-3 is involved in neoplastic transformation and cancer progression, carbohydrate-mediated interference with galectin-3 binding may have therapeutic implications. Monoclonal antibodies and carbohydrate-based compounds (anti-TF antigen, antigalectin-3, modified citrus pectin, and lactulosyl-L-leucine), targeting specifically beta-Gal-mediated tumor–endothelial cell adhesive interactions have been shown to inhibit by >90% the *in vivo* formation of breast and prostate carcinoma metastasis in mouse lung and bones (89).

A modified natural polysaccharide, modified citrus pectin (MCP), is of interest as an anticancer agent (14). There have been several studies of *in vivo* effects of MCP. Colon tumors implanted in Balb/c mice are reduced in size following oral administration of MCP (95). Similar findings have been reported for metastasis of B16-F1 melanoma (90), prostate cancer (57), and breast cancer (14) cells. Although the uptake of the pectin oligosaccharides by colonic epithelium has not been studied in detail, it is likely that MCP that contains shorter, simpler more water-soluble carbohydrate chains (90) would be better absorbed than ordinary long-chain pectins. The smaller complex oligosaccharide units of MCP can combine with the carbohydrate-binding domain of galectin-3 (91) and interfere with its binding to specific cell surface receptors. Besides affecting cell adhesion and tumor embolization that may reduce metastasis, MCP could also affect the noncarbohydrate functions of galectin-3.

Galectin-3 is the sole member of the galectin family that contains the NWGR antideath domain of the Bcl-2 family and can function as an apoptotic inhibitor (92–94) and a critical determinant for anchorage-independent and free radical-resistant cell survival during metastasis.

We have recently established that the oral administration of modified citrus pectin, an inhibitor of the binding of galectin-3 to glycoprotein ligands, can reduce the growth and metastasis of colon tumors orthotopically implanted in mice (Table 10.4). When LS-LiM6 human colon cancer cells were injected into the cecum of nude mice, continuous feeding with modified citrus pectin decreased the weight of the cecal tumors by twofold and caused a complete inhibition of spontaneous liver metastasis (14). While it is not known whether the primary effect of the administration of this galectin-3 inhibitor is the inhibition of growth of the primary tumor or interference with the arrest and survival of metastatic cells in the target organ, these results do provide hope that interference with galectin-3 binding can have therapeutic implications for colon cancer.

TABLE 10.4 Effect of the Galectin-3 Antagonist Modified Citrus Pectin (MCP) on Spontaneous Metastasis of LS-LiM6 Human Colon Cancer Cells After Cecal Wall Implantation

	Control	MCP
Primary tumor weight (95% CI)	1.16 g (1.13–1.19)	0.65 g (0.37–0.93)
Intraabdominal tumor weight	2.00 g (1.94–2.06)	0.88 g (0.37–0.93)
Incidence of lymph node metastasis	100%	25%
Incidence of liver metastasis	60%	0%

Besides MCP and GCS100, a similar pectin derivative (96), other therapeutic strategies involving galectin-3 include dietary polysaccharides (97, 98), synthetic carbohydrates (99), and peptide mimetics (100, 101). Since this is a multifunctional protein, and there are several classes of molecules that could interfere with its activities, galectin-3 remains a promising target for developing novel therapeutic strategies for colon cancer.

REFERENCES

1. Schoeppner HL, Raz A, Ho SB, Bresalier RS. Expression of an endogenous galactose-binding lectin correlates with neoplastic progression in the colon. Cancer 1995;75: 2818–2826.
2. Bresalier RS, Mazurek N, Sternberg LR, Byrd JC, Yunker CK, Nangia-Makker P, Raz A. Metastasis of human colon cancer is altered by modifying expression of the beta-galactoside-binding protein galectin-3. Gastroenterology 1998;115:287–296.
3. Holt PR, Bresalier RS, Ma CK, Liu KF, Lipkin M, Byrd JC, Yang K. Calcium plus vitamin D alters preneoplastic features of colorectal adenomas and rectal mucosa. Cancer 2006;106:287–296.
4. Bresalier RS, Byrd JC, Wang L, Raz A. Colon cancer mucin: a new ligand for the beta-galactoside-binding protein galectin-3. Cancer Res 1996;56:4354–4357.
5. Mazurek N, Conklin J, Byrd JC, Raz A, Bresalier RS. Phosphorylation of the beta-galactoside-binding protein galectin-3 modulates binding to its ligands. J Biol Chem 2000;275:36311–36315.
6. Dudas SP, Yunker CK, Sternberg LR, Byrd JC, Bresalier RS. Expression of human intestinal mucin is modulated by the beta-galactoside binding protein galectin-3 in colon cancer. Gastroenterology 2002;123:817–826.
7. Song S, Byrd JC, Mazurek N, Liu K, Koo JS, Bresalier RS. Galectin-3 modulates MUC2 mucin expression in human colon cancer cells at the level of transcription via AP-1 activation. Gastroenterology 2005;129:1581–1591.
8. Gong HC, Honjo Y, Nangia-Makker P, Hogan V, Mazurak N, Bresalier RS, Raz A. The NH_2 terminus of galectin-3 governs cellular compartmentalization and functions in cancer cells. Cancer Res 1999;59:6239–6245.
9. Takenaka Y, Fukumori T, Yoshii T, Oka N, Inohara H, Kim HR, Bresalier RS, Raz A. Nuclear export of phosphorylated galectin-3 regulates its antiapoptotic activity in response to chemotherapeutic drugs. Mol Cell Biol 2004;24:4395–4406.
10. Mazurek N, Sun YJ, Price JE, Ramdas L, Schober W, Nangia-Makker P, Byrd JC, Raz A, Bresalier RS. Phosphorylation of galectin-3 contributes to malignant transformation of human epithelial cells via modulation of unique sets of genes. Cancer Res 2005;65:10767–10775.
11. Mazurek N, Sun YJ, Liu KF, Gilcrease MZ, Schober W, Nangia-Makker P, Raz A, Bresalier RS. Phosphorylated galectin-3 mediates tumor necrosis factor-related apoptosis-inducing ligand signaling by regulating phosphatase and tensin homologue deleted on chromosome 10 in human breast carcinoma cells. J Biol Chem 2007;282: 21337–21348.

12. Bresalier RS, Byrd JC, Tessler D, Lebel J, Koomen J, Hawke D, Half E, Liu KF, Mazurek N. A circulating ligand for galectin-3 is a haptoglobin-related glycoprotein elevated in individuals with colon cancer. Gastroenterology 2004;127:741–748.
13. Liu K, Byrd JC, Song S, Bresalier RS. Expression of galectin-3 and its binding protein (Mac-2-BP) correlate with progression and metastasis of colorectal adenocarcinoma. Gastroenterology 2006;130:A88.
14. Nangia-Makker P, Hogan V, Honjo Y, Baccarini S, Tait L, Bresalier R, Raz A. Inhibition of human cancer cell growth and metastasis in nude mice by oral intake of modified citrus pectin. J Natl Cancer Inst 2002;94:1854–1862.
15. Cooper DNW. Galectinomics: finding themes in complexity. Biochim Biophys Acta 2002;1572:209–231.
16. Lotz MM, Andrews CW, Korzelius CA, Lee EC, Steele, Clarke A, Mercurio AM. Decreased expression of Mac-2 (carbohydrate binding protein 35) and loss of its nuclear localization are associated with the neoplastic progression of colon carcinoma. Proc Natl Acad Sci USA 1993;90:3466–3470.
17. Sanjuan X, Fernandez PL, Casells A, Casronovo V, van den Brule FA, Liu FT, Cardesa A, Campo E. Differential expression of galectin-3 and galectin-1 in colorectal cancer progression. Gastroenterology 1997;113:1906–1915.
18. Ochieng J, Platt D, Tait L, Hogan V, Raz T, Carmi P, Raz A. Structure–function relationship of a recombinant human galactoside-binding protein. Biochemistry 1993; 32:4455–4460.
19. Raz A, Lotan R. Endogenous galactoside-binding lectins: a new class of functional tumor cell surface molecules related to metastasis. Cancer Metastasis Rev 1987;6: 433–452.
20. Nangia-Makker P, Honjo Y, Sarvis R, Akahani S, Hogan V, Pienta KJ, Raz A. Galectin-3 induces endothelial cell morphogenesis and angiogenesis. Am J Pathol 2000;156: 899–909.
21. Yu F, Finley Jr RL, Raz A, Kim HR. Galectin-3 translocates to the perinuclear membranes and inhibits cytochrome c release from the mitochondria. A role for synexin in galectin-3 translocation. J Biol Chem 2002;277:15819–15827.
22. Irimura T, Matsushita Y, Sutton RC, Carralero D, Ohannesian DW, Cleary KR, Ota DM, Nicolson GL, Lotan R. Increased content of an endogenous lactose-binding lectin in human colorectal carcinoma progressed to metastatic stages. Cancer Res 1991;51: 387–393.
23. Hittelet A, Legendre H, Nagy N, Bronckart Y, Pector JC, Salmon I, Yeaton P, Gabius HJ, Kiss R, Camby I. Upregulation of galectins-1 and -3 in human colon cancer and their role in regulating cell migration. Int J Cancer 2003;103:370–379.
24. Inufusa H, Nakamura M, Adachi T, Aga M, Kurimoto M, Nakatani Y, Wakano T, Miyake M, Okuno K, Shiozaki H, Yasutomi M. Role of galectin-3 in adenocarcinoma liver metastasis. Int J Oncol 2001;19:913–919.
25. Bresalier RS, Hujanen ES, Raper SE, Roll FJ, Itzkowitz SH, Martin GR, Kim YS. An animal model for colon cancer metastasis. Establishment and characterization of murine colon cancer cell lines with enhanced liver-metastazing ability. Cancer Res 1985;47: 249–260.
26. Bresalier RS, Raper SE, Hujanen ES, Kim YS. A new animal model for human colon cancer metastasis. Int J Cancer 1987;39:625–630.

27. Bresalier RS, Rockwell RW, Dahiya R, Duh Q-Y, Kim YS. Cell surface sialoprotein alterations in metastatic murine colon cancer cell lines selected in an animal model for colon cancer metastasis. Cancer Res 1990;50:1299–1307.

28. Bresalier RS, Niv Y, Byrd JC, Duh Q, Toribara NW, Kim YS. Mucin production by human colonic carcinoma cells correlates with their metastatic potential in animal models of colon cancer metastasis. J Clin Invest 1991;87:1037–1045.

29. Qi Y, Moyana T, Matte G, Wilkinson A, Bresalier RS, Xiang J. Immunolocalization of hepatic metastases of human colonic cancer by chimeric anti-AG72 antibody in SCID mice. J Surg Oncol 1995;8:3–9.

30. Qi Y, Moyana T, Bresalier RS, Xiang J. Antibody targeted lymphokine-activated killer cells inhibit liver micrometastases in severe combined immunodeficient mice. Gastroenterology 1995;109:1950–1957.

31. Bresalier RS, Schwartz B, Kim YS, Duh Q-Y, Kleinman HK, Sullam PM. The laminin (α1 chain Ile-Lys-Val-Ala-Val (IKVA)-containing peptide promotes liver colonization by human colon cancer cells. Cancer Res 1995;55:2476–2480.

32. Sternberg LR, Byrd JC, Yunker CK, Dudas S, Hoon V, Bresalier RS. Liver colonization by human colon cancer cells is reduced by antisense inhibition of MUC2 mucin synthesis. Gastroenterology 1999;116:363–371.

33. Byrd JC, Bresalier RS. Mucins and mucin binding proteins in colorectal cancer. Cancer Metastasis Rev 2004;23:77–99.

34. Bresalier RS. Intestinal mucin and colon cancer. Gastroenterology 2002;123:648–649.

35. Andrianifahanana M, Moniaux N, Batra SK. Regulation of mucin expression: mechanistic aspects and implications for cancer and inflammatory diseases. Biochim Biophys Acta 2006;1765:189–222.

36. Hollingsworth MA, Swanson BJ. Mucins in cancer: protection and control of the cell surface. Nat Rev Cancer 2004;4:45–60.

37. Ajioka Y, Allison LJ, Jass JR. Significance of MUC1 and MUC2 mucin expression in colorectal cancer. J Clin Pathol 1996;49:560–564.

38. Matsuda K, Masaki T, Watanabe T, Kitayama J, Nagawa H, Muto T, Ajioka Y. Clinical significance of MUC1 and MUC2 mucin and p53 protein expression in colorectal carcinoma. Jpn J Clin Oncol 2000;30:89–94.

39. Ogata S, Uehara H, Chen A, Itzkowitz SH. Mucin gene expression in colonic tissues and cell lines. Cancer Res 1992;52:5971–5978.

40. Ho SB, Niehans GA, Lyftogt C, Yan PS, Cherwitz DL, Gum ET, Dahiya R, Kim YS. Heterogeneity of mucin gene expression in normal and neoplastic tissues. Cancer Res 1993;53:641–651.

41. Allen A, Hutton DA, Pearson JP. The MUC2 gene product: a human intestinal mucin. Int J Biochem Cell Boil 1998;7:797–801.

42. Chang SK, Dohrman AF, Basbaum CB, Ho SB, Tsuda T, Toribara NW, Gum JR, Kim YS. Localisation of mucin (MUC2 and MUC3) mRNA and peptide expression in normal human intestine and colon cancer. Gastroenterology 1994;107:160–172.

43. Weiss AA, Babyatsky MW, Ogata S, Chen A, Itzkowitz SH. Expression of MUC2 and MUC3 mRNA in human normal, malignant and inflammatory tissues. J Histochem Cytochem 1996;44:1161–1166.

44. Sylvester PA, Myerscough N, Warren BF, Carlstedt I, Corfield AP, Durdey P, Thomas MG. Differential expression of the chromosome 11 mucin genes in colorectal cancer. J Pathol 2001;195:327–335.
45. Blank M, Klussman E, Kruger-Krasagakes S, Schmitt-Graff A, Stolte M, Bornhoeft G, Stein H, Xing PX, McKenzie IFC, Verstijnen CPHJ, Riecken EO, Hanski CJ. Expression of MUC2-mucin in colorectal adenomas and carcinomas of different histological types. Int J Cancer 1994;59:301–306.
46. Biemer-Huttmann AE, Walsh MD, McGuckin MA, Simms LA, Young J, Leggett BA, Jass JR. Mucin core protein expression in colorectal cancers with high levels of microsatellite instability indicates a novel pathway of morphogenesis. Clin Cancer Res 2000;6:1909–1916.
47. Manne U, Weiss HL, Grizzle WE. Racial differences in the prognostic usefulness of MUC1 and MUC2 in colorectal adenocarcinomas. Clin Cancer Res 2000;6:4017–4025.
48. Velcich A, Heyer J, Fragale A, Nicholas C, Viani S, Kucherlapati R, Lipkin M, Yang K, Augenlicht L. Colorectal cancer in mice genetically deficient in the mucin MUC2. Science 2002;295:1726–1729.
49. Schwartz B, Bresalier RS, Kim YS. Role of mucin in colon cancer metastasis. Int J Cancer 1992;52:60–65.
50. Bresalier RS, Ho SB, Schoeppner HL, Kim YS, Sleisenger MH, Brodt P, Byrd JC. Enhanced sialylation of mucin-associated carbohydrate structures in human colon cancer metastasis. Gastroenterology 1996;110:1354–1367.
51. Brodt P, Fallavollita L, Bresalier RS, Norton CR, Wolitzky BA. E-selectin mediates carcinoma cell adhesion and promotes liver metastasis. Int J Cancer 1997;71:612–614.
52. Bresalier RS, Byrd JC, Brodt P, Ogata S, Yunker C. Liver metastasis and adhesion to the sinusoidal endothelium by human colon cancer cells is related to mucin carbohydrate chain length. Int J Cancer 1998;76:556–562.
53. Sternberg LR, Byrd JC, Yunker CK, Hansson GC, Liu K-F, Bresalier RS. Alternative splicing of the human MUC2 gene. Arch Biochem Biophys 2004;421:21–33.
54. Huflejt ME, Turck CW, Lindstedt R, Barondes SH, Leffler H. L-29, a soluble lactose-binding lectin, is phosphorylated on serine 6 and serine 12 *in vivo* and by casein kinase I. J Biol Chem 1993;268:26712–26718.
55. Hamann KK, Cowles EA, Wang JL, Anderson RL. Expression of carbohydrate binding protein 35 in human fibroblasts variations in the levels of mRNA, protein, and isoelectric species as a function of replicative competence. Exp Cell Res 1991;196:82–91.
56. Cowles EA, Agrwal N, Anderson RL, Wang JL. Carbohydrate-binding protein 35. Isoelectric points of the polypeptide and a phosphorylated derivative. J Biol Chem 1990;265:17706–17712.
57. Pienta KJ, Naik H, Akhtar A, Yamazaki K, Repogle TS, Lehr J, Donat TL, Tait L, Hogan V, Raz A. Inhibition of spontaneous metastasis in a rat prostate cancer model by oral administration of modified citrus pectin. J Natl Cancer Inst 1995;87:348–353.
58. Yoshi T, Fukumori T, Honjo Y, Inohara H, Kim HRC, Raz A. Galectin-3 phosphorylation is required for its anti-apoptotic function and cell cycle arrest. J Biol Chem 2002;277:6852–6857.
59. Henrick K, Bawumia S, Barboni EAM, Hughes RC. Evidence for subsites in the galectins involved in sugar binding at the non-reducing end of the central galactose of

oligosaccharide ligands: sequence analysis, homology modeling and mutagenesis studies of hamster galectin-3. Glycobiology 1998;8:45–57.

60. Yang RY, Hill PN, Hsu DK, Liu FT. Role of the carboxyl-terminal domain in self-association of galectin-3. Biochemistry 1998;37:4086–4092.
61. Kim HRC, Lin HM, Biliran H, Raz A. Cell cycle arrest and inhibition of anoikis by galectin-3 in human breast epithelial cells. Cancer Res 1999;59:4148–4154.
62. Nakahara S, Hogan V, Inohara H, Raz A. Importin-mediated nuclear translocation of galectin-3. J Biol Chem 2006;281:39649–39659.
63. Davidson PJ, Li SY, Lohse AG, Vandergaast R, Verde E, Pearson A, Patterson RJ, Wang JL, Arnoys EJ. Transport of galectin-3 between the nucleus and cytoplasm. I. Conditions and signals for nuclear import. Glycobiology 2006;16:602–611.
64. Song S, Byrd JC, Koo JS, Bresalier RS. Bile acids induce MUC2 overexpression in human colon carcinoma cells. Cancer 2005;103:1606–1614.
65. Li JD, Feng WJ, Gallup MG, Gum J, Kim Y, Basbaum C. Activation of NF-kappa-B via a Src-dependent Ras-MAPK-pp90rsk pathway is required for *Pseudomonas aeruginosa*-induced mucin overproduction in epithelial cells. Proc Natl Acad Sci USA 1998;95: 5718–5723.
66. Jono H, Shuto T, Xu H, Kai H, Lim DJ, Gum JR, Kim JS, Yamaoka S, Feng XH, Li JD. Transforming growth factor-beta-Smad signaling pathway cooperates with NF-kappa B to mediate nontypeable *Haemophilus influenzae*-induced MUC2 mucin transcription. J Biol Chem 2002;277:45547–45557.
67. Aslam F, Palumbo L, Augenlicht LH, Velcich A. The Sp family of transcription factors in the regulation of the human and mouse MUC2 gene promoters. Cancer Res 2001;61: 570–576.
68. Perrais M, Pigny P, Copin MC, Aubert JP, Van Seuningen I. Induction of MUC2 and MUC5AC mucins by factors of the epidermal growth factor (EGF) family is mediated by EGF receptor/Ras/Raf/extracellular signal-regulated kinase cascade and Sp1. J Biol Chem 2002;277:32258–32267.
69. Yamamoto H, Bai YQ, Yuasa Y. Homeodomain protein CDX2 regulates goblet-specific MUC2 gene expression. Biochem Biophys Res Commun 2003;300:813–818.
70. Patterson RJ, Wang W, Wang JL. Understanding the biochemical activities of galectin-1 and galectin-3 in the nucleus. Glycoconj J 2004;19:499–506.
71. Shiura T, Takenaka Y, Fukumorio T, Tsutsumi S, Okada K, Hogan V, Kikuchi A, Kuwano H, Raz A. Implication of galectin-3 in Wnt signaling. Cancer Res 2005;65:3535–3537.
72. Oka N, Nakahara S, Takenaka Y, Fukumori T, Hogan V, Kanayama HO, Yanagawa T, Raz A. Galectin-3 inhibits tumor necrosis factor-related apoptosis-inducing ligand-induced apoptosis by activating Akt in human bladder carcinoma cells. Cancer Res 2005;65:7546–7553.
73. Paron I, Scaloni A, Pines A, Bachi A, Liu F-T, Puppin C, Pandolfi M, Ledda L, Di Loretto C, Danante G, Tell G. Nuclear localization of galectin-3 in transformed thyroid cells: a role in transcriptional regulation. Biochem Biophys Res Commun 2003;302:545–553.
74. Akahani S, Nangia-Makker P, Inohara H, Kim HR, Raz A. Galectin-3: a novel antiapoptotic molecule with a functional BH1 (NWGR) domain of Bcl-2 family. Cancer Res 1997;57:5272–5276.
75. Choi JH, Chun KH, Raz A, Lotan R. Inhibition of *N*-(4-hydroxyphenyl)retinamide-induced apoptosis in breast cancer cells by galectin-3. Cancer Biol Ther 2004;3:447–452.

76. Yoshii T, Fukumori T, Honjo Y, Inohara H, Kim HR, Raz A. Galectin-3 phosphorylation is required for its anti-apoptotic function and cell cycle arrest. J Biol Chem 2002;277:6852–6857.

77. Sarafian V, Jadot M, Foidart JM, Letesson JJ, Van den Brule F, Castonovo V, Wattiaux R, Wattiaux-De Coninck S. Expression of Lamp-1 and Lamp-2 and their interactions with galectin-3 in human tumor cells. Int J Cancer 1998;75:105–111.

78. Inohara H, Raz A. Identification of human cellular and secreted ligands for galectin-3. Biochem Biophys Res Commun 1994;201:1366–1375.

79. Ohhannesian DW, Lotan D, Thomas P, Jessup JM, Fukuda M, Gabius HG, Lotan R. Carcinoembryonic antigen and other glycoconjugates act as ligands for galectin-3 in human colon carcinoma cells. Cancer Res 1995;55:2191–2199.

80. Koths K, Taylor E, Halenbeck R, Casipit C, Wang A. Cloning and characterization of a human Mac-2-binding protein, a new member of the superfamily defined by the macrophage scavenger receptor cysteine-rich domain. J Biol Chem 2003;268: 14245–14249.

81. Park JW, Voss PG, Grabski S, Wang JL, Patterson RJ. Association of galectin-1 and galectin-3 with Gemin4 in complexes containing the SMN protein. Nucleic Acids Res 2001;27:3595–3602.

82. Hikita C, Vijayakumar S, Takit J, Erdjument-Bromage H, Tempst P, Al-Awqati Q. Induction of terminal differentiation in epithelial cells requires polymerization of hensin by galectin-3. J Cell Biol 2000;151:1235–1246.

83. Frigeri LG, Liu FT. Surface expression of functional IgE binding protein, an endogenous lectin, on mast cells and macrophages. J Immunol 1992;148:861–867.

84. Cortegano I, del Pozo V, Cardaba B, Arrieta I, Gallardo S, Rojo M, Aceituno E, Takai T, Verbeek S, Palomino P, Liu FT, Lahoz C. Interaction between galectin-3 and FcgRII induces down-regulation of IL-5 gene: implication of the promoter sequence IL-5REIII. Glycobiology 2000;10:237–242.

85. Crider-Pirkle S, Bilingsle P, Faust C, Hardy DM, Lee V, Weitlauf H. Cubulin, a binding partner for galectin-3 in the murine utero-placental complex. J Biol Chem 2002;277:15904–15912.

86. Henrick K, Bawumia S, Barboni EAM, Mehul B, Hughes RC. Evidence for subsites in the galectins involved in sugar binding at the nonreducing end of the central galactose of oligosaccharide ligands: sequence analysis, homology modeling and mutagenesis studies of hamster galectin-3. Glycobiology 1998;8:45–47.

87. Ideo H, Seko A, Ohkura T, Matta KL, Yamashita K. High-affinity binding of recombinant human galectin-4 to SO3-3Gal-GalNAc pyranoside. Glycobiology 2002;12:199–208.

88. Goletz S, Hanisch FG, Karsten U. Novel GalNAc containing glycans on cytokeratins are recognized *in vitro* by galectins with type II carbohydrate recognition domains. J Cell Sci 1997;110:1585–1596.

89. Glinskii OV, Huxley VH, Glinsky GV, Pienta KJ, Raz A, Glinsky VV. Mechanical entrapment is insufficient and intercellular adhesion is essential for metastatic cell arrest in distant organs. Neoplasia 2005;7:522–527.

90. Platt D, Raz A. Modulation of the lung colonization of B16-F1 melanoma cells by citrus pectin. J Natl Cancer Inst 1992;84:438–442.

91. Inohara H, Raz A. Effects of natural complex carbohydrate (citrus pectin) on murine melanoma cell properties related to galectin-3 functions. Glycoconj J 1994;11:527–532.

92. Yang RY, Hsu DK, Liu FT. Expression of galectin-3 modulates T-cell growth and apoptosis. Proc Natl Acad Sci USA 1996;93:6737–6742.
93. Akahani S, Nangia-Makker P, Inohara H, Kim HR, Raz A. Galectin-3: a novel antiapoptotic molecule with a functional BH1 (NWGR) domain of Bcl-2 family. Cancer Res 1997;57:5272–5276.
94. Matarrese P, Fusco O, Tinari N, Natoli C, Liu FT, Semeraro ML, Malorni W, Iacobelli S. Galectin-3 overexpression protects from apoptosis by improving cell adhesion properties. Int J Cancer 2000;85:545–554.
95. Hayashi A, Hayashi A, Gillen AC, Lott JR. Effects of daily oral administration of quercetin chalcone and modified citrus pectin. Altern Med Rev 2000;5:546–552.
96. Chauhan D, Li G, Podar K, Hideshima T, Neri P, He D, Mitsiades N, Richardson P, Chang Y, Schindler J, Carver B, Anderson KC. A novel carbohydrate-based therapeutic GCS-100 overcomes bortezomib resistance and enhances dexamethasone-induced apoptosis in multiple myeloma cells. Cancer Res 2005;65:8350–8358.
97. McCarty MF and Block KI. Toward a core nutraceutical program for cancer management. Integr Cancer Ther 2006;5:150–171.
98. Sathisha UV, Jayaram S, Nayaka MAH, Dharmesh SM. Inhibition of galectin-3 mediated cellular interactions by pectic polysaccharides from dietary sources. Glycoconj J 2007;24:497–507.
99. Rabinovich GA, Cumashi A, Bianco GA, Ciavardelli D, Iurisci I, D'Egidio M, Piccolo E, Tinari N, Nifantiev N, Iacobelli S. Synthetic lactulose amines: novel class of anticancer agents that induce tumor-cell apoptosis and inhibit galectin-mediated homotypic cell aggregation and endothelial cell morphogenesis. Glycobiology 2006;16:210–220.
100. Zou J, Glinsky VV, Landon LA, Matthews L, Deutscher SL. Peptides specific to the galectin-3 carbohydrate recognition domain inhibit metastasis-associated cancer cell adhesion. Carcinogenesis 2005;26:309–318.
101. André S, Maljaars CE, Halkes KM, Gabius HJ, Kamerling JP. Discovery of galectin ligands in fully randomized combinatorial one-bead-one-compound (glyco)peptide libraries. Bioorg Med Chem Lett 2007;17:793–798.

11

GALECTINS IN MALIGNANT GLIOMAS: EXPRESSION, FUNCTIONS, AND POSSIBLE THERAPEUTIC OPTIONS

HERWIG M. STRIK AND ANNA HOFFMANN

Department of Neurology, Medical School, University of Göttingen, Robert Koch Strasse 40, D-37099 Göttingen, Germany

INTRODUCTION

Glioblastoma multiforme is the most malignant human glioma. It is the most common intrinsic malignant brain tumor and one of the most malignant types of human cancer, classified by WHO as grade IV. Despite considerable therapeutic efforts, the median survival of all patients suffering from this disease still does not exceed approximately 12 months (1) in unselected patients. Even in the most successful studies conducted with combined irradiation and chemotherapy, the overall survival does not exceed 14–17 months (2, 3) for selected patients in good clinical condition.

During the past few decades, several reasons for the aggressive growth and poor response of glioblastomas to antineoplastic therapy have been identified. A prominent feature among them is a marked primary resistance against apoptosis-inducing stimuli (4). For example, mutations of wild-type p53 are found in a large part of glioblastomas (5). Their expression allegedly causes a deficient control of cell growth. Overexpression of antiapoptotic proteins of the BCL-2 protein family in conjunction with downregulation of proapoptotic proteins of this group apparently prevents these tumors from extensive damage by radiation or chemotherapy.

Galectins, Edited by Anatole A. Klyosov, Zbigniew J. Witczak, and David Platt

Interestingly, in the development from primary to recurrent glioblastomas, we observed an additional shift of the rheostat toward antiapoptotic functions (6). An improvement of the effects of classical antineoplastic therapies such as irradiation or chemotherapy will have to aim at suppressing the resistance to the induction of apoptosis.

The most widely used and the most effective drugs to treat malignant gliomas are alkylating agents. Nitrosureas such as carmustine (BCNU), lomustine (CCNU), or nimustine (ACNU) are able to pass efficiently the blood–brain barrier and have shown efficacy in numerous studies. The best median overall survival of 17 months was achieved in a large multicenter trial with ACNU (2). Relevant hematotoxicity, however, was experienced with this regimen.

Temozolomide is an orally available imidazotetrazine derivative that also exerts alkylating activity, penetrates well the blood–brain barrier, and has a favorable toxicity profile. With this chemotherapeutic agent, a highly significant effect of chemotherapy as compared to radiation alone has been proven for the first time (3). A retrospective analysis of these data demonstrated that O^6-methyl-DNA-methyl-guanin-transferase (MGMT) is a major predictor of response to chemotherapy (7). MGMT removes the alkyl group from the O^6-guanin residue, thereby preventing the apoptosis-inducing effect of nitrosureas or temozolomide.

Recent efforts aim toward overcoming the effect of this suicide gene by specific inhibitors or—with encouraging results—with prolonged, dose-dense application of temozolomide (8–10). The effect of chemotherapy, though, is expected to be limited even if MGMT activity can be overcome. Therefore, a further improvement of therapy by downregulation of resistance to apoptosis induction would be of major interest.

A marked potential for migration and invasion is another important feature of malignant gliomas (11). Glioma cells are able to migrate up to several centimeters apart from the tumor bulk (12). This makes curative surgery impossible, since resection with a large safety margin would always affect functionally relevant cerebral regions. Other local therapeutic approaches such as radiotherapy or chemotherapy will not be likely able to affect these migrating cells. Systemic chemotherapy is expected to be less efficient for cells apart from the tumor bulk, where the intact blood–brain barrier may protect them more efficiently than the pathological vasculature in the center of the tumor. Furthermore, migrating cells show a decreased proliferation rate, which contributes to a relative resistance against apoptosis and a reduced response to chemotherapy and radiotherapy (13).

Strategies to inhibit migration and invasion of gliomas would therefore be an important addition to other therapeutic options. Matrix metalloproteinases (MMP) such as MMP-2 and -9 and integrins such as $\alpha_v\beta_3$ belong to the best examined proteins with impact on migration and invasion in human gliomas (14).

The need for sufficient blood supply has been recognized to be crucial for tumor growth. Angiogenic factors such as HIF-1, transforming growth factor beta (TGF-β), basic fibroblast growth factor (bFGF), and—most importantly—vascular endothelial growth factor (VEGF) have been shown to be overexpressed in human gliomas and to play an important role for glioma angiogenesis (15). In contrast to the upregulation of angiogenic VEGF, antiangiogenic endostatin is downregulated in the progression from low grade gliomas to glioblastomas (16). Apparently, a shift of the balance between pro- and antiangiogenic factor toward the angiogenic phenotype enables the fast growing tumor to be supplied through an equally fast growing vasculature (17). After first discouraging results with antiangiogenic substances such as thalidomide (18) or with anti-VEGF antisense strategies, numerous promising studies employing compounds for targeted inhibition of the VEGF signaling pathways have followed.

Malignant gliomas are well known for the abundant infiltration with monocytes/ microglial cells up to approximately 30% of the total cell content with only few lymphocytes present (19). The fact that no tumor cell phagocytosis can be observed in those cases indicates a lack of activation of the infiltrating monocytes. Moreover, these infiltrating cells may even promote tumor progression, as Platten demonstrated that after transfection of glioma cells with the chemoattractant MCP-1, more monocytes were attracted into rat gliomas that grew faster, leading to an early death of the animals (20). The infiltrating monocytes were probably modulated toward chronic inflammatory M2 macrophages that could promote tumor growth through production of growth factors. The lack of tumor infiltration by lymphocytes may reflect such a deficient activation toward the M1 phenotype that is able to present antigen and activate lymphocytes. Accordingly, a number of immunosuppressive growth factors such as IL-4, IL-13, and IL-10 are found regularly in human gliomas, while proinflammatory cytokines such as IL-12, TNF-α, or IFN-γ are not found (21, 22). In addition, other immunosuppressive factors such as CD70, HLA-G, or HLA-E contribute to the immune escape of malignant gliomas (23).

It has been demonstrated repeatedly that immune cells are capable of destroying glioma cells when activated *in vitro* (24, 25). Promising results of some independent studies show that immunological approaches can be effective (26, 27), but success seems to be limited by the extensive immunosuppressive environment in malignant gliomas. Circumvention of these immunosuppressive functions would be a promising strategy for the treatment of malignant gliomas.

MALIGNANT GLIOMAS AND GALECTINS

During the past few years, the possible impact of galectins on the biology of malignant gliomas has drawn rising attention. In a comparative study on various tumor cell lines, all analyzed astrocytic tumors expressed galectin-1, -3, and -8, while galectin-2, -4, and -9 RNA was only found in 25–40% of cell lines and galectin-7 in none of these cell lines (28). Of these galectins, especially galectin-1 and -3 have been analyzed for their expression at mRNA and protein levels and for their functional significance in malignant human gliomas (29–31). The following overview will focus on these two galectins.

Galectin-1

Expression of galectin-1 in glioma cells has been previously reported (29, 30, 32). Higher expression of galectin-1 mRNA was found in higher grades of malignancy (29), indicating a tight correlation between galectin-1 and the aggressiveness of these tumors.

Suppression of galectin-1 using an antisense strategy resulted in remarkable changes of morphology and inhibited the growth of 9 L rat gliomas (29). The authors concluded that lower levels of galectin-1 might arrest the growth of 9 L glioma cells. Accordingly, galectin-1 resulted in prolonged survival of nude mice with intracranially implanted U87 or U373 human glioma cells (33). Moreover, low levels of galectin-1 in human malignant astrocytomas or glioblastomas, as detected by immunohistochemistry, were associated with a significantly longer survival of patients (30). No significant effect of exogenous galectin-1, however, was observed on the growth of U87 cells (33).

Pioneer studies carried out by Kiss and colleagues have analyzed the effects of galectins on cell migration and invasion in human gliomas (30, 34). In experimental glioblastoma, a higher level of galectin-1, -3, and -8 was found in the invasive stroma compared to the parenchyma or other sections of the tumor. Galectin-1 promoted migration of U87 glioma cells, as evaluated by computer-assisted microscopy after modulation of the galectin-1 content of the plastic support (34). Further analysis of the molecular basis showed that the influence of galectins on cell migration and invasion results from modifications of the organization of the cytoskeleton through small GTPases such as RhoA (32). This effect seemed to stem from modulation of extracellular matrix (ECM) components and may contribute to the aggressiveness of human gliomas.

We have found the expression of galectin-1 in all of 12 immortalized glioma cell lines at different levels (35). Interestingly, ionizing irradiation induced the expression of galectin-1 detected by immunoblot. While the addition of recombinant galectin-1 caused no significant change of proliferation, suppression by RNA interference resulted in a marked antiproliferative effect in A172 but not in U118 cells. Thus, galectin-1 may have different effects on cell proliferation depending on whether it acts extra- or intracellularly. The effect on cell proliferation may also depend on the susceptibility of different tumor cell types, which is relied upon the balance between different glycosyltransferases creating glycan ligand on the cell surface. The association of different glycan structures with the two different glioma cell lines studied still remains to be ascertained.

A possible explanation for the only mild effects of exogenous recombinant galectin-1 could be that the expression of galectin-1 has already reached maximally high levels, so that incorporation of an excess of this protein to cell cultures can hardly cause an additional effect. Another explanation is that the most relevant effect exerted by galectin-1 would be mainly intracellular, so that exogenous administration of this carbohydrate-binding protein will not be efficient in modulating tumor growth.

We also investigated whether intracellular or extracellular galectin-1 might affect migration of human glioma cell lines. When exogenous galectin-1 was added to cell cultures, a significant effect was observed on cell migration of human glioma cell lines. An increase of about 100% migrating A172 cells was seen after the addition of recombinant galectin-1. However, a bell-shaped effect was observed with U118 cells that migrated significantly better with 1 μg/mL, but not with 2 μg/mL. Surprisingly, migration assays could not be performed when cells were transfected with siRNA before the seeding into the wells. Under this condition, cells remained in suspension and were not able to adhere to the plastic surface, suggesting a close connection between galectin-1 and cellular adhesion molecules. Therefore, we decided to transfect glioma cells with siRNA following adherence of these cells to the migration devices. Using this experimental design, the number of migrating cells dropped by approximately 50% in U118 cells. By contrast, only a slight effect was seen in A172 cells at both concentrations.

Interestingly, galectin-1 downregulation had not the same effect on both cell lines, A172 and U118, and interfered with proliferation only in A172. However, proliferation and migration were noticeable in both cell lines. Regarding addition of exogenous galectin-1, one might speculate that each glioma cell line may express different glycosyltransferases, compromising the susceptibility of these cell lines to galectin-1. However, in the case of intracellular galectin-1, disruption of this protein by siRNA might have different consequences depending on the differentiation state of the cell or its cell cycle progression status. In any case, longer survival associated with lower galectin-1 expression levels in experimental and human gliomas may result not only from a lower migratory phenotype of these cells, but also from a reduced proliferative activity.

Galectin-3

Conflicting results have been reported in the first immunohistochemical studies on the expression of galectin-3 in human gliomas. The first study found a positive correlation with the WHO grade (36), while another group found an inverse correlation (37).

In order to elucidate the reason for the conflicting findings, we studied cellular and regional distribution of galectin-3 and compared with immunoreactivity to the macrophage marker CD68 in 53 gliomas (WHO grades II–IV) by immunohistochemistry. The origin of galectin-3- expressing cells was confirmed by double labeling experiments. Expression of this lectin was seen only in 40% of cells of astrocytic origin, but in 70% of monocytes/macrophages in double labeling immunohistochemistry. In full accordance with the first study above, we found a significant association of galectin-3 expression with the grade of malignancy. Most importantly, we identified tumor-infiltrating macrophages as a substantial source of galectin-3 immunoreactivity.

The tumor parenchyma of human glioblastomas may contain a variety of cell types including macrophages/microglial cells (38), which cannot always be distinguished from glioma cells by pure morphological criteria. As indicated above, we have found a tendency, though not statistically significant, toward higher numbers of tumor-infiltrating macrophages/microglial cells in higher grades of malignancy. In spite of a considerable variability of both results, the positivity for the macrophage/microglial marker CD68 was significantly correlated with positivity for galectin-3, which is known as a macrophage marker itself (39). Thus, galectin-3 immunoreactivity in human gliomas may be considerably influenced by tumor-infiltrating macrophages/microglial cells in a manner that is not significantly correlated with the grade of malignancy. This might explain the conflicting results reported in the literature. In spite of this fact and the lack of clear-cut limits of expression between the WHO grades, galectin-3 has been repeatedly proposed as a marker for glial tumors (40) or for differentiating WHO malignancy grades (41). However, in view of the drawbacks found in our own analysis, galectin-3 appears not to be specific enough for diagnostic purposes.

In accordance with the previous findings (37), we also found less frequent staining of endothelial cells in glioblastomas compared to that in WHO grade II and III astrocytomas. The pathological vasculature of glioblastoma multiforme with tangles of endothelia is thought to develop by increased proliferation of endothelial cells that do not migrate sufficiently, thus forming the characteristic microvascular proliferations (17). Galectin-3 has been shown to induce endothelial cell differentiation (42). One major mechanism seems to be binding to $\alpha_3\beta_1$ integrin on the surface of endothelial cells. This results in a complex formation with NG2 proteoglycan, which may increase transmembrane signaling with effect on endothelial cell motility and morphogenesis (43, 44). The loss of galectin-3 expression may result in inhibited migration of glioblastoma endothelial cells. Subsequently, continuing the proliferation of the immobilized endothelial cells and a lack of differentiation signals may result in the typical pathological glomerulum-like vascular deformation. Functional studies have to elucidate if and how galectin-3 plays a role in tumor angiogenesis in human gliomas.

In functional studies, we found only a weak influence of exogenous galectin-3 protein on the proliferation of glioma cells: a bell-shaped effect was observed with a significantly increased proliferation of U118 cells with 1 and 2 μg/mL and a reduction to almost baseline with 4 μg/mL. In A172 cells, however, no consistent effect was seen (unpublished data). Migration in the transwell assay was induced in both cell lines by exogenous galectin-3 protein with a significant increase on top of that observed earlier with 1 μg/mL in A172 and with 2 μg/mL in both cell lines. This exogenous effect, however, is in contrast to a previous study by Debray et al. (45) that found an increased motility of U373 glioma cells and upregulation of α_6- and β_1-integrins after galectin-3 suppression with an antisense construct. These contradictory results may be caused either by a concentration-dependent effect or a difference between extra- and intracellular functions of galectins.

THERAPEUTIC CONSIDERATIONS

The data available to date indicate a prominent role of galectin-1 and-3, possibly also of galectin-8 and other glycoprotein-binding molecules in glioma biology.

Galectin-1 is the functionally best examined galectin in gliomas. Its influence on glioma cell proliferation raises hope to achieve growth inhibition in human gliomas. In addition to a reduced proliferation rate, an effect on resistance to apoptosis induction has to be examined. Since targeted therapies with inhibition of signaling pathways often are not efficient in monotherapy, but produce positive results in the combination with conventional chemotherapy, combination of galectin-1 inhibition with temozolomide, aiming at an improved apoptosis-inducing effect, would be a promising approach. Such an effect has already been shown with experimental melanomas (46).

With regard to galectin-3, less impressive data have been collected during the past few years. Our own unpublished data also point toward an induction of cell growth by recombinant galectin, which was effective only in moderate doses. In previous studies, however, recombinant galectin-3 has been shown to act as a mitogen in lung fibroblast cells (47). Furthermore, galectin-3 has been demonstrated to possess antiapoptotic effects (48) that could also be targeted by galectin interference to be combined with antineoplastic chemotherapy.

Inhibition of migration and invasion is another functional target of glioma biology. Growing evidence of a promotion of migration makes galectin-1 inhibition interesting also in this regard. Restraining cells from migrating far away from the tumor bulk may render tumors more susceptible to local therapies such as radiotherapy or **local** chemotherapy. Following the data collected by Camby et al. (32) and our own unpublished results, galectin-3 seems to promote glioma cell migration in a similar manner.

To date, one principle of cell biology apparently is to either go or grow, which may result in a more disseminated growth pattern when cell proliferation is impaired (49). One of the molecules regulating the balance between proliferation and migration is the membrane-spanning proteoglycan NG2. This molecule promotes both functions depending on whether it interacts with extracellular or intracellular-binding partners (50). Further elucidation of the interaction between galectin-1 and -3 may help to assess the differential effect of these molecules on migration and proliferation in gliomas.

After first discouraging results of targeting angiogenesis in brain tumors, recent studies with newer compounds achieved promising results. One advantage of aiming toward the tumor vasculature in gliomas is that the blood–brain barrier has not to be overcome, since the target lies at the endovascular side of the blood vessel. By such, even therapies with large molecular antibodies are feasible. Galectin-1 has been shown to be expressed in endothelial cells of different tumors and to be involved in angiogenesis. Galectin-1 knockdown has been shown recently to result in defective vasculature and reduced tumor growth (51). Accordingly, our own data of a distinct galectin-3 expression in endothelial cells of different WHO grades of human gliomas (31) indicate that this galectin may also be involved in glioma angiogenesis. This assumption is supported by reports that galectin-3 induces endothelial cell differentiation (42). Furthermore, it has been shown to influence endothelial cell motility and morphogenesis (43). Targeting both galectin-1 and -3 may result in inhibited tumor angiogenesis with a reduced growth potential of gliomas.

Finally, the overexpression of galectins may contribute to the immunosuppressive microenvironment of human gliomas. Galectin-1 has been shown to suppress the production of proinflammatory cytokines like interleukin (IL)-12 in T cells and macrophages, but not of immunosuppressive IL-10 (52). It induces apoptosis in T lymphocytes (53) and enhanced tumor cell rejection (54). In contrast to galectin-1, galectin-3 has no clear functions in immune responses. Recruitment of phagocytic cells is lacking in galectin-3-deficient mice, indicating a possible role of galectin-3 as a chemoattractant of monocytes in human gliomas.

Proinflammatory functions are promoted by galectin-3, such as phagocytosis by macrophages and adhesion between dendritic cells and lymphocytes. However, T2 allergic reactions are suppressed by galectin-3. Taken together, galectin-1 suppression clearly seems to be a promising strategy to revert the immunosuppressive environment of gliomas, whereas the functions of galectin-3 have to be further elucidated and probable distinct functions dependent on dosage and extra- or intracellular interaction clarified. The data available to date indicate that an upregulation of the immune responses may induce regression of gliomas. The extent of inducing inflammatory response, however, will have to be handled with caution, as autoimmune responses have to be avoided.

Our findings of different effectiveness of extra- or intracellular galectin modification point toward the importance of the mode of modulation. Extracellular interference with galectin-1 may be easier to achieve than intracellular modulation. The effect, however, may be minor in comparison with intracellular interference. The delivery of a gene therapeutic approach, however, will be difficult since the efficiency of transfection rates *in vivo* is very low, as demonstrated by a large gene therapy trial in human gliomas (55).

Our findings that ionizing irradiation upregulates galectin-1 expression may be of particular relevance to the treatment of glioma patients, since external beam irradiation is one of the most important elements of malignant glioma therapy. Migratory capacity of glioma cells has been shown to be enhanced by ionizing irradiation (56). This may in part be explained by our findings that galectin-1 expression is induced by irradiation and can promote migration of human glioma cells.

In conclusion, the data presented here support a concept that galectin-1 has an impact not only on migration, but also on proliferation of human glioma cells. This seems to be primarily an intracellular effect. In the light of these findings, galectin-1 seems to be an interesting target to treat malignant gliomas by interfering with proliferation and migration. Since ionizing irradiation induces the expression of galectin-1, downregulation of this molecule would be feasible before radiotherapy to avoid undesired side effects of this standard therapy. Taken together, these observations in conjunction with the novel role of galectin-1 in modulating angiogenesis (51) and tumor cell evasion of immune responses (54, 57), we conclude that blocking galectin-1 may help overcome several steps of tumorigenesis, including cell proliferation, migration, angiogenesis, and tumor immune escape. Thus, it can be predicted that inhibitors of galectin-1 (58–60) will find their way into cancer clinical trials, aiming at delaying the tumor progression and improving overall survival.

The wide distribution of galectins throughout cell compartment and organism and the multifaceted roles of galectins that may depend not only on extra- and intracellular functions, but also on concentration, make prediction of clinical effects difficult. Careful preclinical studies using cell cultures and animal models will be required to assess the appropriate mode of interference, application, and dosage before the expected enormous potential of this therapeutic approach can be used for the benefit of glioma patients.

REFERENCES

1. DeAngelis LM, Burger PC, Green SB, Cairncross JG, Malignant glioma. who benefits from adjuvant chemotherapy?Ann Neurol 1998;44:691–695.
2. Weller M, Muller B, Koch R, Bamberg M, Krauseneck P. Neuro-Oncology Working Group 01 trial of nimustine plus teniposide versus nimustine plus cytarabine chemotherapy in addition to involved-field radiotherapy in the first-line treatment of malignant glioma. J Clin Oncol 2003;21:3276–3284.
3. Stupp R, Mason WP, van den Bent MJ, Weller M, Fisher B, Taphoorn MJ, Belanger K, Brandes AA, Marosi C, Bogdahn U, Curschmann J, Janzer RC, Ludwin SK, Gorlia T, Allgeier A, Lacombe D, Cairncross JG, Eisenhauer E, Mirimanoff RO. Radiotherapy plus concomitant and adjuvant temozolomide for glioblastoma. N Engl J Med 2005;352:987–996.
4. Bogler O, Weller M. Apoptosis in gliomas, and its role in their current and future treatment. Front Biosci 2002;7:339–353.
5. Ohgaki H, Dessen P, Jourde B, Horstmann S, Nishikawa T, Di Patre PL, Burkhard C, Schuler D, Probst-Hensch NM, Maiorka PC, Baeza N, Pisani P, Yonekawa Y, Yasargil MG, Lutolf UM, Kleihues P. Genetic pathways to glioblastoma: a population-based study. Cancer Res 2004;64:6892–6899.
6. Strik H, Deininger M, Streffer J, Grote E, Wickboldt J, Dichgans J, Weller M, Meyermann R. Bcl-2 family protein expression in initial and recurrent glioblastomas: modulation by radiochemotherapy. J Neurol Neurosurg Psychiatry 1999;67:763–768.
7. Hegi ME, Diserens AC, Gorlia T, Hamou MF, de Tribolet N, Weller M, Kros JM, Hainfellner JA, Mason W, Mariani L, Bromberg JE, Hau P, Mirimanoff RO, Cairncross JG, Janzer RC, Stupp R. MGMT gene silencing and benefit from temozolomide in glioblastoma. N Engl J Med 2005;352:997–1003.
8. Wick A, Felsberg J, Steinbach JP, Herrlinger U, Platten M, Blaschke B, Meyermann R, Reifenberger G, Weller M, Wick W. Efficacy and tolerability of temozolomide in an alternating weekly regimen in patients with recurrent glioma. J Clin Oncol 2007;25:3357–3361.
9. Brandes AA, Tosoni A, Cavallo G, Bertorelle R, Gioia V, Franceschi E, Biscuola M, Blatt V, Crino L, Ermani M. Temozolomide 3 weeks on and 1 week off as first-line therapy for recurrent glioblastoma: phase II study from gruppo italiano cooperativo di neuro-oncologia (GICNO). Br J Cancer 2006;95:1155–1160.
10. Strik HM, Buhk JH, Nitsche M, Giese A, Bock C, Bähr M, Rechallenge of malignant gliomas with temozolomide—can it be effective? Abstract ASCO Meeting Chicago; 2007.
11. Bolteus AJ, Berens ME, Pilkington GJ. Migration and invasion in brain neoplasms. Curr Neurol Neurosci Rep 2001;1:225–232.
12. Kelly PJ, Daumas-Duport C, Kispert DB, Kall BA, Scheithauer BW, Illig JJ. Imaging-based stereotaxic serial biopsies in untreated intracranial glial neoplasms. J Neurosurg 1987;66:865–874.
13. Evan G, Littlewood T. A matter of life and cell death. Science 1998;281:1317–1322.
14. Wild-Bode C, Weller M, Wick W. Molecular determinants of glioma cell migration and invasion. J Neurosurg 2001;94:978–984.
15. Fischer I, Gagner JP, Law M, Newcomb EW, Zagzag D. Angiogenesis in gliomas: biology and molecular pathophysiology. Brain Pathol 2005;15:297–310.

16. Strik HM, Schluesener HJ, Seid K, Meyermann R, Deininger MH. Localization of endostatin in rat and human gliomas. Cancer 2001;91:1013–1019.
17. Wesseling P, Ruiter DJ, Burger PC. Angiogenesis in brain tumors; pathobiological and clinical aspects. J Neurooncol 1997;32:253–265.
18. Riva M, Imbesi F, Beghi E, Galli C, Citterio A, Trapani P, Sterzi R, Collice M. Temozolomide and thalidomide in the treatment of glioblastoma multiforme. Anticancer Res 2007;27:1067–1071.
19. Graeber MB, Scheithauer BW, Kreutzberg GW. Microglia in brain tumors. Glia 2002;40:252–259.
20. Platten M, Kretz A, Naumann U, Aulwurm S, Egashira K, Isenmann S, Weller M. Monocyte chemoattractant protein-1 increases microglial infiltration and aggressiveness of gliomas. Ann Neurol 2003;54:388–392.
21. Parney IF, Hao C, Petruk KC. Glioma immunology and immunotherapy. Neurosurgery 2000;46:778–791.
22. Hao C, Parney IF, Roa WH, Turner J, Petruk KC, Ramsay DA. Cytokine and cytokine receptor mRNA expression in human glioblastomas: evidence of Th1, Th2 and Th3 cytokine dysregulation. Acta Neuropathol (Berl) 2002;103:171–178.
23. Wischhusen J, Jung G, Radovanovic I, Beier C, Steinbach JP, Rimner A, Huang H, Schulz JB, Ohgaki H, Aguzzi A, Rammensee HG, Weller M. Identification of CD70-mediated apoptosis of immune effector cells as a novel immune escape pathway of human glioblastoma. Cancer Res 2002;62:2592–2599.
24. Kito T, Kuroda E, Yokota A, Yamashita U. Enhancement of macrophage cytotoxicity against murine gliomas by interferon beta: increase in nitric oxide production in response to glioma-derived soluble factors. J Neurosurg 2002;97:619–626.
25. Kito T, Kuroda E, Yokota A, Yamashita U. Cytotoxicity in glioma cells due to interleukin-12 and interleukin-18-stimulated macrophages mediated by interferon-gamma-regulated nitric oxide. J Neurosurg 2003;98:385–392.
26. Yamanaka R, Homma J, Yajima N, Tsuchiya N, Sano M, Kobayashi T, Yoshida S, Abe T, Narita M, Takahashi M, Tanaka R. Clinical evaluation of dendritic cell vaccination for patients with recurrent glioma: results of a clinical phase I/II trial. Clin Cancer Res 2005;11:4160–4167.
27. Steiner HH, Bonsanto MM, Beckhove P, Brysch M, Geletneky K, Ahmadi R, Schuele-Freyer R, Kremer P, Ranaie G, Matejic D, Bauer H, Kiessling M, Kunze S, Schirrmacher V, Herold-Mende C. Antitumor vaccination of patients with glioblastoma multiforme: a pilot study to assess feasibility, safety, and clinical benefit. J Clin Oncol 2004;22:4272–4281.
28. Lahm H, Andre S, Hoeflich A, Kaltner H, Siebert HC, Sordat B, der Lieth CW, Wolf E, Gabius HJ. Tumor galectinology: insights into the complex network of a family of endogenous lectins. Glycoconj J 2004;20:227–238.
29. Yamaoka K, Mishima K, Nagashima Y, Asai A, Sanai Y, Kirino T. Expression of galectin-1 mRNA correlates with the malignant potential of human gliomas and expression of antisense galectin-1 inhibits the growth of 9 glioma cells. J Neurosci Res 2000;59:722–730.
30. Rorive S, Belot N, Decaestecker C, Lefranc F, Gordower L, Micik S, Maurage C, Kaltner H, Ruchoux MM, Danguy A, Gabius HJ, Salmon I, Kiss R, Camby I. Galectin-1 is highly expressed in human gliomas with relevance for modulation of invasion of tumor astrocytes into the brain parenchyma. Glia 2001;33:241–255.

31. Strik HM, Deininger MH, Frank B, Schluesener HJ, Meyermann R. Galectin-3: cellular distribution and correlation with WHO-grade in human gliomas. J Neurooncol 2001; 53:13–20.

32. Camby I, Decaestecker C, Gordower L, DeDecker R, Kacem Y, Ledmmers A, Siebert H, Bovin N, Wesseling P, Danguy A, Salmon I, Gabius HJ, Kiss R. Distinct differences in binding capacity to saccharide epitopes in supratentorial pilocytic astrocytomas, astrocytomas, anaplastic astrocytomas and glioblastomas. J Neuropathol Exp Neurol 2001;60:75–84.

33. Camby I, Belot N, Lefranc F, Sadeghi N, de Launoit Y, Kaltner H, Musette S, Darro F, Danguy A, Salmon I, Gabius HJ, Kiss R. Galectin-1 modulates human glioblastoma cell migration into the brain through modifications to the actin cytoskeleton and levels of expression of small GTPases. J Neuropathol Exp Neurobiol 2002;61:585–596.

34. Camby I, Belot, Rorive S, Lefranc F, Lahm H, Kaltner H, Hadari Y, Ruchoux MM, Brotchi J, Zick Y, Salmon I, Gabius HJ, Kiss R. Galectins are differentially expressed in supratentorial pilocytic astrocytomas, astrocytomas, anaplastic astrocytomas and glioblastomas, and significantly modulate tumor astrocyte migration. Brain Pathol 2001;11:12–26.

35. Strik HM, Schmidt K, Lingor P, Tonges L, Kugler W, Nitsche M, Rabinovich GA, Bähr M. Galectin-1 expression in human glioma cells: Modulation by ionizing radiation and effects on tumor cell proliferation and migration. Oncol Rep 2007;18:483–488.

36. Bresalier RS, Yan PS, Byrd JC, Lotan R, Raz A. Expression of the endogenous galactose-binding protein galectin-3 correlates with the malignant potential of tumors in the central nervous system. Cancer 1997;80:776–787.

37. Gordower L, Decaestecker Y, Kacem Y, Lemmers J, Gusman J, Burchert M, Danguy A, Gabius HJ, Salmon R, Camby I. Galectin-3 and galectin-3-binding site expression in human adult astrocytic tumours and related angiogenesis. Neuropathol Appl Neurobiol 1999;25:319–330.

38. Roggendorf W, Strupp S, Paulus W. Distribution and characterization of microglia/ macrophages in human brain tumors. Acta Neuropathol Berl 1996;92:288–293.

39. Liu FT, Hsu DK, Zuberi RI, Kuwabara I, Chi EY, Henderson-WR J. Expression and function of galectin-3, a beta-galactoside-binding lectin, in human monocytes and macrophages. Am J Pathol 1995;147:1016–1028.

40. Kuklinski S, Pesheva P, Heimann C, Urschel S, Gloor S, Graeber S, Herzog V, Pietsch T, Wiestler OD, Probstmeier R. Expression pattern of galectin-3 in neural tumor cell lines. J Neurosci Res 2000;60:45–57.

41. Neder L, Marie SK, Carlotti CG Jr, Gabbai AA, Rosemberg S, Malheiros SM, Siqueira RP, Oba-Shinjo SM, Uno M, Aguiar PH, Miura F, Chammas R, Colli BO, Silva WA Jr, Zago MA. Galectin-3 as an immunohistochemical tool to distinguish pilocytic astrocytomas from diffuse astrocytomas, and glioblastomas from anaplastic oligodendrogliomas. Brain Pathol 2004;14:399–405.

42. Nangia MP, Honjo Y, Sarvis R, Akahani S, Hogan V, Pienta KJ, Raz A. Galectin-3 induces endothelial cell morphogenesis and angiogenesis. Am J Pathol 2000;156:899–909.

43. Fukushi J, Makagiansar IT, Stallcup WB. NG2 proteoglycan promotes endothelial cell motility and angiogenesis via engagement of galectin-3 and alphabeta1 integrin. Mol Biol Cell 2004;15:3580–3590.

44. Wen Y, Makagiansar IT, Fukushi J, Liu FT, Fukuda MN, Stallcup WB. Molecular basis of interaction between NG2 proteoglycan and galectin-3. J Cell Biochem 2006;98:115–127.

45. Debray C, Vereecken P, Belot N, Teillard P, Brion JP, Pandolfo M, Pochet R. Multifaceted role of galectin-3 on human glioblastoma cell motility. Biochem Biophys Res Commun 2004;325:1393–1398.

46. Mathieu V, Le Mercier M, De Neve N, Sauvage S, Gras T, Roland I, Lefranc F, Kiss R. Galectin-1 knockdown increases sensitivity to temozolomide in a B16F10 mouse metastatic melanoma model. J Invest Dermatol 2007;127:2399–2410.

47. Inohara H, Akahani S, Raz A. Galectin-3 stimulates cell proliferation. Exp Cell Res 1998;245:294–302.

48. Akahani S, Nangia MP, Inohara H, Kim HR, Raz A. Galectin-3: a novel antiapoptotic molecule with a functional BH1 (NWGR) domain of Bcl-2 family. Cancer Res 1997;57: 5272–5276.

49. Giese A, Bjerkvig R, Berens ME, Westphal M, Cost of migration. invasion of malignant gliomas and implications for treatment. J Clin Oncol 2003;21:1624–1636.

50. Makagiansar IT, Williams S, Mustelin T, Stallcup WB. Differential phosphorylation of NG2 proteoglycan by ERK and PKCalpha helps balance cell proliferation and migration. J Cell Biol 2007;178:155–165.

51. Thijssen VL, Postel R, Brandwijk RJ, Dings RP, Nesmelova I, Satijn S, Verhofstad N, Nakabeppu Y, Baum LG, Bakkers J, Mayo KH, Poirier F, Griffioen AW. Galectin-1 is essential in tumor angiogenesis and is a target for antiangiogenesis therapy. Proc Natl Acad Sci USA 2006;103:15975–15980.

52. Rubinstein N, Ilarregui JM, Toscano MA, Rabinovich GA. The role if galectins in the initiation, amplification and resolution of inflammatory response. Tissue Antigens 2004;64:1–12.

53. Perillo NL, Pace KE, Seilhamer JJ, Baum LG, Sato S, Hughes RC. Apoptosis of T cells mediated by galectin-1. Nature 1995;378:736–739.

54. Rubinstein N, Alvarez M, Zwirner NW, Toscano MA, Ilarregui JM, Bravo A, Mordoh J, Fainboim L, Podhajcer OL, Rabinovich GA. Targeted inhibition of galectin-1 gene expression in tumor cells results in heightened T cell-mediated rejection: a potential mechanism of tumor-immune privilege. Cancer Cell 2004;5:241–251.

55. Rainov NG. A phase III clinical evaluation of herpes simplex virus type 1 thymidine kinase and ganciclovir gene therapy as an adjuvant to surgical resection and radiation in adults with previously untreated glioblastoma multiforme. Hum Gene Ther 2000;11: 2389–2401.

56. Wild-Bode C, Weller M, Rimner A, Dichgans J, Wick W. Sublethal irradiation promotes migration and invasiveness of glioma cells: implications for radiotherapy of human glioblastoma. Cancer Res 2001;61:2744–2750.

57. Le QT, Shi G, Cao H, Nelson DW, Wang Y, Chen EY, Zhao S, Kong C, Richardson D, O'Byrne KJ, Giaccia AJ, Koong AC. Galectin-1: a link between tumor hypoxia and tumor immune privilege. J Clin Oncol 2005;23:8932–8941.

58. Rabinovich GA, Cumashi A, Bianco GA, Ciavardelli D, Iurisci I, D'Egidio M, Piccolo E, Tinari N, Nifantiev N, Iacobelli S. Synthetic lactulose amines: novel class of anticancer agents that induce tumor-cell apoptosis and inhibit galectin-mediated homotypic cell aggregation and endothelial cell morphogenesis. Glycobiology 2006;16:210–220.

59. Sorme P, Kahl-Knutsson B, Wellmar U, Magnusson BG, Leffler H, Nilsson UJ. Design and synthesis of galectin inhibitors. Methods Enzymol 2003;363:157–169.

60. Andre S, Pieters RJ, Vrasidas I, Kaltner H, Kuwabara I, Liu FT, Liskamp RM, Gabius HJ. Wedgelike glycodendrimers as inhibitors of binding of mammalian galectins to glycoproteins, lactose maxiclusters, and cell surface glycoconjugates. Chem Bio Chem 2001;2:822–830.

12

FOOD-RELATED CARBOHYDRATE LIGANDS FOR GALECTINS

Valeri V. Mossine, Vladislav V. Glinsky, and Thomas P. Mawhinney
Department of Biochemistry, University of Missouri, Columbia, MO 65211, USA

INTRODUCTION

Carbohydrates constitute over 50% of the dry mass in the human diet. More than 100 structures of mono-, oligo-, or polysaccharides have been identified as natural food ingredients or food additives. Sugars and starches (e.g., monosaccharides such as glucose and fructose, disaccharides such as sucrose and lactose, and polysaccharides such as amylose and amylopectin) constitute the main energy source from food and are readily digested and absorbed, in the form of monosaccharides, in the small intestine (1). In the Western diet, carbohydrates account for approximately 50% of human energy intake, and the daily consumption of main monosaccharides totals about 240 g of glucose, 65 g of fructose, and 15 g of galactose. Water-insoluble nondigestible carbohydrate fibers (mainly cellulose) pass through the digestive system with little change and make up the main bulk of undigested fecal mass (2). Nevertheless, these fibers are used by plants as a major structural material of cell walls and thus provide for structural integrity of many plant foods. They are also important for colon protection and the normal functioning of microflora in the gut.

Water-soluble nondigestible carbohydrates (WSNC) pass, with little change, through the upper intestine but are mostly fermented by gut bacteria into short fatty acids and gases. Small WSNCs, such as sweeteners and milk oligosaccharides, may be partially absorbed into the bloodstream, without being metabolized by the human

Galectins, Edited by Anatole A. Klyosov, Zbigniew J. Witczak, and David Platt

organism. The bulk of dietary WSNCs, however, is made up of polysaccharides, also referred to as "soluble fibers" (3). Though this term is firmly established in general use, it is somewhat misleading, as it also includes many insoluble, gel-forming polysaccharides.

Known types of nondigestible carbohydrates, with alleged health-beneficial effects, include polysaccharides and low molecular weight carbohydrates and their conjugates, derived mostly from fructose and galactose. The fructose-containing oligo- and polysaccharides are represented by fructans (found in chicory), levans, and inulin (4). Galactose-containing short oligosaccharides are found in legumes (raffinose and stachyose) or milk. Many polysaccharides from plant cell walls, exudates, and storage material are abundant in galactoside residues, such as galactomannans, agar, or pectic substances.

Emerging evidence shows the importance of many dietary nondigestible oligo- and polysaccharides for protection against diabetes, cancer, cardiovascular, and infectious diseases. Thus, oligosaccharides from human milk appear to be essential for the protection of newborns from pathogens and may act as anti-inflammatory agents (5). The risk of cardiovascular disease may be lowered by diets rich in prebiotic carbohydrates (6). In diabetics, better control over postprandial glucose concentration was achieved by ingestion of soluble fibers (7). A number of studies also showed that in populations with enhanced consumption of fruits and vegetables, rates of colorectal, prostate, and other cancers are significantly lower, and this beneficial effect was also consistently related to diets rich in WSNCs (8, 9). These and many other examples of beneficiary effects of nondigestible saccharides for human health have stimulated an interest in the details of their biological activities at the molecular level. One can, however, expect the mechanistic studies to be very complicated because of the structural diversity and heterogeneity of WSNC.

As of today, the exact mechanisms of the immunomodulatory or cancer protective effects of dietary carbohydrates are not established. In the previous chapters of this book, the role of galectins in the signaling of the immune cells, tumor cell proliferation, and adhesion along with other physiologically important cellular functions has been extensively reviewed. Since galectins have been recognized as novel therapeutic targets in the fields of immunology and oncology (10, 11), a search for galectin inhibitors, aimed at developing potential anticancer drugs, is under way (12, 13). Considering many hurdles that traditionally accompany the development of carbohydrate-based drugs, galactose-rich nondigestible saccharides from food might provide an attractive natural alternative to synthetic antigalectin agents. Indeed, in the majority of *in vivo* studies, involving galectin inhibition experiments, the lectin blockers were of food-related origin.

In this chapter, we have compiled available data on two types of food-related β-galactoside WSNCs: (1) pectic saccharides and (2) lactulosamines (LAs), which have been validated as potential galectin ligands and tumor growth inhibitors in the initial studies.

PECTIN AND ARABINOGALACTAN

Structure and Sources

Pectin, as defined, is a group of acidic heteropolysaccharides from plant and fungal cell walls, exhibiting significant heterogeneity with respect to both chemical structure and molecular weight (14, 15). The pectic "backbone" mainly consists of galacturonan, a homopolymer of (1 → 4)-α-D-galactopyranuronic acid with varying proportions of methyl-esterified carboxyl groups, and rhamnogalacturonan I, a heteropolymer of repeating (1 → 2)-α-L-rhamnopyranosyl-(1 → 4)-α-D-galactopyranuronic acid disaccharide units. Both galacturonan and rhamnogalacturonan I contain side substituents, but the level of the substitution and the size and the diversity of the substituents are dramatically different. Galacturonan is only sparsely populated with short side chains that are, nevertheless, very diverse and are represented by D-apiose, D-xylose monosaccharides, or short branches in the rhamnogalacturonan II region. The branches within this region are notable for their heterogeneity and for containing saccharide residues such as D-glucose, D-galactose (predominantly as α-anomer), D-xylose, L-rhamnose, L-arabinose, L-fucose, D-glucuronic acid, 3-deoxy-D-manno-2-octulosonic acid (Kdo), and 3-deoxy-D-lyxo-2-heptulosonic acid (Dha) along with others.

Within rhamnogalacturonan I, however, the side chains are attached to the backbone exclusively through rhamnose residues and consist of three types of branched (1–20 residues) oligosaccharides: arabinan and type I or type II arabinogalactans. The structure of the arabinogalactans itself is of the main interest for further discussion in this chapter due to the presence of β-D-galactopyranosyl residues. Type I arabinogalactan (AG-I) is the most abundant neutral pectic structure in nature and is typically found in fruits such as citrus (16) or apple (17). AG-I is composed of (1 → 4)-β-D-galactopyranose chains with α-L-arabinofuranosyl units attached to O-3 of the galactosyl residues (18). A small proportion of galactoside residues is mutually bound by β-(1 → 3) bonds (19). A typical primary structure is shown in Fig. 12.1. The type II arabinogalactan (AG-II) has considerably more branched galactan chains. Its main chain is composed of (1 → 3)-β-D-galactopyranose residues and is substituted by short branches of (1 → 6)-β-D-galactopyranose chains, which in turn carry (1 → 3) or (1 → 6)-linked α-L-arabinofuranosyl residues (Fig. 12.1).

According to the "canonical" model (14, 15, 20), pectin adopts an extended and curved "worm-like" tertiary structure that exhibits a large amount of flexibility. It is often described in terms of alternating "smooth" and "hairy" regions, which correspond to primary structures of the galacturonan and rhamnogalacturonan I. The "smooth" regions are defined by stretches of homogalacturonan, about 25–50 residues in length, with acetylated and/or methyl-esterified GalA residues as the only modifications to the α-D-galactopyranuronic acid. The macroscopic properties of pectins such as solubility or gelation depend on the extent of the methyl ester residues and the proportion of the homogalacturonan in a particular pectin sample.

The "hairy" regions include all pectic structures containing neutral sugars: non-branched xylo- and apiogalacturonans and branched rhamnogalacturonans I and II,

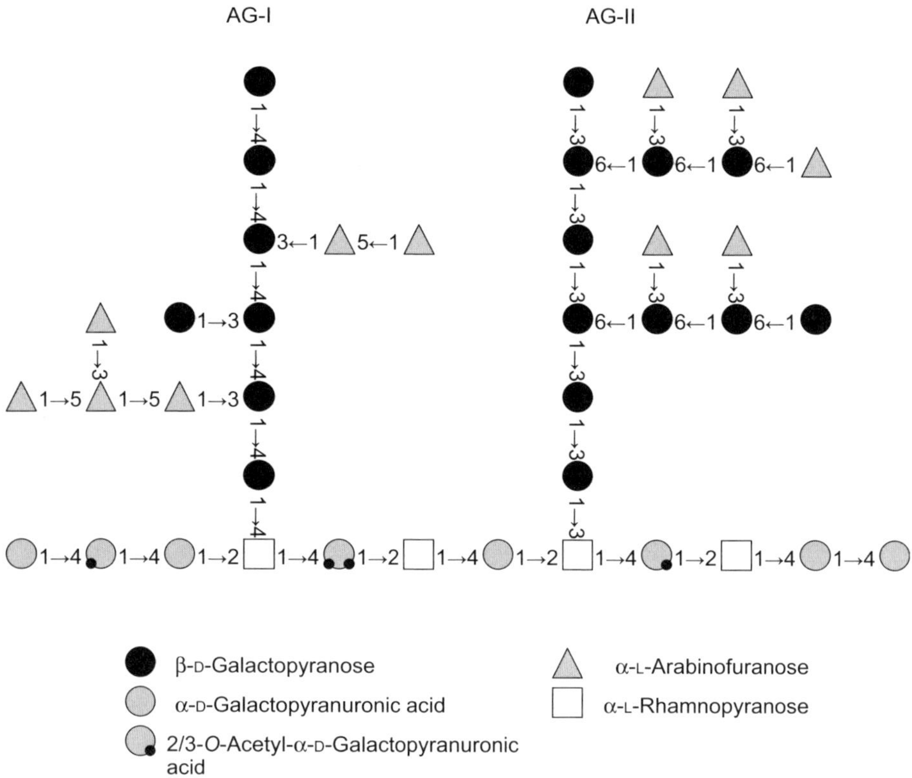

FIGURE 12.1 Primary structure of rhamnogalacturonan I (RG-I) from pectins. Side chains of type I and type II arabinogalactans (AG-I and AG-II), as well as arabinan (not shown), are attached to the backbone of RG-I exclusively through L-rhamnose residues.

with AG-I and AG-II providing the bulk of the "hair." Molecular modeling of AG-I tertiary structure showed (21) that its lateral chains are orthogonal to the rhamnogalacturonan backbone and thus create ample unoccupied space that is available for attachment of other molecules to the β-galactoside residues. The modeling of AG-II predicts that the lateral (1 → 3)-β-galactan chain has a fairly open helical structure, with linear (1 → 6)-β-galactan side chains that stretch away from both the pectin backbone and the (1 → 3)-β-galactan (21). Such spatial arrangement of the β-Gal residues in arabinogalactan II also makes them sterically relaxed and available for both efficient packing and exogenous interactions.

Plant food is the main source of pectic substances in the human diet. The texture of fruits and vegetables is strongly influenced by the amount and type of pectic polymers in cell walls. The pectin content in edible portions of some plants is listed in Table 12.1. In industry, pectins are mostly prepared from waste citrus peel and apple pomace (23). Because of the excellent gelling properties of pectins, the modern food industry utilizes them as important additives in jellies, jams, marmalade, yogurt, and so on. Pectins are used in medical preparations as well, because of their beneficial effects as antidiarrheals, detoxicants, and protectants of the gastrointestinal tract.

TABLE 12.1 Average Pectin Content in Edible Portions of Fresh Fruits and Vegetables

Food	Pectin Content, %
Apples	0.6
Apricots	1.0
Bananas	0.7
Beans	0.5
Carrots	0.8
Grapes	0.7
Grapefruit	0.3
Oranges	0.6
Squash	0.7
Watermelon	0.02

Adapted from Reference [22].

Commercial processing of plant products and extraction cause degradation of pectins. The optimal conditions for pectin stability require a weakly acidic medium, pH 3–4, since methylated galacturonate carboxylic groups and neutral carbohydrate side chains are readily hydrolyzed upon both acidic and basic catalysis, as well as at elevated temperatures and upon catalysis by endogenous pectin methyl esterases (15). The saponification of galacturonan is accompanied, even at neutral pH and low temperatures, by a β-eliminative depolymerization. The backbone fragmentation is also aided by a number of endogenous hydrolases and lyases. Therefore, commercial extraction of pectins brings about a decrease in neutral sugar content and degree of esterification, and an increase in galacturonic acid content along with some depolymerization of the backbone.

The structural aspects of two pectic preparations, larch arabinogalactan (LAG) and modified citrus pectin (MCP), deserve special interest, since both have demonstrated the ability to bind β-galactoside-specific lectins and inhibit the lectin-dependent tumor metastasis *in vivo*, as discussed below.

Arabinogalactans are found in a wide range of plants and often accompany pectin molecules in cell walls. They are most abundant in plants of the genus *Larix*, with Western larch being the main source of commercially prepared LAG. The primary structure of LAG is close to AG-II from pectin and consists of the (1 → 3)-β-D-galactan backbone, with extensive branching by (1 → 6)-β-D-galactan side chains that are in turn substituted by terminal and 3-linked L-Ara*f* and L-Ara*p* residues. A linkage analysis (24) of purified LAG gives the following distribution of linkages for galactose (mol%): 3,4,6-(2.2), 2,3,6-(1.6), 3,6-(30.7), 3,4-(2.3), 6-(19.4), 3-(1.6), and terminal-(26.8), with the molar ratio Gal/Ara in LAG equal to about 11:2. The authors estimated molecular weight of their LAG preparation by light scattering and by gel chromatography to be 40 and 16 kDa, respectively. Other sources assign varying molecular weights for LAG in the 10–120 kDa interval.

Arabinogalactan proteins (AGPs) are plant glycoproteins rich in hydroxyproline, to which arabinogalactan is attached. AGPs are commonly located on cell surface and are believed to play roles in plant cell adhesion, signaling, and growth regulation (25).

Certain AGPs are present in food grade *Acacia* plant gums and thus represent an important class of WSNCs in human diet (26). The carbohydrate portion of AGP largely resembles type II arabinogalactan, with the (1 → 3) backbone links and the (1 → 6) branching links between β-D-galactopyranose residues. This branched β-galactan core is decorated with L-arabinose and, to a smaller extent, L-rhamnose and D-glucuronic acid (27).

The name "modified citrus pectin" has been originally assigned by David Platt and Avraham Raz to a therapeutically active preparation of partially hydrolyzed and depolymerized reagent-grade citrus pectin (28, 29). The starting material had 70–100 kDa molecular weight, a very low content of methyl ester groups (10%), and a relatively high portion of arabinogalactan, 35% of the combined Gal + Ara molar content. After brief degradation at pH 10, the average molecular weight of the preparation dropped to 10 kDa, as estimated viscosimetrically by the authors. Although no detailed speciation analysis of MCP was provided, it is conceivable to suggest that, after the alkaline hydrolysis, the citrus pectin sample degraded mainly into a mixture of neutral arabinogalactan, oligogalacturonic acid, and a variety of smaller entities.

Recently, another therapeutically active preparation of partially hydrolyzed and depolymerized citrus pectin has been termed "modified pectin" (30). Although the starting material has been treated essentially as in the MCP preparation procedure, the resulting hydrolysis mixture was fractionated and the most active fraction with a molecular weight of 90 kDa was isolated. In this preparation, one would expect all structural features of citrus pectin to be retained to a certain extent.

Pectin depolymerization may also be achieved by treating its aqueous suspensions at 100–130°C (31). Commercial preparations referred to as modified citrus pectin under such brand names as PectaSol® (EcoNugenics), "Modified Citrus Pectin" (NutriCology, MW 30 kDa), "Fractionated Pectin" (Thorne), and others are sold by many distributors of dietary supplements and may have variations in molecular weight, structure, and composition of pectic substances. To avoid any confusion, the abbreviation MCP in this chapter is related only to the preparation of modified citrus pectin described in the original publication of Platt and Raz (28) and following works of the authors.

Binding of Pectic Substances to β-Galactoside-Specific Lectins

The availability of multiple terminal β-galactoside residues in arabinogalactan makes it and all AG-containing pectic polysaccharides potential ligands for such physiologically important mammalian β-galactoside-specific lectins as hepatic ASG receptors and galectins.

Characterization of interactions between individual lectins and the oligosaccharide preparations by direct methods, such as calorimetry or spectroscopy, would not be particularly informative, owing to the high level of molecular heterogeneity in the pectic substances. Because of this, indirect binding and competition assays were used as convenient alternatives to evaluate the lectin–pectic ligand interactions.

In one study (24), a purified arabinogalactan was radiolabeled and its binding to asialoglycoprotein receptor (ASGR) from rat liver was evaluated in a competitive binding assay. The relative affinities of some ligands to the ASGR were estimated as follows: asialofetuin > lactosylated BSA > arabinogalactan ≫ galactan > dextran ≫ galactose. The purified AG sample was also completely retained by a *Ricinus communis* agglutinin affinity column.

Interaction of galectin-3 with MCP *in vitro* was studied by Inohara and Raz (32). They demonstrated binding of gal3 to immobilized MCP by immunodetection of the lectin. Lactose at 50 mM completely blocked gal3 in this experiment. MCP, but not original citrus pectin, inhibited, apparently β-Gal-dependent, adhesion of B16-F1 murine melanoma cells to immobilized laminin. In addition, MCP, but not original citrus pectin, inhibited asialofetuin-induced homotypic aggregation of the cells. Finally, MCP was significantly better inhibiting, in a concentration-dependent manner, the colony formation by B16-F1 cells in agarose, as compared to the original citrus pectin.

Since highly metastatic MAT-LyLu rat prostate adenocarcinoma cells express galectin-3 on their surface, it was hypothesized that metastasis of the malignant tumors to distant organs may be initiated by galectin-3-mediated adhesion of circulating cancer cells to blood vessel endothelial cells. Pienta et al. (29) demonstrated the ability of MCP to block, dose dependently, adhesion of MAT-LyLu cells to rat aortic endothelial cells, as well as to inhibit MAT-LyLu colony formation in agarose. Similarly, interaction of recombinant galectin-3 or galectin-3-expressing MDA-MB-435 human breast carcinoma cells with monolayers of human umbilical vein endothelial cells was dose dependently inhibited by MCP, while nonmodified citrus pectin was not active (33).

Interestingly, blood vessel endothelial cells were also affected by MCP, *in vitro*. Thus, MCP, but not the original citrus pectin, inhibited capillary tube formation by HUVECs on Matrigel, as well as chemotaxis of HUVECs toward positive gradients of galectin-3, fibronectin, or basic fibroblast growth factor (33).

The striking difference between the lectin-binding abilities of MCP and the nonhydrolyzed "source" citrus pectin requires explanation, given that structural analysis suggests that terminal β-galactoside residues are sterically accessible in both pectic substances. It appears that a lack of significant interaction with native pectin is common for both galectin-3 and other β-Gal-specific lectins, regardless of their origin (34). However, native pectin readily interacts with galacturonic acid-specific lectins, such as *Aplysia depilans* gonad lectin (35). Treating pectin with weak alkali at room temperature or heat treatment in neutral medium leads to a dramatic increase in its reactivity toward β-Gal-specific lectins (34, 36). Therefore, depolymerization, rather than deesterification of pectin, appears to be essential for the access to its β-Gal residues. Because depolymerization of pectin is mostly associated with partial or complete dissociation of oligogalacturonan from the neutral sugar clusters, it may be hypothesized that oligogalacturonan strands in underivatized pectin shield arabinogalactan from an external access. The new structural data in the pectin field have led to calls to revise the canonical "smooth–hairy region" macrostructure model of pectin, by replacing it with, for example, a model (37) where oligogalacturonan is not a backbone but rather is attached as branches to rhamnogalacturonan I along with AG-I and AG-II (Fig. 12.2).

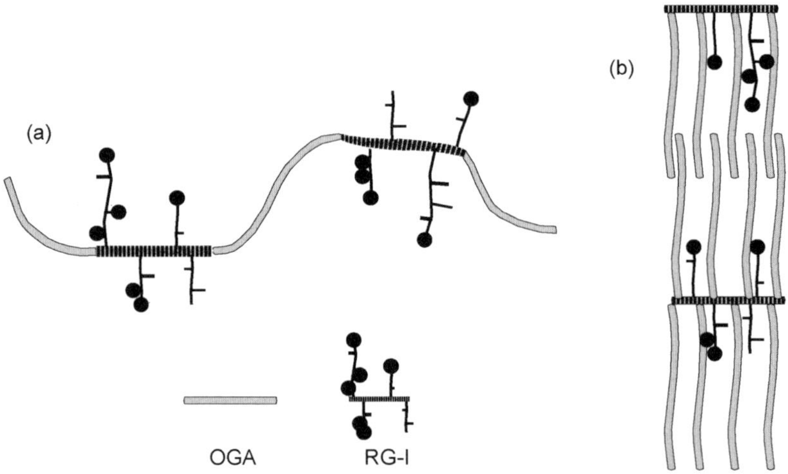

FIGURE 12.2 Tertiary structure of pectin. (a) "Canonical" model of pectin, with its backbone made of alternating regions of oligogalacturonic acid (OGA) and rhamnogalacturonan I (RG-I). Stretches of OGA mainly consist of "smooth" regions of homogalacturonan and also include xylogalacturonan, apiogalacturonan, and rhamnogalacturonan II. "Hairy" blocks of RG-I contain side chains of arabinan, type I and type II arabinogalactans. Terminal β-galactoside residues in AG-I and AG-II are depicted as black circles. (b) A hypothetical alternative structure of pectin (37). In this model, OGA is shown attached to RG-I as side chains, which noncovalently bind to each other through ionic interactions (calcium galacturonates), hydrogen bonding, boronate complexes (with RG-II), and so on. The association bonds between OGA chains are strong enough to survive gel chromatography, which may give overestimates for pectin molecular size. Arabinan and arabinogalactan chains fill voids within the OGA network and may not be readily available for lectin probes.

Modified citrus pectin may modulate galectin-related tumor cell proliferation not only by directly blocking the lectin, but also by inhibiting its expression in tumor cells. In one study, Chauhan et al. investigated effects of a modified citrus pectin preparation, GCS-100, on apoptosis in several human multiple myeloma cell lines (38). GCS-100 triggered apoptosis, at least partially, through intrinsic caspase-8-mediated pathway and overcame Bcl-2-mediated cytoprotective effects. The antiproliferative effect of GCS-100 was synergistically enhanced by dexamethasone. As galectin-3 has been identified as an antiapoptotic factor, the authors examined its expression in MM.1S cells treated with the GCS-100/dexamethasone and found that, indeed, galectin-3 was significantly downregulated in the cells by the combination, rather than by separate agents.

Bioavailability

Despite a large practical interest in health effects of dietary pectic oligosaccharides, there is a surprising lack of information regarding their delivery and transformations *in vivo*. In part, it may be explained by the complexity of pectic structures and difficulties with obtaining preparations of reasonable homogeneity and purity. LAG is one of a few pectic substances with relatively simple structure and fair resistance that make its tracking easier. Pharmacokinetics of a radiolabeled arabinogalactan conjugate was investigated in rats (24). Ninety minutes postintravenous injection, more than a half of the labeled AG was localized in liver, with about one third of the initial radiolabeled dose excreted in urine. In contrast, coinjection of the AG with asialofetuin dramatically decreased the amount of radioactivity retained in the liver and prolonged the circulation time of the AG in the bloodstream.

Whether arabinogalactan or its oligosaccharide fragments can be absorbed into the circulation after oral ingestion remains unanswered. Pectic substances are generally regarded as nondigestible in the upper intestine. In rats, pectin is depolymerized and deesterified only to a small extent after passing through the small intestine and is excreted nearly unchanged in germ-free rats (39). Fermentation of pectic polysaccharides by gut microflora, however, may produce short intermediates that could be absorbed in the colon. Interestingly, larch arabinogalactan better resisted fermentation by human fecal bacteria when compared to underivatized or rhamnogalacturonan-enriched apple pectins, which were fermented rapidly and completely in similar conditions (40).

Beneficial Therapeutic Effects of Pectic Substances *In Vivo*

Health-beneficial effects of dietary pectic substances, including native pectins, arabinogalactan, and their modifications, have been a subject for numerous studies. The structural complexity and variability of pectic saccharides provide for a variety of physicochemical properties of these substances, which, in turn, define a multitude of nutritional and physiological effects reported for this class of dietary agents.

Glycemic control in diabetics. Three decades ago, it was discovered that pectin attenuated the postprandial rise in blood glucose and insulin concentrations in patients with both type I and type II diabetes (41). The effect was attributed to the ability of pectin and other soluble fibers to decrease the rate of carbohydrate absorption within the gastrointestinal tract. Pectins gel readily at low pH and may increase viscosity of digesta, thus contributing to slower degradation and movement of starch (42).

Lipid-lowering effects of dietary pectin had been consistently demonstrated in a wide variety of subjects and experimental conditions since 1960s (43). Incorporating pectic substances in diet lowers serum cholesterol and low density lipoprotein and, in turn, lowers the risk of cardiovascular disease. The mechanisms of this effect are uncertain, but may include binding the lipids and reducing their absorption (44).

Immunostimulation and immunoprotection. Pectins that activate the complement system have been isolated from several plants used in traditional oriental medicines (45). Interestingly, a structural investigation of a fractionated pectin revealed that its biological activity was determined by the galactose side chains in arabinogalactan II (46). The immunoadjuvant effect of dietary pectins such as lemnan may result from an enhanced antigen ingestion and stimulation of macrophage activity (47). Stimulation of the innate immune system by pectin was determined as a possible mechanism of the polysaccharide protection against systemic pneumococcal infection in mice (48).

Heavy metal detoxification. Oligogalacturonic acid in pectins is a potential binding site for metal ions, as a result of the presence of negatively charged carboxylic groups in its structure. In plant cell walls, these binding sites are occupied with calcium ions, which are removed during industrial extraction and processing of pectin, or dissociated in the stomach acidic medium upon ingestion. In addition, hydrolysis of pectin increases its metal-binding capacity (49). Heavy metal ions then can be absorbed by the oligogalacturonate or mobilized into complexes with galacturonic acid. As a proof of principle, modified citrus pectin, given to patients in one small clinical study, increased urinary secretion of lead, mercury, cadmium, and arsenic (50).

Antioxidant activity of pectin may also be related to its ability to bind metal ions catalyzing redox reactions, such as copper or iron. This explains a phenomenon of increased antioxidant activity of pectins with lower level of glucuronate esterification (51).

Colon cancer prevention. Significant experimental and epidemiological evidence suggests that diets rich in soluble fiber, including pectic substances, reduce the risk of colorectal cancer (8, 52). Pectin preparations from apple or citrus were examined for the prevention of chemically induced colon carcinogenesis in a number of studies (53, 54). Although most published data showed the protective activities of pectins, this effect is not consistent throughout all pectic preparations or carcinogenesis models. The possible anticarcinogenic mechanisms are still a subject of debates; the ones most often mentioned in the literature, however, are the suppression of procarcinogenic bile acids by the polysaccharides and the production of the protective short-chain fatty acids (SCFAs) during pectin fermentation in the gut (55).

The ability of larch arabinogalactan and modified citrus pectin to block β-galactoside-specific lectins stimulated their investigation as *potential antimetastatic agents*. Two European groups studied the effects of commercial arabinogalactan on metastasis, primarily to liver, in murine models. In one study, Beuth et al. (56) treated murine sarcoma L-1 cells with neuraminidase to render β-Gal residues on the cell wall glycoproteins accessible for interactions with lectins. The cells were then injected i.v. into experimental animals treated with LAG, D-galactose, gum arabic galactan, dextran, and mannan. The carbohydrates were given i.p. to the animals 1 h before and for several days after the injection of the L-1 cell suspensions. After 14 days, tumor nodules in lungs and livers were counted. While there was no difference in lung colonization between any of the treated and control groups, livers of LAG-treated animals were completely free of any metastases, while D-galactose-treated animals had significantly less tumors in livers, as compared to the controls or other carbohydrate-treated groups. LAG and, to a lesser extent, D-galactose inhibited liver colonization with i.v. injected murine lymphoma cells in a similar experiment conducted by these researchers (57).

Hagmar et al. (58) showed that murine sarcoma MCG1-SS cells, pretreated with LAG and i.v. injected, produced significantly less metastases in lungs, livers, and other organs, as compared to the untreated controls or animals that received arabinogalactan before and after injection of the untreated cells. In addition, pretreatment of MCG1-SS cells with arabinogalactan led to a dramatic improvement in the animal survival. The same authors have demonstrated that survival of experimental animals, injected i.v. with murine hepatoma cells, had significantly improved, when the cells were pretreated with the arabinogalactan or when the animals obtained LAG intraperitoneally shortly before and after the tumor cell injection (59). The ability of both sarcoma and hepatoma cells to form metastases in these experiments, however, was not affected by treatments with such small Gal-derived substances as D-galactose, D-lactose, α-Gal(1 → 3)Gal, or lacto-*N*-fucopentaose, implying an importance of the carbohydrate multivalency for successful blockage of the circulating tumor cells from lodging onto vessel walls.

In 1992, Platt and Raz introduced MCP that abrogated lung colonization by B16-F1 tumor cells when coinjected with the cell suspension (28). Highly metastatic human breast and prostate carcinoma cells failed to colonize lungs in SCID mice, when the suspensions were pretreated by MCP or antigalectin-3 (60). These experiments established LAG and MCP as potential therapeutic agents for the prevention of advanced solid tumor dissemination and provided insights for significance of the β-Gal-lectin interactions in the process of metastasis.

Pienta et al. (29) demonstrated that orally administered MCP retained its ability to suppress spontaneous metastasis of galectin-3-expressing MAT-LyLu cells in the Dunning rat prostate adenocarcinoma model. Oral administration of a preparation of thermally treated citrus pectin significantly inhibited growth of the implanted human colon-25 tumors in Balb/c mice (61). Primary tumor growth inhibition by MCP was also documented for MDA-MB-435 human breast and LS-LiM6 human colon carcinomas in nude mice (33). As a result of this and other studies, the antiangiogenic activity of MCP was offered as an additional possible explanation of the observed therapeutic effects.

In a small pilot study, a modified citrus pectin preparation, PectaSol®, taken orally at 14.4 g/day as a nutritional supplement, benefited patients with prostate cancer by increasing the prostate-specific antigen doubling time (62).

As the exact composition, intestinal adsorbability, and pharmacokinetics of MCP have not yet been established, and owing to a variety of nutritional and physiological effects reported for pectic substances in literature, it would be very premature to assign the observed *in vivo* effects of modified citrus pectin solely to the blockage of β-Gal–lectin interactions. Nevertheless, the antitumor effects of orally delivered MCP justify its further investigation as a potential chemopreventive agent, regardless of the exact *in vivo* mechanisms.

LACTULOSAMINES

Lactulosamines belong to a reasonably well-studied group of food-related glycoaminoconjugates, investigated owing primarily to their relatively high content in many processed dairy products. Synthetic lactulosamines have been prepared since over 50 years ago (63, 64) by employing a two-step reaction between lactose and an

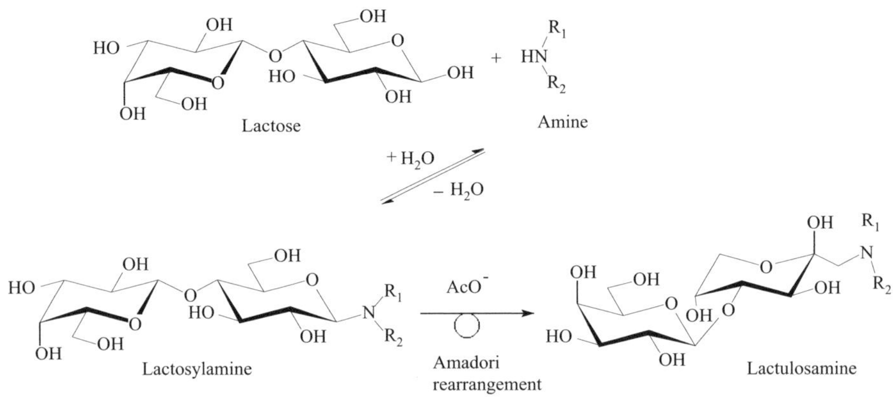

FIGURE 12.3 Formation of lactulosamines from D-lactose and primary or secondary amines. β-Pyranose anomers shown here are the most abundant tautomers detected for lactose, lactosylamines, and lactulosamines in equilibrated aqueous solutions.

aromatic amine. Initial proton-catalyzed condensation of the aldose and a primary or secondary amine results in the formation of lactosylamine (Fig. 12.3). Although many lactosylamines have been prepared and isolated in pure form, the glycosylamine bond is labile, especially in acidic media, and readily undergoes hydrolysis back to the original amine and lactose.

In the presence of nucleophilic species, such as carboxylate, phosphate, hydroxide, and so on, glycosylamines undergo a vinilogous Amadori rearrangement (Fig. 12.3). This is thought to proceed through the enamine formation that is promoted by the nucleophilic catalyst. The amine basicity is essential, since no Amadori rearrangement could be achieved for glycosylamines with $pK_a < 3$. In our experience, acetic acid or sodium acetate is the most convenient catalyst for preparative synthesis of LAs (65). Unfortunately, the reaction between lactose and amine does not stop at the lactulosamine formation step, but proceeds further to a host of the Maillard reaction products. Many of these products are high molecular weight, strongly absorbing compounds, commonly referred to as melanoidins or nonenzymatic browning products (66, 67). As LAs do not crystallize, the presence of the Maillard by-products greatly complicates their isolation. Nevertheless, by using ion-exchange chromatography, we have succeeded in preparing and purifying many synthetic LAs, with one through eight lactulosamine groups per molecule (Table 12.3), of high purity and gram quantities.

Structure, Origin, and Analysis in Dairy Products

Lactulosamine is a disaccharide, consisting of terminal D-galactopyranose and D-fructosamine at the reducing end. The glycosidic bond is $\beta(1 \rightarrow 4)$, and the structure, therefore, can be regarded as a D-lactulose derivative, with OH group at the fructose carbon-1 being replaced by the nitrogen of an amino group. *Lactulosamine* should not be confused with more familiar *lactosamine*, as the latter represents a

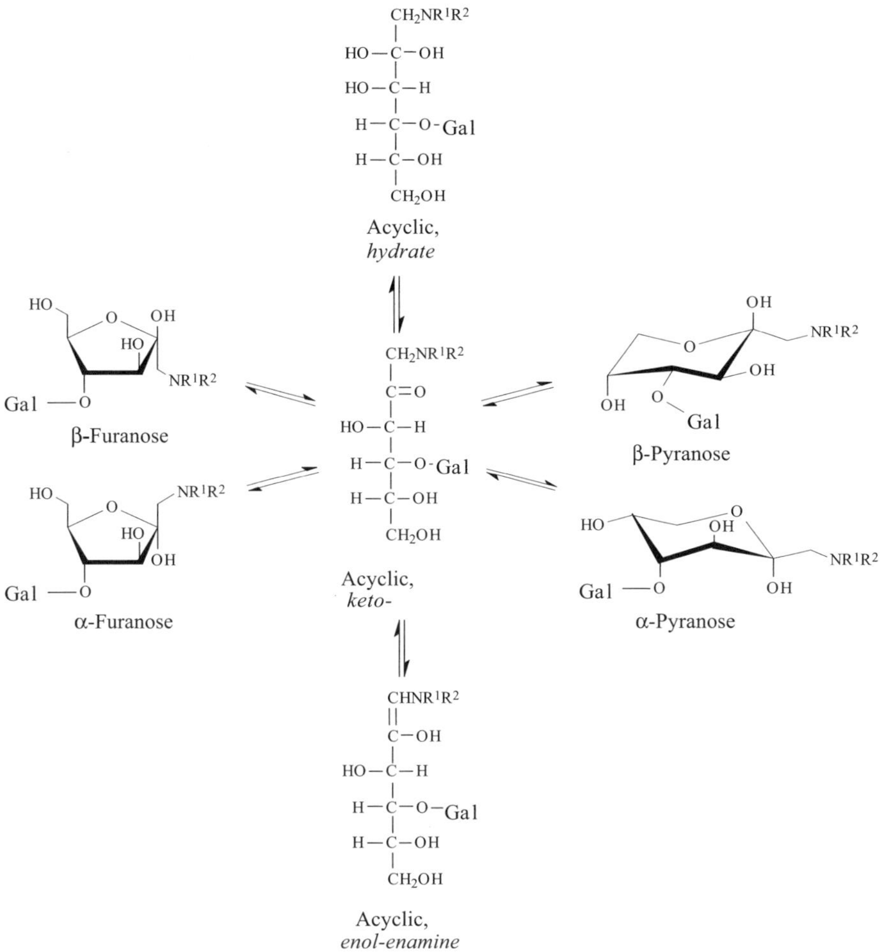

FIGURE 12.4 A general tautomerization scheme for lactulosamines. β-Pyranose anomer is the most thermodynamically stable. Acyclic enol-enamine is the most reactive tautomer.

derivative of lactose, with amino group attached at carbon-2. In the absence of crystallographic studies, lactulosamines have been studied in solution. In an aqueous environment, lactulosamines are observed to tautomerize (65, 68) in the same fashion as lactulose or fructose (Fig. 12.4). In equilibrium, the β-fructopyranose form predominates at 60–70% of the equilibrium mixture, the α-pyranose content does not exceed 2–3%, and the remainder is divided between the α- and β-furanose anomers. The percentage of acyclic *keto* and hydrated tautomers has not been determined but is lower than 1%.

Lactulosamines derived from primary amines are, themselves, secondary amines and can further react with a second lactose molecule to produce *N,N*-di(lactulos) amines. Tautomerization of these carbohydrates appears to be more complicated,

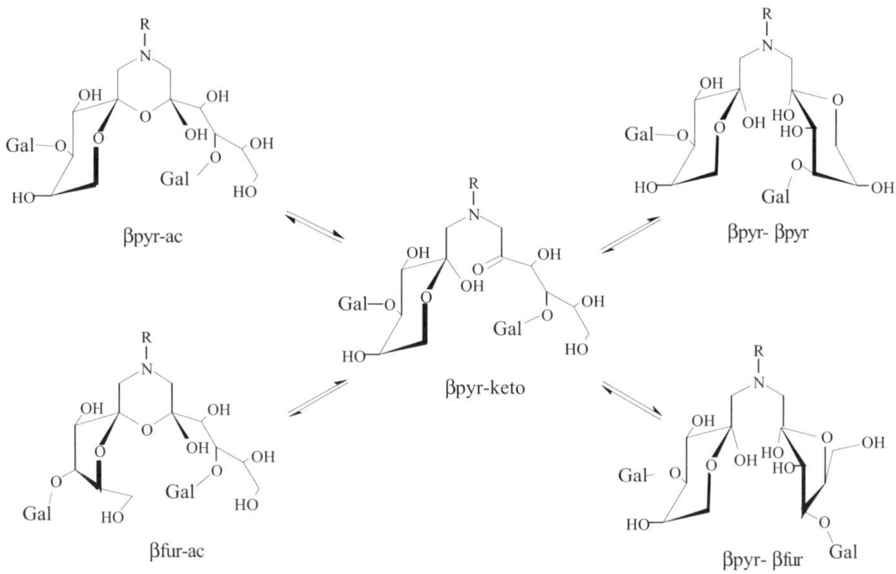

FIGURE 12.5 Tautomerization of *N*, *N*-di(lactulos)amines in aqueous media. Only a few representative structures are shown. In equilibrium, the βpyr-ac isomer is the most abundant.

owing to the additional formation of *spiro*-bicyclic structures (Fig. 12.5), which is driven by the anomeric effect. For example, out of about 30 theoretical tautomers, we have been able to identify the signals of at least 12 tautomers in C-13 NMR spectrum of *N*,*N*-di(fructos)amine (69).

The molecular structure of lactulosamines suggests that these compounds may possess some antioxidant properties, by acting as reducing agents (70) or metal chelators. Although there are no published data on LAs, their structural analogues, fructose-amino acids, have demonstrated a superior ability to protect oxidative DNA fragmentation as compared to other water-soluble antioxidants (71). Both fructosamines and LAs tautomerize into the enol-enamine form (Fig. 12.4), which is a potent reducing agent (72). Fructose-amino acids can strongly chelate redox-active metals such as copper and iron, and stability of such complexes is higher as compared to parent amino acids (71, 73).

Milk is the single major source of lactose in human nutrition. The dairy industry widely uses milk dehydration to manufacture powdered milk as a base for numerous dairy products, such as infant formulas, confectionaries, reconstituted milk, and so on. During the process of heating, drying, and storage, the lactose in milk can readily interact with amino compounds that are naturally present. Primarily, lysine residues in proteins and other available amino acids would react to form, consecutively, lactosylamine, then lactulosamine structures, followed by multiple products of their degradation. Because of the importance of this reaction in the food industry and human nutrition (74–77), a number of analytical methods have been developed for detection and measurement of lactulosamine. The most popular method to estimate the extent of the lactose-lysine conjugation is a convenient indirect procedure that takes advantage of the formation of UV-absorbing compound furosine from lactulose-lysine that occurs during the acid hydrolysis of milk proteins (67, 74). Alternatively, a number or polyclonal and monoclonal antibodies against lactulosamine-modified proteins and polylysine have been developed, which dramatically increased sensitivity of LA analysis by the use of immunodetection techniques (78–80). More recently, efforts to develop direct instrumental analysis methods for determining lactulosamine modification in dairy products have succeeded in applying the liquid chromatography–mass spectrometry to enzymatic hydrolyzates of milk proteins (81, 82).

With a host of detection methods currently available, lactulosamine formation has been confirmed in UHT milk (74, 78, 83), dried milk (74, 78, 83, 84), infant formulas (83, 84), whey (83, 85, 86), chocolate (83, 87), cheese (84), soy milk (87), and drug formulations containing powdered lactose (88, 89). These same analytical techniques have been indispensable for tracking lactulose-lysine in blood, urine, and feces to establish metabolic fates of ingested lactulosamine in LA-containing foods.

The extent of LA formation in processed dairy foods depends on many parameters, including temperature, water content, pH, storage time, and so on. Estimated contents of LA in selected products are compiled in Table 12.2. Commercial dairy products may contain, therefore, up to 40% of protein lysine in the form of lactulose-lysine.

Interaction of Lactulosamines with Galactose-Specific Lectins

The first report on the interaction between lactulosamines and galactose-specific lectins, we are aware of, appeared in 1989. Kolb and colleagues established (90) that both lactitol- and lactulosamine-conjugated human serum albumin (HSA) was able to efficiently inhibit adhesion of desialylated erythrocytes to hepatocytes and Kupffer cells, in a concentration-dependent manner. Modification of HSA with galactose or glucose, which ended up with the formation of, respectively, tagatosyl- or fructosyl-HSA conjugates, did not show any significant inhibition of the cell adhesion. In

TABLE 12.2 Estimated Contents of Lactulose-Lysine (g per 100 g) in Dairy Products

Product	Per Dry Matter (83)	Per Protein (84)
Milk UHT	0.15	0.5
Condensed milk	0.9	1.2
Spray dried milk	0.6	0.2
Roller dried milk	1.1	
Spray dried whey	0.3	
Roller dried whey	1.7	
Whey concentrates	1.2	
Baby food	0.3	0.3–0.7
Chocolate	0.45	
Cheese		0.6–1.1

contrast, samples of HSA conjugated with increasing number of lactitol or lactulosamine residues demonstrated increasing blocking of the liver lectins. A notation made in this work was that lactulosamine-conjugated peptides, entering bloodstream after intestinal digestion of the modified milk proteins, would be subjected to efficient clearance by liver (90).

Our initial studies involving synthetic lactulosamines had a specific focus on testing the hypothesis that carbohydrate-mediated homo- and heterotypic cancer cell aggregation and adhesion is one of the key events in the metastatic process (91). If β-galactoside participated in mediating such interactions, then adding carbohydrates with terminal β-Gal to the medium could potentially block both interactions of galactose-specific lectins with the cancer cell glycoproteins and interactions of terminal β-galactoside glycoproteins with the cancer cell lectins. Indeed, lactulose-L-leucine (LL) inhibited asialofetuin-induced aggregation of B16-F1 murine melanoma cells, as well as adhesion of peanut agglutinin (PNA lectin) to human breast carcinoma MDA-MB-435 cells (92).

Galectin-1 readily interacts with lactulosamines, as estimated by both indirect and direct methods. Employing immobilized laminin, a putative ligand for galectins in the extracellular matrix, we have evaluated the ability of mono- and multivalent LAs to inhibit binding of human recombinant galectin-1 to this glycoprotein (Table 12.3). In a similar fashion, the ability of LAs to bind to galectin-3 has been probed in a competition with immobilized immunoglobulin E (Table 12.4). All but one tested LAs performed better than lactose as inhibitors of the laminin–galectin interactions, and most LAs were better than "standard" galectin ligands LacNAc and β-thiodigalactoside in this assay.

As seen in Table 12.3, a few structural factors apparently influence the affinity of LAs to galectin-1. First, the basicity of the amine nitrogen positively correlates with the LA–galectin affinity. Thus, LAs derived from highly basic aliphatic amines are better galectin-1 blockers than less basic lactulose-*p*-aminophenylacetic acid (Lct-pApa) and α-amino acid derivatives. Second, in a series of divalent LAs, the distance between the lactulosamine residues appears to influence the LA affinity to galectin-1 in a nonlinear fashion. For *N,N*-dilactulosamines, the distance is zero atoms, and the affinity does not differ from that for monovalent LA. With increased numbers of atoms bridging the lactulosamine residues, the inhibitory potential of the LA increases, reaching its maximum at $n = 12$. With longer bridges, the blocking ability of LA falls, possibly because of the long hydrocarbon chains coiling in aqueous media, which, in effect, results in the decreasing distance between the lactulosamine residues.

Next, with the increase in the number of lactulosamine residues per molecule, or lactulose valency, the LA inhibitory potential increases significantly. This is not surprising, since carbohydrate valency has been recognized long ago as an important parameter for carbohydrate–lectin affinity (93). It is likely, however, that a net positive charge of the LA molecules also plays a role in stabilizing the lactulosamine–galectin-1 complexes. Results published recently by Rabinovich et al. (94) lend support in favor of this possibility. These researchers found no difference in galectin-1 or -3-binding abilities between monovalent Lct-NH-$(CH_2)_8$-NH_2 and

TABLE 12.3 Inhibition of hrGalectin-1 Binding to Laminin by Synthetic Lactulosamines

Compound	Valency	Lct–Lct Distance, Atoms	IC_{50}, M
Lct-NH_2	1		100
Lct-NH-$(CH_2)_3$-NH-Lct	2	3	30
Lct-NH-$(CH_2)_6$-NH-Lct	2	6	15
Lct-NH-$(CH_2)_9$-NH-Lct	2	9	10
Lct-NH-$(CH_2)_{12}$-NH-Lct	2	12	8
$(Lct\text{-}Ava)_2$-1,9-diamine	2	21	20
$(Lct\text{-}Ava)_2$-1,12-diamine	2	24	40
Lct-pApa-morpholine	1		500
$(Lct\text{-}pApa)_2$-piperazine	2	16	150
$(Lct\text{-}pApa)_3$-cyclo[2,2,2]	3	16	80
$(Lct\text{-}pApa)_4$-cyclo[2,2,2,2]	4	16	80
Lct_2Aca-NH$(CH_2)_3$OH	2	0	100
$(Lct_2Aca)_2$-1,3-diamine	4	0/19	40
$(Lct_2Aca)_3$-[3,3]triamine	6	0/19	20
$(Lct_2Aca)_4$-[3,3,3]tetramine	8	0/19	10
Lct-Gly	1		800
Lct-L-Leu	1		450
Lct-L-Pro	1		300
LacNAc	1		200
β-Thiodigalactoside	1		200
Lactose	1		800

Estimated IC_{50} values are shown for lactulosamines and representative carbohydrates in competitive ELISA.

divalent Lct-NH-$(CH_2)_8$-NH-Lct molecules. Both compounds inhibited binding of the galectins to the immobilized glycoprotein 90 K equally well, with IC_{50} values of about 30 μM (gal1) and 60 μM (gal3). Finally, a comparison of the IC_{50} values for lactulose-amino acids in Table 12.4 suggests an influence of the aglycon structure on the interaction of LAs with galectin-1. Indeed, thermodynamic dissociation constants K_D measured for the interactions between galectin-1 and lactulose-amino acids by the surface plasmon resonance (SPR) (95) varied in the range of 70 μM (Lct-L-Pro) through 380 μM (Lct-L-Thr).

The SPR technique has also allowed for estimating the affinities of lactulose-amino acids to galectin-4 (95). On average, the K_D values obtained for individual lactulose-amino acid/galectin-4 complexes are about 10 times higher (that is the binding weaker) than the respective values for the lactulose-amino acid/galectin-1 complexes, with a similar pattern within the set of the LAs: from the lowest 530 μM (Lct-L-Pro) to the highest 3800 μM (Lct-L-Thr).

Interestingly, dehydration and oxidative degradation of lactulosamines lead to an array of so-called advanced glycation end products (AGEs), including protein

TABLE 12.4 Inhibition of hrGalectin-3 Binding to Immunoglobulin E by Synthetic Lactulosamines

Compound	IC_{50}, μM
Lct-NH-$(CH_2)_6$-NH-Lct	90
Lct-Gly	210
Lct-L-Ala	240
Lct-L-Val	110
Lct-L-Leu	130
Lct-L-Phe	110
β-Thiodigalactoside	240
Lactose	450

Estimated IC_{50} values for LAs and representative carbohydrates in competitive ELISA.

cross-links, browning, and fluorescent structures (96). The AGEs, which also form as a result of the nonenzymatic modification of proteins by glucose in humans, have been identified as ligands for galectin-3 (97), although the exact galectin-3-binding AGE species is yet to be established.

Cellular Responses to Lactulosamines *In Vitro*

The ability of lactulosamines to bind to galectins and inhibit interactions of galectins with their physiological glycoprotein ligands suggests that these carbohydrates might modulate some galectin-mediated events in mammalian cells. Since the discovery of tumor-associated Gal-binding lectins in the early 1980s (98), it has been thought that the homotypic cancer cell–cell adhesion and related clonogenic cell survival were partly dependent on β-galactoside–galectin interactions. Following the pioneering work of Platt and Raz with MCP (28), we have established the ability of LAs to also modify cancer cell adhesion and viability. Most of the work was performed with lactulose-L-leucine, which was the first LA that demonstrated potentially beneficial effects in various *in vitro* and *in vivo* cancer models.

It was determined that LL inhibited homotypic cell aggregation in highly metastatic B16F10 murine melanoma cell line, with $IC_{50} = 20$ μM (65, 99). The spontaneous aggregation of human MDA-MB-435 breast carcinoma cells was reduced by half when LL was added to the medium. In addition, the reduction was accompanied with a fivefold increase in the level of apoptosis of the MDA-MB-435 cells (92) after 2 days of cultivation on plastic. In 0.9% agarose, however, LL caused more than fivefold decrease in colony formation of this same cell line (92, 100).

Dissemination of metastatic cancer requires entrapment of circulating blood-borne tumor cells in blood vessels for it to successfully form secondary tumors. In the light of this, we have been testing the hypothesis that interactions between β-galactoside glycoproteins and their lectin receptors are essential for initial adhesion of the circulating tumor cells to blood vessel endothelium. Specifically, aberrant glycosylation of mucins, a class of cell wall glycoproteins, in tumor cells leads to exposure of terminal β-galactoside residues in these glycoproteins. The most common modification is disaccharide β-D-Gal*p*(1 → 3)-α-D-GalNAc, also known as Thomsen–Friedenreich antigen, or TF antigen. This pancarcinoma tumor-associated antigen was found in almost all major types of tumors and was once considered as one of the most promising targets for cancer immunotherapy (101).

Our studies have demonstrated participation of TF antigen and galectin-3, which is translocated to the surface of vascular endothelial cells, in aiding the formation of the initial endothelial cell–cancer cell contacts in flow (102, 103). If the hypothesis is correct, such glycoprotein–lectin interactions might be targeted by blocking galectin-3 or TF antigen with the appropriate antibodies or other galectin ligands, such as lactulosamines. Our data showed that LL, along with lactose and an antigalectin-3 monoclonal antibody, significantly inhibited rolling of suspended MDA-MB-435 cells on monolayers of HUVECs and abrogated adhesion of the tumor cells to the endothelial cells under conditions of laminar flow (103).

In a head-to-head comparison experiment (104), MCP and LL demonstrated similarity in both induction of apoptosis and dose-dependent inhibition of clonogenic growth of murine hemangiosarcoma SVR cell line, with IC_{50} about 0.08% for MCP and 350 μM (0.016%) for LL. Notably, both MCP and LL synergized, in a similar manner, with chemotherapeutic drug doxorubicin. This provided further clues for mechanistic similarity of their therapeutic activity.

Rabinovich et al. (94) have recently confirmed the ability of synthetic LAs to inhibit homotypic tumor cell–cell adhesion by employing A375 human melanoma line, to induce apoptosis in tumor cells such as H69 small-cell lung carcinoma line, and to disrupt in vitro tube-like formation by HUVECs.

In 1995, galectin-1 was identified as an inducer of apoptosis in activated cytotoxic T lymphocytes (105), possibly due to cross-linking and aggregation of CD45 glycoproteins on the cell surface. Since then, evidence has accumulated in support of the hypothesis that solid tumors may use galectin-1 to inactivate infiltrating CTLs and escape the immune surveillance. Blocking galectin-1, therefore, could potentially aid cancer immunotherapies. For example, divalent Lct-NH-$(CH_2)_6$-NH-Lct successfully protected TF antigen activated CTLs from galectin-1-induced apoptosis and restored the ability of the cells to produce γ-interferon (95). Consequently, this LA has improved efficiency of dendritic cell immunotherapy in a transgenic murine model of breast cancer.

Bioavailability of Dietary Lactulosamines

Since the discovery of lactulosamine modification of lysine in milk proteins, nutritionists have established that such a lysine glycoconjugate was no longer available to mammals as a nutrient (106, 107). Analyses of urine samples of adults fed LA-rich milk showed (107, 108), however, measurable amounts of lactulose-lysine. This clearly demonstrated that the lysine glycoconjugates were partially absorbed into the bloodstream after milk protein digestion and were excreted unchanged. The level of LA in urine returned to its normal level within one day after the LA ingestion (109), indicating rapid renal clearance of LA from blood. As only the traces of LA were detected in feces, the majority of ingested lactulose-lysine was probably fermented in the gut.

In infants fed heat-processed formulas, lactulose-lysine is also excreted in urine, reaching up to 4% of the ingested amount (110). Measurements of LA content in pig blood after an LA-enriched meal showed (77) that intestinal absorption of the lactulose-lysine is rather slow and reaches maximal flux rate about 6 h after the meal. Even after 12 h, its absorption was not complete and the amount of the LA absorbed between 4th and 12th h was about 8% of the ingested quantity. These results suggest that lactulosamines are most likely passively absorbed from ingested food. Of interest, the lactulosamine modification has been used to improve the bioavailability and stability of some drugs (111, 112).

Antimetastatic Properties of Lactulosamines

Establishment of LA antiadhesion and proapoptotic activities has justified a number of *in vivo* experiments aimed at validation of the ability of these nontoxic carbohydrate derivatives to inhibit metastasis in melanoma, breast, and prostate cancer models. Murine melanoma B16 cells, upon treatment with LL and intravenous injection into Balb/c mice, produced 80% less lung colonies than the untreated controls (99). Highly metastatic MDA-MB-435 human breast carcinoma xenografts in athymic nude mice readily form primary tumors that metastasize into the lungs with about 80% incidence. Treatment of the animals with lactulosamine LL decreased the lung metastasis incidence almost threefold and reduced the average number of the pulmonary metastases from 37 in the control to 0.9 in the treated mice (92). In another experiment (113), MDA-MB-435 or human prostate adenocarcinoma DU145 cell suspensions, pretreated with LL or other galectin-3 and TF antigen blockers, were injected intravenously in SCID mice. Both LL and other tested inhibitors of the galectin-3-dependent cancer cell–endothelium adhesion reduced the number of breast and prostate carcinoma subpleural metastatic deposits by more than 90%.

There are at least two important conclusions that may be drawn from these experiments. First, despite its relatively modest affinity to galectins, lactulosamine LL appears to successfully block *in vivo* interactions involving galectin and/or β-Gal-glycoprotein ligands and acts as a therapeutic antimetastatic agent. Second, these results provide a new perspective toward the development of novel adjuvant anticancer therapies based on glycoaminoconjugates that are nontoxic and target dissemination of malignant tumors—the primary cause of death in the advanced cancer disease.

OTHER POTENTIAL GALECTIN BINDERS FROM FOOD

Humans are an omnivorous species and consume a large variety of animal and plant products that contain both digestible and nondigestible saccharides capable of interacting with galectins.

Interest in the physiological significance of galectins in humans has led to the identification of many galectin ligands in tissues of both humans and other species of the animal kingdom. A list of endogenous glycoproteins that are proven binders for mammalian galectins is constantly growing (114, 115) and may eventually contain hundreds of items. Some of the best known are laminin, fibronectin, 90 K, IgE, CD45, mucin, and so on. The chemical transformations of these glycoproteins during processing of meat or other animal products are not clear. The native glycoproteins are efficiently digested in the upper intestine by hydrolytic enzymes to amino acids, short peptides, and monosaccharides and short oligosaccharides. Nonenzymatic modifications involving carbohydrates and other food components during food processing, however, may bring about, as in the case of LAs, some modified carbohydrate structures that are decorated with β-Gal residues but are resistant to intestinal digestion.

Several known galactose-containing saccharides that originate from plant and dairy foods and that resist digestion and could at least theoretically act as galectin ligands, owing to the presence of β-galactoside in their structure, are listed in Table 12.5.

Plant storage material in seeds, grains, and tubers consists mostly of starch granules, which provide a bulk of nutritional and energy value for the human diet. In many seeds, however, the cell walls of storage tissues are thick and contain deposits of storage polysaccharides that are used up after germination, and in the growing plant are replaced with other types of cell wall polysaccharides such as pectins or hemicelluloses. Cell wall storage polysaccharides are represented by a variety of "gum" structures, some of which have found wide use in food industry. For example, a principal structure of gums from locust beans, guar, tara, fenugreek, or cassia is galactomannan, with (1 → 4)-β-D-mannan backbone, which carries (1 → 6)-α-D-galactopyranosyl substituents (116). *Tamarind gum*, which is obtained from the flour of tamarind tree seeds, is represented by a polysaccharide known as the only xyloglucan used in the food industry (117). Its backbone is (1 → 4)-β-D-glucan, with short side chains of (1 → 6)-α-linked D-xylopyranose monosaccharide or β-D-Gal*p*-(1 → 2)-D-Xyl*p* disaccharide (Fig. 12.6). Tamarind gels are popular in Asia as prebiotic food additives and thickeners used for confectionary and dairy products.

Psyllium gum is present in seed coat and husks of *Plantago* species and can be extracted with hot water, yielding a gel. Its principal structure consists of main chain with (1 → 4)-β-linked D-xylopyranosyl residues, densely substituted with (1 → 2) or (1 → 3)-linked branches of D-xylopyranose of α-L-Ara*f*-(1 → 3)-β-D-Xyl*p*-(1 → 3)-α-L-Ara*f*, which are occasionally capped with D-galactose (118). Psyllium is regarded as a valuable soluble dietary fiber and was credited for multiple beneficial effects including lowering cholesterol, risk of colon cancer, and cardiovascular disease (119). *Yellow mustard* and *flaxseed mucilages* are readily obtained from hydrated ground seeds. Yellow mustard gum is a mixture of several polysaccharides; one of the two major constituents is a pectic polysaccharide composed of (1 → 2)-α-L-Rha*p*-(1 → 4)-α-D-GalA*p* rhamnogalacturonan backbone and short β-(1 → 4)-linked branches of (1 → 6)-β-D-Gal*p* oligosaccharides, capped with residues of 4-*O*-methyl-β-D-glucuronic acid (120). The main fraction in flaxseed gum consists of

TABLE 12.5 Major Nondigestible β-Galactoside Saccharides in Human Food

Plant Polysaccharides	Core Chain	Terminal Monosaccharides
Plant cell wall polysaccharides		
Larch arabinogalactan	β-D-Galactan	β-D-Gal*p*, α-L-Ara*f*
Pectins	Galacturonan, rhamnogalacturonan	β/α-D-Gal*p*, α/β-L-Ara*f*, α-L-Rha*p*, β-D-GalA*p*, α-L-Fuc*p*, β-D-Xyl*p*, β-D-Api*f*, and so on
Plant storage polysaccharides		
Tamarind gum (xyloglucan)	Glucan	α-D-Xyl*p*, β-D-Gal*p*
Psyllium gum	Xylan	α-L-Ara*f*, β-D-Xyl*p*, β-D-Gal*p*
Yellow mustard mucilage	Rhamnogalacturonan	β-D-Gal*p*, 4-*O*-Me-β-D-GlcA*p*
Plant exudates		
Gum Arabic	β-D-galactan	β-D-Gal*p*, α-L-Ara*f*, α-L-Rha*p*, β-D-GlcA*p*
Gum tragacanth	Galacturonan, β-D-galactan	β-D-Gal*p*, α-L-Fuc*p*, β-D-Xyl*p*, α-L-Ara*f*, β-D-Glc*p*
Gum karaya	Rhamnogalacturonan	β-D-Gal*p*, β-D-GlcA*p*
Gum ghatti	Mannoglucuronan	β-D-Gal*p*, α-L-Ara*f*, β-D-GlcA*p*
Okra mucilage	Rhamnogalacturonan	β-D-Gal*p*, α-L-Ara*f*
Algal polysaccharides		
Agar	α-L-/β-D-Galactan	None
Carrageenans	α-D-/β-D-Galactan	None

Short Saccharides	Structure
Processed milk, prebiotic, and sweetener saccharides	
Lactulose	βGal(1-4)Fru
Lactulosamines	βGal(1-4)FruNR
Lactitol	βGal(1-4)Glucitol
Lactosucrose	βGal(1-4)αGlc(1-2)αFru
Galacto-oligosaccharides, $n = 1-4$	[βGal(1-4/6)]$_n$βGal(1-4)Glc
Human milk oligosaccharides	
2′-Fucosyl-lactose	αFuc(1-2)βGal(1-4)Glc
3-Fucosyl-lactose	βGal(1-4)Glc αFuc(1-3)/
Lacto-*N*-tetraose	βGal(1-3)βGlcNAc(1-3)βGal(1-4)Glc
Lacto-*N*-neotetraose	βGal(1-4)βGlcNAc(1-3)βGal(1-4)Glc
Lacto-*N*-fucopentaose I	αFuc(1-2)βGal(1-3)βGlcNAc(1-3)βGal(1-4)Glc
Lacto-*N*-fucopentaose II	βGal(1-3)βGlcNAc(1-3)βGal(1-4)Glc αFuc(1-4)/
Sialylα(2-3)lactose	αNeuAc(2-3)βGal(1-4)Glc
Sialylα(2-6)lactose	αNeuAc(2-6)βGal(1-4)Glc
Sialyl-lacto-*N*-tetraose b	βGal(1-3)βGlcNAc(1-3)βGal(1-4)Glc αNeuAc(2-6)/

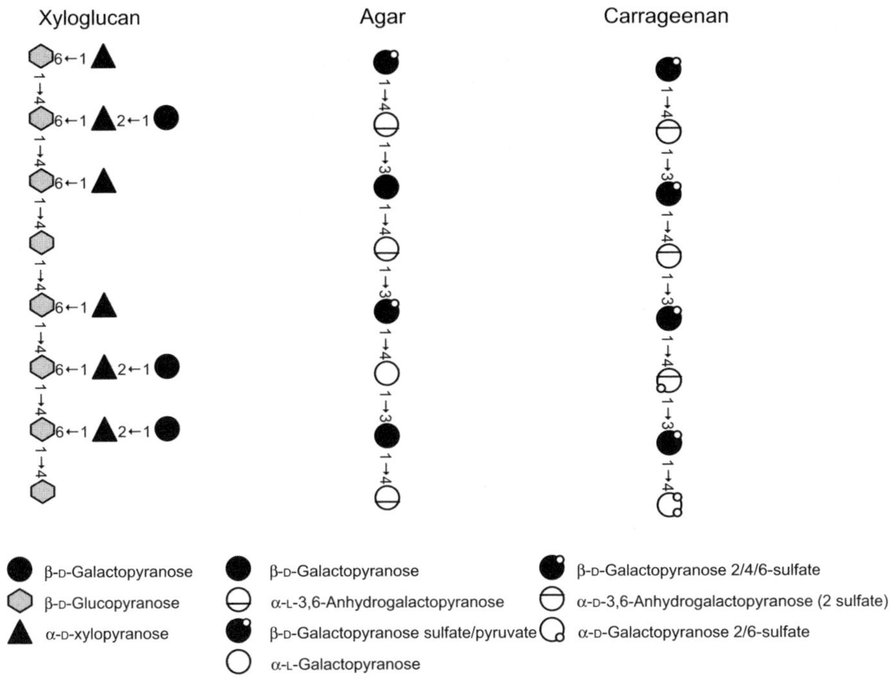

FIGURE 12.6 Fragments of primary structures of xyloglucan, agar, and carrageenan polysaccharides. Two main fractions of agar, agarose and agaropectin, differ in, respectively, lower and higher levels of esterification with sulfate/pyruvate in the polysaccharide molecules. Carrageenan types are also recognized by the sulfation levels, with the most common β-, ι-, κ-, λ, μ-, and ν-carrageenan fractions.

three arabinoxylan populations with varying amounts of D-Gal and L-Fuc residues in side chains (121). Yellow mustard gum is valued for synergistic effects with galactomannans and starches (122). Supplementing a diet with flaxseed improved glycemic control and provided protection against breast and prostate cancers, among other beneficial effects (123).

Plant exudates usually consist of mixtures of both low and high molecular weight substances released from plant tissues in conditions of normal growth or in response to environmental stress or damage. *Gum arabic* is by far the most popular plant exudate in the food industry. The gum, commercially obtained from *Acacia* trees, contains mainly arabinogalactan protein, with the polysaccharide attached to hydroxyproline-rich protein (see section "Pectin and Arabinogalactan: Structure and Sources"). *Gum tragacanth*, the exudate from *Astragalus* trees, contains two major saccharide components, a neutral branched arabinogalactan and tragacanthic acid. The arabinogalactan structure is complex and not completely understood (124). Its core resembles AG-II, with (1 → 3)-β-D-Gal*p* galactan main chain and (1 → 6)-β-D-Gal*p* branches, decorated with (1 → 2)-, (1 → 3)-, and (1 → 5)-linked α-L-Ara*f* arabinan (125). Also present are (1 → 4)-β-D-Gal*p*, (1 → 4)-L-Ara*p*, and β-D-Glc*p* residues. The tragacanthic acid has (1 → 4)-α-D-GalA*p* galacturonan backbone, with β-(1 → 3)-attached D-xylopyranose residues. The latter are occasionally capped at *O*-2 by β-D-Gal*p* and α-L-Fuc*p* (126). Arabinogalactan is water soluble and has a spherical molecular shape, while the water-swellable tragacanthic acid tertiary structure is rod-like.

Gum karaya from the bark exudates of the *Sterculia* trees and *okra mucilage* from immature *Hibiscus esculentis* seed pods share the basic structure of the main polysaccharide chains with pectinic rhamnogalacturonan I. Both have (1 → 2)-α-L-Rha*p*-(1 → 4)-α-D-GalA*p* repeats in the backbone. Unlike RG-I, the side chains in karaya gum are short and composed of β-D-Gal*p* or β-D-GlcA*p* residues attached to the main chain at O-2 or O-3 of the galacturonic acid. In addition, about half of the rhamnose units of the main chain are (1 → 4)-linked to (1 → 4)-β-D-Gal*p* saccharides (127). *Gum ghatti* from an exudate of *Anogeissus latifolia* is an example of a polysaccharide with mannoglucuronan backbone, composed of (1 → 2)-α-D-Man*p*-(1 → 4)-β-D-GlcA*p* repeats. The mannose units are further substituted at O-3 and O-6 by both arabinan and arabinogalactan branches, some of which are capped with β-D-GlcA*p* (126). D-Galactopyranose residues in the arabinogalactan branches are β-(1 → 6)-linked.

Algal polysaccharides keep the biggest share of the world markets for nondigestible food hydrocolloids and are dominated by alginate and carrageenan from seaweeds. *Carrageenans* are a family of sulfated polymers with a linear chain of D-galactose linked alternately α-(1 → 3) and β-(1 → 4), as shown in Fig. 12.6. There are no carbohydrate side chains in carrageenan. Some of the residues are in the form of 3,6-anhydrogalactose, and this feature provides for a helical tertiary structure of the polysaccharide chain (128). There are several types of carrageenans, which differ considerably in gelling properties, depending on the sulfation level, and this diversity explains their popularity in numerous food applications, where they are used as gelling agents. *Agars* are obtained from cell walls of red seaweeds. Agars differ from carrageenans in that the (1 → 4)-linked galactose is in the L-configuration and D-galactose residues may be alternatively esterified by pyruvate (Fig. 12.6). As in carrageenans, the formation of anhydrogalactose favors helical macrostructure for the polysaccharide chain. Agar gels are generally too hard for many food applications and are mostly used in confectionary.

Human milk contains about 7% lactose, 1% neutral oligosaccharides, and 0.1% acidic oligosaccharides. The core structure of human milk oligosaccharides (HMOs) can be expressed by the formula: $[\beta\text{Gal}(1 \rightarrow 3/4)\beta\text{GlcNAc}(1 \rightarrow 3/6)]_n$ βGal(1 → 4) Glc, where $n = 0{-}25$. Branching of the core structure occurs at βGal residues, and further structural variability is achieved by attaching L-fucose (in neutral HMOs) or sialic acid, NeuAc, (in acidic HMOs) residues at GlcNAc or terminal Gal and Glc. So far, between 150 and 200 neutral and acidic oligosaccharides have been isolated from human milk and milk of farm animals (129). A few examples of the most abundant structures are listed in Table 12.5. Thus, concentrations of the most common lacto-*N*-tetraose and lacto-*N*-fucopentaose I in human milk are 0.5–1.5 and 1.2–1.7 g/L, respectively (5).

Ingested HMOs resist acid hydrolysis and enzymatic digestion in the upper intestine (130). Importantly, large quantities of HMOs were found in feces and, to a smaller extent, in urine of breast-fed infants (131). This fact implies that milk oligosaccharides pass the gastrointestinal tract with little change, even in the gut, and that HMOs are partially absorbed into the blood circulation. The physiological significance of HMOs as immunoprotectors has been suggested on the basis of experiments that established the ability of these oligosaccharides to inhibit adhesion of various pathogens to epithelia in gastrointestinal and urine tracts (132, 133), as well as in the upper respiratory tract (134). The mechanism of protection may be related to the ability of HMOs to block lectins that are critical for the attachment of the bacteria to the epithelial mucosa (135). Although there are fewer types of oligosaccharides in bovine milk, they also have demonstrated the ability to block pathogen adhesion to epithelia (132, 136).

Binding of milk oligosaccharides to galectins was a subject of several studies. Selected published data on relative affinities of HMOs to human galectins are given in Table 12.6. It is obvious from the data that HMOs represent an important pool of galectin binders; whether their affinity to galectins has any physiological significance for breast-fed infants remains unexplored.

In recent decades, a growing demand for functional foods that contain WSNCs, or prebiotic saccharides with bifidogenic properties, has led to the development of a number of dietary carbohydrates based on chemical transformations of lactose, sucrose, and other saccharides (143). *Lactulose*, βGal(1 → 4)Fru, is a leader among these in production volume, which amounts to over 20,000 tons per year. It is manufactured by alkaline isomerization of lactose (144). *Lactitol* is produced by the catalytic reduction of lactose with hydrogen (145); formally, it is not a true disaccharide but a conjugate of β-galactose and glucitol. When lactose is treated by bacterial β-galactosidases in the presence of sucrose, a transgalactosydation reaction occurs, resulting in the production of trisaccharide βGal(1 → 4)αGlc(1 → 2)αFru, also referred to as *lactosucrose* (146).

On a larger industrial scale, enzymes are employed for the transgalactosydation of lactose itself in the process of production of *galacto-oligosaccharides* (GOSs), with tri- to hexasaccharides of the general formula $[\beta Gal(1 \rightarrow 4/6)]_n\beta Gal(1 \rightarrow 4)Glc$ being the main species of the reaction output (143). All of the above-listed lactose-derived semisynthetic saccharides possess a sweet taste, although not as strong as sucrose, and are used as low calorie sweeteners in the food industry. Lactulose, lactosucrose, and GOSs are bifidogenic, and for this property are also used as prebiotic functional additives in infant formulas and many dairy products. Additional health-beneficiary effects of these dietary carbohydrates include promotion of intestinal calcium absorption and anti-inflammatory effects in the bowel (147). Owing to their small size, these saccharides, when ingested, can slowly penetrate into the circulation by passive intestinal and colonic absorption and can be detected in urine (148, 149). As follows from Table 12.6, lactulose is a better ligand for galectins as compared to "standard" lactose, while lactitol and GOS βGal(1 → 4)βGal(1 → 4)Glc appear to be weak binders for galectins.

TABLE 12.6 Relative Affinities of Some Lactose-Derived Saccharides to Galectins

	Human Galectin-1 (137)	Bovine Galectin-1 (138)	Human Galectin-3 (139)	Human Galectin-3 (140)	Human Galectin-4 (140)	Human Galectin-7 (141)	Human Galectin-8 (142)
Lactose	1	1.0	1	1.0	1.0	1.0	1.0
N-Acetyllactosamine	7.8	5.5	11.3	4.1	1.0	1.1	0.3
Lactulose	1.1	1.9	5.3				
Lactitol				<0.1	<1		<0.1
GOS mucotriose		<0.05					
2′-Fucosyl-lactose	0.3	0.5	1.8	2.8	8.3		1.3
3-Fucosyl-lactose	<0.04	<0.04	<0.01	<0.1	<1		<0.1
Lacto-*N*-tetraose	1.9		9.4	6.7	5.0	1.3	1.2
Lacto-*N*-neotetraose				10.0	4.3	1.7	4.8
Lacto-*N*-fucopentaose I	1.9		10.0	12.7	8.3		2.5
Lacto-*N*-fucopentaose II				10.8	4.3		6.5
Sialylα(2-3)lactose	0.6	0.4	1.2	1.8	<1	0.3	54
Sialylα(2-6)lactose	<0.03	<0.1	<0.05	<0.1	<1	<0.1	<0.1

CONCLUSIONS

The human diet contains significant amounts of carbohydrates, both low and high molecular weight, that are potential ligands and blockers for mammalian galectins. Two principal sources of food oligo- and polysaccharides that carry terminal β-galactose residues are (1) plant gums and pectic substances and (2) milk. Initial studies suggest that these compounds may be partially absorbed into circulation and that they can modify cellular adhesion, proliferation, and signaling. Some of these food-related galectin blockers demonstrated antimetastatic effects in animal solid tumor models in the lab. Data obtained with modified citrus pectin and lactulosamines indicate that targeting galectins with dietary saccharides may have a therapeutic potential in cancer and other health disorders. Validating these beneficial therapeutic effects in humans and clarifying the *in vivo* mechanisms for biologically active galectin blockers remain the major practical challenges for the field of galectinology.

ABBREVIATIONS

AG	arabinogalactan
AG-I	type I arabinogalactan
AG-II	type II arabinogalactan
AGP	arabinogalactan protein
ASGR	asialoglycoprotein receptor
HMO	human milk oligosaccharides
LA	lactulosamine
LAG	larch arabinogalactan
LL	lactulose-L-leucine
MCP	modified citrus pectin (related to specific preparations by D. Platt and A. Raz)
RG	rhamnogalacturonan
WSNCs	water-soluble nondigestible carbohydrates

REFERENCES

1. Kushak RI, Winter HS. Dietary carbohydrates: digestion and absorption. In: Landlow MV, editor. Trends in Dietary Fats Research. Nova Science Publishers; 2005. p 1–30.
2. Brownlee IA, Dettmar PW, Strugala V, Pearson JP. The interaction of dietary fibres with the colon. Curr Nutr Food Sci 2006;2: 243–264.
3. Meyer D, Tungland B. Non-digestible oligosaccharides and polysaccharides: their physiological effects and health implications. In: McCleary BV, Prosky L, editors. Advanced Dietary Fibre Technology. Oxford, UK: Blackwell Science Ltd; 2001. p 455–470.
4. Tungland BC. Fructooligosaccharides and other fructans: structures and occurrence, production, regulatory aspects, food applications, and nutritional health significance. ACS Symp Ser 2003;849:135–152.

5. Kunz C, Rudloff S, Baier W, Klein N, Strobel S. Oligosaccharides in human milk: structural, functional, and metabolic aspects. Annu Rev Nutr 2000;20:699–722.
6. Lairon D. Dietary fibre, lipid metabolism and cardiovascular disease. In: Van der Kamp JW, editor. Dietary Fibre. Wageningen, Netherlands: Wageningen Academic Publishers; 2004. p 183–195.
7. Nuttall FQ. Dietary fiber in the management of diabetes. Diabetes 1993;42:503–508.
8. Moore MA. Dietary factors for cancer prevention: focus on insoluble and soluble fiber in cereals and vegetables. Foods Food Ingredients J Jpn 2002;196:20–33.
9. Pelucchi C, Talamini R, Galeone C, Negri E, Franceschi S, Dal Maso L, Montella M, Conti E, La Vecchia C. Fibre intake and prostate cancer risk. Int J Cancer 2004;109:278–280.
10. Takenaka Y, Fukumori T, Raz A. Galectin-3 and metastasis. Glycoconj J 2004;19: 543–549.
11. Parsonage G, Trebilcock E, Toscano MA, Bianco GA, Ilarregui JM, Buckley CD, Rabinovich GA. Roles of galectins in chronic inflammatory microenvironments. Future Rheumatol 2006;1:441–454.
12. Soerme P, Kahl-Knutsson B, Wellmar U, Magnusson B-G, Leffler H, Nilsson UJ. Design and synthesis of galectin inhibitors. Methods Enzymol 2003;363:157–169.
13. Ingrassia L, Camby I, Lefranc F, Mathieu V, Nshimyumukiza P, Darro F, Kiss R. Anti-galectin compounds as potential anti-cancer drugs. Curr Med Chem 2006;13: 3513–3527.
14. Ridley BL, O'Neill MA, Mohnen D. Pectins: structure, biosynthesis, and oligogalacturonide-related signaling. Phytochemistry 2001;57:929–967.
15. Voragen AGJ, Pilnik W, Thibault J-F, Axelos MAV, Renard CMGC. Pectins. Food Sci Technol 1996;67:287–339.
16. de Vries JA, Rombouts FM, Voragen AGJ, Pilnik W. Comparison of the structural features of apple and citrus pectic substances. Carbohydr Polym 1984;4:89–101.
17. Rouau X, Thibault J-F. Apple juice pectic substances. Carbohydr Polym 1984;4: 111–125.
18. Dey PM, Brinson K. Plant cell-walls. Adv Carbohydr Chem Biochem 1984;42:265–382.
19. Hinz SWA, Verhoef R, Schols HA, Vincken J-P, Voragen AGJ. Type I arabinogalactan contains β-D-Gal*p*-(1→3)-β-D-Gal*p* structural elements. Carbohydr Res 2005;340: 2135–2143.
20. Pérez S, Mazeau K, Hervé du Penhoat C. The three-dimensional structures of the pectic polysaccharides. Plant Physiol Biochem 2000;38:37–55.
21. Engelsen SB, Cros S, Mackie W, Pérez S. A molecular builder for carbohydrates: application to polysaccharides and complex carbohydrates. Biopolymers 1996;39: 417–433.
22. Baker RA. Reassessment of some fruit and vegetable pectin levels. J Food Sci 1997;62:225–229.
23. Rolin C. Commercial pectin preparations. In: Seymour GB, Knox JP, editors. Pectins and Their Manipulation. Oxford, UK: Blackwell Publishing Ltd; 2002. p 222–241.
24. Groman EV, Enriquez PM, Jung C, Josephson L. Arabinogalactan for hepatic drug delivery. Bioconjug Chem 1994;5:547–556.

25. Gaspar Y, Johnson KL, McKenna JA, Bacic A, Schultz CJ. The complex structures of arabinogalactan-proteins and the journey towards understanding function. Plant Mol Biol 2001;47:161–176.
26. Williams PA, Phillips GO. Gum arabic. In: Phillips GO, Williams PA, editors. Handbook of Hydrocolloids. Cambridge, UK: Woodhead Publishing Ltd.; 2000. p 155–168.
27. Street CA, Anderson DMW. Refinement of structures previously proposed for gum arabic and other *Acacia* tree exudates. Talanta 1983;30:887–895.
28. Platt D, Raz A. Modulation of the lung colonization of B16-F1 melanoma cells by citrus pectin. J Natl Cancer Inst 1992;84:438–442.
29. Pienta KJ, Naik H, Akhtar A, Yamazaki K, Replogle TS, Lehr J, Donat TL, Tait L, Hogan V, Raz A. Inhibition of spontaneous metastasis in a rat prostate cancer model by oral administration of modified citrus pectin. J Natl Cancer Inst 1995;87: 348–353.
30. Staples M, Rolke J. Modified pectins, compositions and methods related thereto. WO 2005/095463.2005.
31. Mitchell CR, Mitchell PR. Treated pectinic acid process and product. US Patent 5,498,702, 1996.
32. Inohara H, Raz A. Effects of natural complex carbohydrate (citrus pectin) on murine melanoma cell properties related to galectin-3 functions. Glycoconj J 1994;11:527–532.
33. Nangia-Makker P, Hogan V, Honjo Y, Baccarini S, Tait L, Bresalier R, Raz A. Inhibition of human cancer cell growth and metastasis in nude mice by oral intake of modified citrus pectin. J Natl Cancer Inst 2002;94:1854–1862.
34. Luck K, Ehwald R, Ziska P, Koppitz H. Affinity sorbent for galactose specific lectins consisting of deproteinized plant cell wall ghosts. Plant Sci 1992;82:29–35.
35. Benhamou N, Gilboa-Garber N, Trudel J, Asselin A. A new lectin–gold complex for ultrastructural localization of galacturonic acids. J Histochem Cytochem 1988;36: 1403–1411.
36. Koppitz H, Woehlecke H, Ehwald R. Crosslinking treatment for increasing lectin binding capacity of a vesicular affinity sorbent consisting of higher plant cells walls. Phytochemistry 1994;35:1205–1213.
37. Vincken J-P, Schols HA, Oomen RJFJ, McCann MC, Ulvskov P, Voragen AGJ, Visser RGF. If homogalacturonan were a side chain of rhamnogalacturonan I. Implications for cell wall architecture. Plant Physiol 2003;132:1781–1789.
38. Chauhan D, Li G, Podar K, Hideshima T, Neri P, He D, Mitsiades N, Richardson P, Chang Y, Schindler J, Carver B, Anderson KC. A novel carbohydrate-based therapeutic GCS-100 overcomes bortezomib resistance and enhances dexamethasone-induced apoptosis in multiple myeloma cells. Cancer Res 2005;65: 8350–8358.
39. Dongowski G, Anger H. Metabolism of pectin in the gastrointestinal tract. Prog Biotechnol 1996;14 (Pectins and Pectinases):659–666.
40. Gulfi M, Arrigoni E, Amado R. *In vitro* fermentability of a pectin fraction rich in hairy regions. Carbohydr Polym 2007;67:410–416.
41. Jenkins DJA, Leeds AR, Gassull MA, Wolever TMS, Goff DV, Alberti KGMM, Hockaday TDR. Unabsorbable carbohydrates and diabetes: decreased postprandial hyperglycemia. Lancet 1976;2:172–174.
42. Ellis PR, Rayment P, Wang Q. A physico-chemical perspective of plant polysaccharides in relation to glucose absorption, insulin secretion and the entero-insular axis. Proc Nutr Soc 1996;55:881–898.

43. Keys A, Grande F, Anderson JT. Fiber and pectin in the diet and serum cholesterol concentration in man. Proc Soc Exp Biol Med 1961;106:555–558.

44. Furda I. Interaction of dietary fiber with lipids—mechanistic theories and their limitations. Adv Exp Med Biol 1990;270:67–82.

45. Hirano M, Matsumoto T, Kiyohara H, Yamada H. Lipopolysaccharide-independent *Limulus* amebocyte lysate activating, mitogenic and anti-complementary activities of pectic polysaccharides from Chinese herbs. Planta Med 1994;60:248–252.

46. Samuelsen AB, Paulsen BS, Wold JK, Otsuka H, Kiyohara H, Yamada H, Knutsen SH. Characterization of a biologically active pectin from *Plantago major* L. Carbohydr Polym 1996;30:37–44.

47. Popov SV, Golovchenko VV, Ovodova RG, Smirnov VV, Khramova DS, Popova GY, Ovodov YS. Characterisation of the oral adjuvant effect of lemnan, a pectic polysaccharide of *Lemna minor* L. Vaccine 2006;24:5413–5419.

48. Hetland G, Samuelsen AB, Lovik M, Paulsen BS, Aaberge IS, Groeng E-C, Michaelsen TE. Protective effect of *Plantago major* L. pectin polysaccharide against systemic *Streptococcus pneumoniae* infection in mice. Scand J Immunol 2000;52:348–355.

49. Kupchik LA, Kartel' NT, Bogdanov ES, Bogdanova OV, Kupchik MP. Chemical modification of pectin to improve its sorption properties. Russ J Appl Chem 2006;79:457–460.

50. Eliaz I, Hotchkiss AT, Fishman ML, Rode D. The effect of modified citrus pectin on urinary excretion of toxic elements. Phytother Res 2006;20:859–864.

51. Motomura Y, Yoshida Y. Antioxidative ability of cell wall components in fruits against ascorbic acid oxidation. Acta Hort 2003;628:669–676.

52. Harris PJ, Ferguson LR. Dietary fibre: its composition and role in protection against colorectal cancer. Mutat Res 1993;290:97–110.

53. Ohkami H, Tazawa K, Yamashita I, Shimizu T, Murai K, Kobashi K, Fujimaki M. Effects of apple pectin on fecal bacterial enzymes in azoxymethane-induced rat colon carcinogenesis. Jpn J Cancer Res 1995;86:523–529.

54. Ohno K, Narushima S, Takeuchi S, Itoh K, Mitsuoka T, Nakayama H, Itoh T, Hioki K, Nomura T. Inhibitory effect of apple pectin and culture condensate of *Bifidobacterium longum* on colorectal tumors induced by 1,2-dimethylhydrazine in transgenic mice harboring human prototype c-Ha-*ras* genes. Exp Anim 2000;49: 305–307.

55. Klurfeld DM. Dietary fiber-mediated mechanisms in carcinogenesis. Cancer Res 1992;52 (7 Suppl):2055s–2059s.

56. Beuth J, Ko HL, Oette K, Pulverer G, Roszkowski K, Uhlenbruck G. Inhibition of liver metastasis in mice by blocking hepatocyte lectins with arabinogalactan infusions and D-galactose. J Cancer Res Clin Oncol 1987;113:51–55.

57. Beuth J, Ko HL, Schirrmacher V, Uhlenbruck G, Pulverer G. Inhibition of liver tumor cell colonization in two animal models by lectin blocking with D-galactose and arabinogalactan. Clin Exp Metastasis 1988;6:115–120.

58. Hagmar B, Erkell LJ, Ryd W, Skomedal H. Effect of arabinogalactan and other glycoconjugates on experimental metastases. Cell Pharmacol 1994;1:87–90.

59. Hagmar B, Ryd W, Skomedal H. Arabinogalactan blockade of experimental metastases to liver by murine hepatoma. Invasion Metastasis 1991;11:348–355.

60. Glinskii OV, Huxley VH, Glinsky GV, Pienta KJ, Raz A, Glinsky VV. Mechanical entrapment is insufficient and intercellular adhesion is essential for metastatic cell arrest in distant organs. Neoplasia 2005;7:522–527.

61. Hayashi A, Gillen AC, Lott JR. Effects of daily oral administration of quercetin chalcone and modified citrus pectin in implanted colon-25 tumor growth in Balb-c mice. Altern Med Rev 2000;5:546–552.

62. Guess BW, Scholz MC, Strum SB, Lam RY, Johnson HJ, Jennrich RI. Modified citrus pectin (MCP) increases the prostate-specific antigen doubling time in men with prostate cancer: a phase II pilot study. Prostate Cancer Prostatic Dis 2003;6:301–304.

63. Adachi S. *N-p*-Tolyl-lactosylamine and its Amadori rearrangement product. Chem Ind 1957;956.

64. Kuhn R, Krüger G. Interaction of Amadori compounds with diazonium salts. Liebigs Ann Chem 1959;628:240–255.

65. Glinsky GV, Mossine VV, Price JE, Bieleuberg D, Glinsky VV, Ananthaswamy H, Feather MS. Inhibition of colony formation in agarose of metastatic human breast carcinoma and melanoma cells by synthetic glycoamine analogs. Clin Exp Metastasis 1996;14:253–267.

66. Finot PA, Mauron J. The blockage of lysine in the Maillard reaction. II. Chemical properties of *N*-(deoxy-1-D-fructosyl-1) and *N*-(deoxy-1-D-lactulosyl-1) derivatives of lysine. Helv Chim Acta 1972;55:1153–1164.

67. Krause R, Knoll K, Henle T. Studies on the formation of furosine and pyridosine during acid hydrolysis of different Amadori products of lysine. Eur Food Res Technol 2003;216:277–283.

68. Olano A, Santa-Maria G, Corzo N, Calvo MM, Martinez-Castro I, Jimeno ML. Determination of free carbohydrates and Amadori compounds formed at the early stages of non-enzymic browning. Food Chem 1992;43:351–358.

69. Mossine VV, Barnes CL, Glinsky GV, Feather MS. Interaction between two C(1)-X-C′(1) branched ketoses: observation of an unprecedented crystalline *spiro*-bicyclic hemiketal tautomer of *N,N*-di(1-deoxy-D-fructos-1-yl)-glycine ("difructose glycine") having open chain carbohydrate. Carbohydr Lett 1995;1:355–362.

70. Hashiba H. The browning reaction of Amadori compounds derived from various sugars. Agric Biol Chem 1982;46:547–548.

71. Mossine VV, Mawhinney TP. D-Fructose-L-histidine: a potent copper chelator from tomato powder. J Agric Food Chem 2007;55:10373–10381.

72. Mossine VV, Linetsky M, Glinsky GV, Ortwerth BJ, Feather MS. Superoxide free radical generation by Amadori compounds: the role of acyclic forms and metal ions. Chem Res Toxicol 1999;12:230–236.

73. Mossine VV, Feather MS. Metal binding properties of glycated amino acids and proteins. In: The Maillard Reaction in Foods and Medicine. London: The Royal Society of Chemistry;1998. p 429.

74. Finot PA, Deutsch R, Bujard E. The extent of the Maillard reaction during the processing of milk. Prog Food Nutr Sci 1981;5:345–355.

75. Erbersdobler HF. Protein reactions during food processing and storage—their relevance to human nutrition. In: Somogyi JC, Müller HR, editors. Nutritional Impact of Food Processing. Karger: Basel;1989. p 140–155.

76. Berg HE, van Boekel MAJS. Degradation of lactose during heating of milk. 1. Reaction pathways. Neth Milk Dairy J 1994;48:157–175.

77. Rérat A, Calmes R, Vaissade P, Finot P-A. Nutritional and metabolic consequences of the early Maillard reaction of heat treated milk in pig. Significance for man. Eur J Nutr 2002;41:1–11.

78. Matsuda T, Kato Y. Haptenic sugar antigens as immunochemical markers for the Maillard reaction of processed and stored milk products. ACS Symp Ser 1996;631:221–226.

79. Pizzano R, Nicolai MA, Siciliano R, Addeo F. Specific detection of the Amadori compounds in milk by using polyclonal antibodies raised against a lactosylated peptide. J Agric Food Chem 1998;46:5373–5379.

80. Pallini M, Compagnone D, Palleschi G, Di Stefano S, Marini S, Coletta M. Immunodetection of lactosylated proteins as a useful tool to determine heat treatment in milk samples. Analyst 2001;126:66–70.

81. Siciliano R, Rega B, Amoresano A, Pucci P. Modern mass spectrometric methodologies in monitoring milk quality. Anal Chem 2000;72:408–415.

82. Czerwenka C, Maier I, Pittner F, Lindner W. Investigation of the lactosylation of whey proteins by liquid chromatography–mass spectrometry. J Agric Food Chem 2006;54: 8874–8882.

83. Erbersdobler HF. The biological significance of carbohydrate–lysine cross-linking during heat-treatment of food proteins. Adv Exp Med Biol 1977;86B: 367–378.

84. Dworschák E, Tarján V, Turos S. Characteristics of some new flavoring materials produced by the Maillard reaction. ACS Symp Ser 1983;215:159–168.

85. Desrosiers T, Savoie L, Bergeron G, Parent G. Estimation of lysine damage in heated whey proteins by furosine determinations in conjunction with the digestion cell technique. J Agric Food Chem 1989;37:1385–1391.

86. Holt C, McPhail D, Nevison I, Nylander T, Otte J, Ipsen RH, Bauer R, Ogendal L, Olieman K, De Kruif KG, Leonil J, Molle D, Henry G, Maubois JL, Perez MD, Puyol P, Calvo M, Bury SM, Kontopidis G, McNae I, Sawyer L, Ragona L, Zetta L, Molinari H, Klarenbeek B, Jonkman MJ, Moulin J, Chatterton D. Apparent chemical composition of nine commercial or semi-commercial whey protein concentrates, isolates and fractions. Int J Food Sci Technol 1999;34:543–556.

87. Oak J-H, Nakagawa K, Miyazawa T. UV analysis of Amadori-glycated phosphatidylethanolamine in foods and biological samples. J Lipid Res 2002;43:523–529.

88. Kumar V, Banker GS. Maillard reaction and drug stability. Spec Pub R Soc Chem 1994;151 (*Maillard Reactions in Chemistry Food and Health*): 20–27.

89. Wirth JJ, Olsen BA, Hallenbeck DK, Lake ME, Gregg SM, Perry FM. Screening methods for impurities in multi-sourced fluoxetine hydrochloride drug substances and formulations. Chromatographia 1997;46:511–523.

90. Aring J, Schlepper-Schaefer J, Burkart V, Kolb H. Nonenzymatically glycated serum albumin: interaction with galactose-specific liver lectins. Biochim Biophys Acta 1989;1010:140–144.

91. Glinsky GV. Glycodeterminants of melanoma cell adhesion. A model for antimetastatic drug design. Crit Rev Oncol Hematol 1993;14:1–13.

92. Glinsky GV, Price JE, Glinsky VV, Mossine VV, Kiriakova G, Metcalf JB. Inhibition of human breast cancer metastasis in nude mice by synthetic glycoamines. Cancer Res 1996;56:5319–5324.

93. Weis WI, Drickamer K. Structural basis of lectin–carbohydrate recognition. Annu Rev Biochem 1996;65:441–473.

94. Rabinovich GA, Cumashi A, Bianco GA, Ciavardelli D, Iurisci I, D'Egidio M, Piccolo E, Tinari N, Nifantiev N, Iacobelli S. Synthetic lactulose-amines: novel class of anticancer agents that induce tumor-cell apoptosis and inhibit galectin-mediated homotypic cell aggregation and endothelial cell morphogenesis. Glycobiology 2006;16:210–220.

95. Huflejt ME, Mossine VV, Croft M. Galectin-1 and galectin-4 in tumor development and use in prognosis and treatment of breast cancer. PCT Int Appl WO 0326494 A2.2003.

96. Pellegrino L, De Noni I, Cattaneo S. Formation of protein bound lysine-derived galactosyl and glucosyl pyrroles in heated model systems. Nahrung 2000;44:193–200.

97. Pricci F, Leto G, Amadio L, Iacobini C, Romeo G, Cordone S, Gradini R, Barsotti P, Liu F-T, Di Mario U, Pugliese G. Role of galectin-3 as a receptor for advanced glycosylation end products. Kidney Int Suppl 2000;77:S31–S39.

98. Raz A, Lotan R. Lectin-like activities associated with human and murine neoplastic cells. Cancer Res 1981;41:3642–3647.

99. Denisevich TV, Semyonova-Kobzar RA, Glinsky VV, Mossine VV. The influence of synthetic aminoglycoconjugates on the aggregation ability and metastatic potential of tumor cells. Exp Oncol 1995;17:111–117.

100. Glinsky VV, Glinsky GV, Glinskii OV, Huxley VH, Turk JR, Mossine VV, Deutscher SL, Pienta KJ, Quinn TP. Intravascular metastatic cancer cell homotypic aggregation at the sites of primary attachment to the endothelium. Cancer Res 2003;63:3805–3811.

101. Springer GF. Immunoreactive T and Tn epitopes in cancer diagnosis, prognosis, and immunotherapy. J Mol Med 1997;75:594–602.

102. Glinsky VV, Glinsky GV, Rittenhouse-Olson K, Huflejt ME, Glinskii OV, Deutscher SL, Quinn TP. The role of Thomsen–Friedenreich antigen in adhesion of human breast and prostate cancer cells to the endothelium. Cancer Res 2001;61:4851–4857.

103. Khaldoyanidi SK, Glinsky VV, Sikora L, Glinskii AB, Mossine VV, Quinn TP, Glinsky GV, Sriramarao P. MDA-MB-435 human breast carcinoma cell homo- and heterotypic adhesion under flow conditions is mediated in part by Thomsen–Friedenreich antigen–galectin-3 interactions. J Biol Chem 2003;278:4127–4134.

104. Johnson KD, Glinskii OV, Mossine VV, Turk JR, Mawhinney TP, Anthony DC, Henry CJ, Huxley VH, Glinsky GV, Pienta KJ, Raz A, Glinsky VV. Galectin-3 as a potential therapeutic target in tumors arising from malignant endothelium.Neoplasia 2007;9:662–670.

105. Perillo NL, Pace KE, Seilhamer JJ, Baum LG. Apoptosis of T cells mediated by galectin-1. Nature 1995;378:736–739.

106. Hurrell RF, Finot PA, Ford JE. Storage of milk powders under adverse conditions. 1. Losses of lysine and of other essential amino acids as determined by chemical and microbiological methods. Br J Nutr 1983;49:343–354.

107. Hurrell RF. Influence of the Maillard reaction on the nutritional value of foods. In: Finot PA, Aeschbacher HU, Hurrell RF, Liardon R, editors. The Maillard Reaction in Food Processing, Human Nutrition and Physiology. Basel: Birkhäuser Verlag;1990. p 245–258.

108. Henle T, Schwenger V, Ritz E. Preliminary studies on the renal handling of lactuloselysine from milk products. Czech J Food Sci 2000;18:101–102.

109. Schwenger V, Morath C, Schönfelder K, Klein W, Weigel K, Deppisch R, Henle T, Ritz E, Zeier M. An oral load of the early glycation compound lactuloselysine fails to accumulate in the serum of uraemic patients. Nephrol Dial Transplant 2006;21: 383–388.

110. Langhendries JP, Hurrell RF, Furniss DE, Hischenhuber C, Finot PA, Bernard A, Battisti O, Bertrand JM, Senterre J. Maillard reaction products and lysinoalanine: urinary excretion and the effects on kidney function of preterm infants fed heat-processed milk formula. J Pediatr Gastroenterol Nutr 1992;14:62–70.

111. Albert R, Marbach P, Bauer W, Briner U, Fricker G, Bruns C, Pless J. SDZ CO 611: a highly potent glycated analog of somatostatin with improved oral activity. Life Sci 1993;53:517–525.

112. Hurley TR, Lovdahl MJ, Tobias B. Pregabalin lactose conjugates. PCT Int Appl WO 078747 A3.2002.

113. Glinskii OV, Huxley VH, Glinsky GV, Pienta KJ, Raz A, Glinsky VV. Mechanical entrapment is insufficient and intercellular adhesion is essential for metastatic cell arrest in distant organs. Neoplasia 2005;7:522–527.

114. Elola MT, Chiesa ME, Alberti AF, Mordoh J, Fink NE. Galectin-1 receptors in different cell types. J Biomed Sci 2005;12:13–29.

115. Dumic J, Dabelic S, Flögel M. Galectin-3: an open-ended story. Biochim Biophys Acta 2006;1760:616–635.

116. Wielinga WC. Galactomannans. In: Phillips GO, Williams PA, editors. Handbook of Hydrocolloids. Cambridge, UK: Woodhead Publishing Ltd.;2000. p 137–154.

117. Nishinari K, Yamatoya K, Shirakawa M. Xyloglucan. In: Phillips GO, Williams PA, editors. Handbook of Hydrocolloids. Cambridge, UK: Woodhead Publishing Ltd.; 2000. p 247–267.

118. Fischer MH, Yu N, Gray GR, Ralph J, Anderson L, Marlett JA. The gel-forming polysaccharide of psyllum husk (*Plantago ovata* Forsk). Carbohydr Res 2004;339: 2009–2017.

119. Singh B. Psyllium as therapeutic and drug delivery agent. Int J Pharm 2007;334:1–14.

120. Cui W, Eskin MNA, Biliaderis CG, Marat K. NMR characterization of a 4-*O*-methyl-β-D-glucuronic acid-containing rhamnogalacturonan from yellow mustard (*Sinapis alba* L.) mucilage. Carbohydr Res 1996;292:173–183.

121. Warrand J, Michaud P, Picton L, Muller G, Courtois B, Ralainirina R, Courtois J. Structural investigations of the neutral polysaccharide of *Linum usitatissimum* L. seeds mucilage. Int J Biol Macromol 2005;35:121–125.

122. Cui SW, Eskin MAN, Wu Y, Ding S. Synergisms between yellow mustard mucilage and galactomannans and applications in food products—a mini review. Adv Colloid Interface Sci 2006;128–130,249–256.

123. Tarpila A, Wennberg T, Tarpila S. Flaxseed as a functional food. Curr Top Nutraceut Res 2005;3:167–188.

124. Tischer CA, Iacomini M, Gorin PAJ. Structure of the arabinogalactan from gum tragacanth (*Astragalus gummifer*). Carbohydr Res 2002;337:1647–1655.

125. Aspinall GO, Baillie J. Gum tragacanth. Part II. The arabinogalactan. J Chem Soc 1963;1714–1721.

126. Aspinall GO. Gums and mucilages. Adv Carbohydr Chem 1969;24:333–379.

127. Wang W. Tragacanth and karaya. In: Phillips GO, Williams PA, editors. Handbook of Hydrocolloids. Cambridge, UK: Woodhead Publishing Ltd.; 2000. p 231–246.

128. Stortz CA. Carrageenans: structural and conformational studies. In:Yarema KJ, editor. Handbook of Carbohydrate Engineering. Boca Raton, FL: CRC Press LLC; 2005. p 211–245.

129. Mehra R, Kelly P. Milk oligosaccharides: structural and technological aspects. Int Dairy J 2006;16:1334–1340.

130. Engfer MB, Stahl B, Finke B, Sawatzki G, Daniel H. Human milk oligosaccharides are resistant to enzymatic hydrolysis in the upper gastrointestinal tract. Am J Clin Nutr 2000;71:1589–1596.

131. Chaturvedi P, Warren CD, Buescher CR, Pickering LK, Newburg DS. Survival of human milk oligosaccharides in the intestine of infants. Adv Exp Med Biol 2001;501:315–323.

132. Martin-Sosa S, Martin M-J, Hueso P. The sialylated fraction of milk oligosaccharides is partially responsible for binding to enterotoxigenic and uropathogenic *Escherichia coli* human strains. J Nutr 2002;132:3067–3072.

133. Coppa GV, Zampini L, Galeazzi T, Facinelli B, Ferrante L, Capretti R, Orazio G. Human milk oligosaccharides inhibit the adhesion to Caco-2 cells of diarrheal pathogens: *Escherichia coli*, *Vibrio cholerae*, and *Salmonella fyris*. Pediatr Res 2006;59:377–382.

134. Newburg DS. Human milk glycoconjugates that inhibit pathogens. Curr Med Chem 1999;6:117–127.

135. Perret S, Sabin C, Dumon C, Pokorna M, Gautier C, Galanina O, Shahov I, Bovin N, Nicaise M, Desmadril M, Gilboa-Garber N, Wimmerov M, Mitchell EP, Imberty A. Structural basis for the interaction between human milk oligosaccharides and the bacterial lectin PA-IIL of *Pseudomonas aeruginosa*. Biochem J 2005;389:325–332.

136. Gopal PK, Gill HS. Oligosaccharides and glycoconjugates in bovine milk and colostrum. Br J Nutr 2000;84 (Suppl. 1):S69–S74.

137. Sparrow CP, Leffler H, Barondes SH. Multiple soluble β-galactoside-binding lectins from human lung. J Biol Chem 1987;262:7383–7390.

138. Ahmed H, Fink NE, Pohl J, Vasta GR. Galectin-1 from bovine spleen: biochemical characterization, carbohydrate specificity and tissue-specific isoform profiles. J Biochem 1996;120:1007–1019.

139. Leffler H, Barondes SH. Specificity of binding of three soluble rat lung lectins to substituted and unsubstituted mammalian β-galactosides. J Biol Chem 1986;261: 10119–10126.

140. Ideo H, Seko A, Ohkura T, Matta KL, Yamashita K. High-affinity binding of recombinant human galectin-4 to $SO_3^- \rightarrow$ 3Galβ1 → 3GalNAc pyranoside. Glycobiology 2002;12: 199–208.

141. Ahmad N, Gabius H-J, Kaltner H, Andre S, Kuwabara I, Liu F-T, Oscarson S, Norberg T, Brewer CF. Thermodynamic binding studies of cell surface carbohydrate epitopes to galectins-1, -3, and -7: evidence for differential binding specificities. Can J Chem 2002;80:1096–1104.

142. Ideo H, Seko A, Ishizuka I, Yamashita K. The N-terminal carbohydrate recognition domain of galectin-8 recognizes specific glycosphingolipids with high affinity. Glycobiology 2003;13:713–723.

143. Sako T, Matsumoto K, Tanaka R. Recent progress on research and applications of non-digestible galacto-oligosaccharides. Int Dairy J 1999;9:69–80.

144. Aider M, de Halleux D. Isomerization of lactose and lactulose production: review. Trends Food Sci Technol 2007;18:356–364.

145. Mesters PHJ, Van Velthuijsen JA, Brokx S. Lactitol: a new reduced-calorie sweetener. Food Sci Technol 2001;112:297–315.

146. Arakawa K, Aoyama Y, Ikeda H, Mikuni K, Fujita K, Hara K. The development of lactosucrose production and its applications in foods for specified health use. J Appl Glycosci 2002;49:63–72.

147. Singh A, Porwal O, Sharma N, Singh A, Kumar S, Sharma PK. Effects of prebiotics on gut and human health: a review. J Pure Appl Microbiol 2007;1:69–82.

148. Kalach N, Rocchiccioli F, de Boissieu D, Benhamou PH, Dupont C. Intestinal permeability in children: variation with age and reliability in the diagnosis of cow's milk allergy. Acta Paediatr 2001;90:499–504.

149. Anderson ADG, Jain PK, Fleming S, Poon P, Mitchell CJ, MacFie J. Evaluation of a triple sugar test of colonic permeability in humans. Acta Physiol Scand 2004;182:171–177.

INDEX